THE WRONG KIND OF SNOW

THE
WRONG
KIND
OF SNOW

How the
Weather Made Britain

**Antony Woodward
and Robert Penn**

HODDER

First published in Great Britain in 2007 by Hodder & Stoughton
An Hachette Livre UK company

First published in paperback in 2008

1

A CIP catalogue record for this title is available from the British Library

ISBN 978 0 340 93788 4

Typeset in Garamond by Palimpsest Book Production Limited,
Grangemouth, Stirlingshire

Printed and bound by Clays Ltd, St Ives plc

Hodder & Stoughton policy is to use papers that are natural, renewable
and recyclable products and made from wood grown in sustainable forests.
The logging and manufacturing processes are expected to conform to the
environmental regulations of the country of origin.

Hodder & Stoughton Ltd
338 Euston Road
London NW1 3BH

www.hodder.co.uk

To Ivo, Kitty, Lucas, Maya, Scarlett and Storm

You can spend your whole life trying to be popular but, at the end of the day, the size of the crowd at your funeral will be largely dictated by the weather.

Frank Skinner

INTRODUCTION

Rain gave us *Inspector Morse* and the sliding tackle. Fog gave us the Cat's Eye and chains on front doors. Wind brought a Protestant monarchy and the greatest turnaround in Ashes history – Headingley 1981. A hail-storm led to the Norwich Union insurance company. A sunny summer afternoon gave us *Alice in Wonderland* and the 1992 Conservative government. A blizzard gave us Elizabeth David's cook books. Storms gave us the pencil, the Norfolk Broads, the lifeboat and the weather forecast. And cold, damp days? Penicillin and the Industrial Revolution.

In Britain, what isn't affected by the weather? Since the first chilly Roman on Hadrian's Wall pulled socks on before his sandals (yes, they're the culprits), British life and British weather have been inseparable. Our collection of islands, sandwiched between an ocean and a continent, where nowhere is more than 100 miles from the sea, lies at a point where four competing air masses meet. This combination gives our weather two key characteristics – mildness and changeability. Bizarrely, together, they make the British Isles the most weather-affected place on earth.

Mild dampness, of course, with its lack of extremes, is what our weather is famous for. 'You will probably not like the almost continual rains and mists and the absence of snow and crisp cold,' warns the *Instructions for American Servicemen in Britain*, issued in 1942. 'Actually, London has less rain for the whole year than many places in the United States, but the rain falls in frequent drizzles. Most people get used to the British climate eventually.' In Britain, we witness only three or four extreme weather events in a lifetime (over the sweep of history, of course, these soon add up: 85% of the worst UK disasters, by death toll, are somehow weather-linked).

But this same drizzly dampness, which so fails to impress foreigners, is our most precious natural resource. Beneath our overcast skies grows the best grass in the world – the reason why our land is so green and pleasant, and why the wool industry made England a great trading nation. The wealth from wool – damp weather, converted for export – gave us the most advanced economy in the world, bankrolling the Industrial Revolution, and helping create the largest empire of modern times. Dampness gave us the roast beef of olde Englande, the best hard cheeses in the world, and it still means we are the only country with fresh milk in every corner shop and filling station. It gave us turf so perfect we had to invent and codify endless things to do on it – football, bowls, golf, rugby, croquet, lawn tennis and hockey.

'The Calmness of the Air doth mollify Men's Minds, not corrupting them with venereal Lusts, but preserving them from savage and rude Behaviour,' observed William Fitzstephen in 1180, some nine hundred years before the first sociologist mentioned 'environmental determinism'. 'Has a mild and gentle climate, rarely too hot and rarely extremely cold, played a role in producing a moderate, pragmatic people?' asks Jeremy Paxman in *The English, A Portrait of a People*. Into that calm, mild dampness, generations of children have been ordered to 'get some fresh air' – resulting in a national obsession with the outdoors, from sport, to dog and hill walking, to gardening. Just as often that same weather keeps us indoors, of course: thinking, inventing, writing, practising darts and forming pop groups.

What about that second crucial consequence of those ever-competing air masses – changeability? 'In our island,' said Dr Johnson, 'every man goes to bed unable to guess whether he shall behold in the morning a bright or cloudy atmosphere.' And that's the crux of our vexed relationship with our weather: the uncertainty. This is the reason we're forever caught out in the rain (or by the wrong kind of snow), the reason we're always taken by surprise and always wearing the wrong kind of clothes. It's why the weather is the one safe topic of conversation, whatever the circumstances. 'Stick to the weather and your health,' Professor Higgins advises Eliza Doolittle in *My Fair Lady*, before her first foray into society.

The constant uncertainty meant we pioneered the science of meteorology. We classified the wind and the clouds. We invented the weather journal, the tipping rain gauge, the cup anemometer, the storm warning, the forecast, the domestic barometer, the isobar and the Stevenson Screen weather station. It also meant we led the way in waterproofing, from the *birrus Britannicus*, the hooded woollen cloak exported by the Romans, to the folding umbrella, the Mackintosh and the Wellington boot.

Yet, when we started researching this book, despite living all our lives under British skies, breathing British air, despite hearing at least 100,000 weather forecasts between us, we began to realise that, really, we knew almost nothing about the British weather. There are local winds and giant waves we had never heard of. There are offshore rocks and sand banks that, in bad weather, have claimed more lives than wars. Where, we wondered, is the wettest place and what is the wettest month? When's the best time to find mushrooms? And the worst time for midges? When's autumn leaf colour at its best? How much can it rain in a day? What's the strongest wind? Why don't (and can't) we get hurricanes? Why is hail more likely on hot summer days? What day does spring begin? When, on average, do we see the first swallow? Or get the last frost? Do we really get more tornadoes than anywhere else in the world? Where's our bleakest, iciest road? When's the coldest and hottest day of the year likely to be? When's the best time to plan a barbecue? We wanted answers – and found some of them surprisingly hard to come by.

The aim of this book is fourfold. First, to show how our weather affects our history, politics, literature, science, industry, art, film, television, music, fashion, sport, health, leisure, shopping, advertising and food. Then, to give a practical general guide to the kind of weather that may be expected on any given day of the year, with hard facts about extremes and likely average conditions. To show how, plunging back over two millennia of recorded history, every day of the year has made its mark – well, every day but 18 March, but let's not be churlish. And, finally, to give a flavour, through diary comments and reportage, of how weather has affected us as humans – showing that, however surprised we are at the weather, millions have been just as surprised before us.

THE BEST WEATHER IN THE WORLD

'Not another f***ing beautiful day' complains Lady Diana Broughton, the Greta Scacchi character in the film *White Mischief*, as she opens the curtains on another cloudless, cobalt blue, equatorial African sky. Not, perhaps, a sentiment with which every Brit will immediately empathise, but it goes to the heart of why British weather is the best in the world. Over the last year of writing, we've witnessed the hottest July ever, the warmest autumn, the wettest December and the worst floods for sixty years. Not bad for twelve months. But even in Britain, with all its variety, we only see, on average, ten lightning shows a year. Just eight days will be properly foggy. We may get as few as twenty frosts, these days, and perhaps fifteen days (and falling) of snow. Seventy-seven days will be predominantly sunny. So next time you have to scrape ice off your car windscreen in the morning, or turn off the garden sprinkler, or slow down because of fog, or a thunderstorm cuts off your mobile signal, or a centimetre of snow brings Britain to a halt, don't complain – treasure the experience. Because, beneath our restless skies, we know something that others don't.

In Britain, there's no such thing as a dull day.

TECHNICAL STUFF

This book is not – nor could it attempt to be, without being very dull indeed – a comprehensive survey of all the recorded weather events in the history of the British Isles. For that, you'll have to read through the entire contents of the National Meteorological Library and Archive in Exeter. Good luck. Storms alone could fill several volumes. Floods another. Blizzards yet another.

Many of the major weather events and all of the significant climate changes of the last thousand years are included. But scale alone does not warrant a place. This is not a list of freak or extreme weather: there are many stories about average or ordinary weather.

We have selected events for each day using the following criteria: Did the weather have *consequences* that resonate through the centuries? Or was it exceptional in some way – for the century, for the season, or even the day? Our intention is to offer a raking glance, 365 snapshots, of our relationship with our weather, rather than a history of the weather itself.

Where possible, we've dated events to the critical moment – the eye of the storm, the day the snow fell thickest, the moment the sea-surge breached; but just occasionally there have been monumental date clashes and we've approached an event from a different angle – the beginning of a prolonged cold snap, the end of a drought, the middle of London's foggiest ever month.

We have tried to allow the meteorological shape of the year to come through. Of course, heat waves happen in summer and snow storms in winter, but the prevalence of, for example, gales in January, frosts in May, hail and thunderstorms in July and August, Indian Summers in October and fog in November and December are the details that make the British weather calendar fascinating.

HOW THE BOOK WORKS

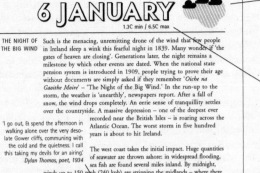

6 JANUARY

1.3C min / 6.5C max

THE NIGHT OF THE BIG WIND

Such is the menacing, unremitting drone of the wind that few people in Ireland sleep a wink this fearful night in 1839. Many wonder if 'the gates of heaven are closing'. Generations later, the night remains a milestone by which other events are dated. When the national state pension system is introduced in 1909, people trying to prove their age without documents are simply asked if they remember '*Oíche na Gaoithe Moire*' – 'The Night of the Big Wind.' In the run-up to the storm, the weather is 'unearthly', newspapers report. After a fall of snow, the wind drops completely. An eerie sense of tranquillity settles over the countryside. A massive depression – one of the deepest ever recorded near the British Isles – is roaring across the Atlantic Ocean. The worst storm in five hundred years is about to hit Ireland.

'I go out, & spend the afternoon in walking alone over the very desolate Gower cliffs, communing with the cold and the quietness. I call this taking my devils for an airing.'
Dylan Thomas, poet, 1934

The west coast takes the initial impact. Huge quantities of seawater are thrown ashore: in widespread flooding, sea fish are found several miles inland. By midnight, winds up to 150 mph (240 kph) are stripping the midlands – where there are practically no natural windbreaks – of all structures. Because the storm arrives at night, only three hundred are killed – but the damage is unprecedented. In Carlow, the tower of the cathedral is torn off. Many large country houses are de-roofed. Windmills, factories, trees, ancient monuments and barns are blown down. Tombstones are lifted from the ground and found up to a mile away. Tens of thousands are left homeless; many more lose their savings, hidden in the thatched roofs. In spring, birdsong is absent – scarcely any birds survive. In an age before comprehensive relief measures and insurance, the nation is kneecapped. Devout Christians believe it's Judgement Day. But for one man at least, it is simply a colossal storm: the Reverend Romney Robinson, Director of the Armagh Observatory, is inspired to develop his cup-anemometer, a breakthrough in the accurate measurement of wind speed.

71.1C West Unites 1841

17.3C May 1955

In 1947 (*4 March*), wintry weather divides the British Isles today. In London skiers practise parallel turns on Hampstead Heath and ice slabs wash on to Southend foreshore in Essex, while in Torquay in Devon 'coatless promenaders' enjoy nearly seven hours of sunshine and primroses are being picked in Pembrokeshire.

Guide to likely general conditions, averaged from the last 50 years.

Daily maximum and minimum mean temperatures, averaged from the Central England Temperature (CET) series daily data, 1878-2006.

Dates in italics refer to a related story on that date.

Extremes and map references
☼ – Highest recorded temperatures for each day of the year, 1875-2006.
❅ – Lowest recorded temperatures for each day of the year (October-May), 1875-2006, at altitudes below 500 m.
🖋 – Notable daily rainfall extremes.
🌀 – Notable wind extremes at altitudes below 500m.
♀ ♛ ✗ ✿ – Refer to stories in the prose.

Eye-witness accounts: quirky, funny, charming, apposite or telling firsthand accounts of weather events that happened on this day, culled from official records, private letters, diaries, newspapers or literature – by artists, sportsmen, naturalists, painters, poets, engineers, historians, gardeners, generals, inventors, politicians and, of course, whingeing foreigners.

OUR 'FORECAST' – REALLY A 'BACKCAST'

In the British Isles, almost any weather can happen on almost any day: so our weather graphic is NOT intended as a forecast. It is merely intended as a guide to the *kind of weather conditions* you can expect at that time of year. To compile our 'backcast', we have consulted: (1) Central England Temperature (CET) series – the longest available instrumental record of temperature in the world, devised by Professor Gordon Manley, comprising monthly means from 1659 to the present and mean daily data from 1772; (2) 'HadUKP' regional precipitation series – the longest running rainfall series, including the England and Wales Precipitation (EWP) series which begins in 1766; (3) Daily sunshine data from Met Office weather stations at Ringway/Woodford, Filton/Yeovilton and Heathrow. Our data, from 1957 to the end of 2006, approximately covers the area defined by CET (a roughly triangular area enclosed by Preston, Bristol and London).

After analysing the relative presence (or not) of sunshine, cloud and precipitation we have presented the results using the familiar idiom of weather symbols (see 'Backcast Key'). To illustrate the meteorological shape of the year, we have distinguished between winter and summer sunshine by using differently-sized sun symbols.

Inevitably, choosing thresholds to distinguish the different weather typically experienced at different times of year, from data averaged over long periods, has a certain arbitrariness. For example, for rainfall, we know, from Met Office statistics, that on average there are approximately 115 'heavy' days in Central England (≥1mm) and forty-five 'light' rain days (0.2-1.0mm) each year. By ranking each day of the year by its associated average rainfall (based on the HadUKP regional precipitation series), we have then set thresholds to determine the choice of symbol.

Relative amounts of sun/cloud have been calculated by taking the maximum amount of sunshine (as a proportion of daylight hours) possible for each day. Taking these figures, we divided the year into equal proportions of 'overcast', 'broken', 'scattered' and 'few' in terms of cloud cover (i.e. relative proportions of cloud) based on the sunshine data. We have also paid attention to weather 'singularities' as defined by the climatologist Professor Hubert Lamb and others. Singularities are points in the year when certain types of weather may show a tendency to recur.

In no way is the weather symbol intended to be a forecast – it is purely for interest.

WEATHER EXTREMES

The Royal Meteorological Society established a network of Climatological Weather Stations, using verified instruments that were systematically inspected, in 1875, standardising thermometer exposure (in a Stevenson Screen) across Britain. Reliable highest and lowest temperatures date from this time.

EYE-WITNESS ACCOUNTS

Chronicles and newspaper quotes are corrected to appear on the day of the events they describe rather than the date on which they were published. If we particularly liked quotes that were not precisely dated – or easily datable by the events mentioned – but were datable by the season or the seasonal event referred to, then we have occasionally included them in the appropriate place, eg. 'end of March', 'early spring', 'the second cut of hay', or 'Michaelmas Term lately over'.

There is no significance to the order of the quotes on the page: we took each quote on its merits. Because we have quoted verbatim, there are some stylistic inconsistencies. Dates in brackets refer to the year of publication, if a literary work. Dates following the author biographies refer to the year that that particular extract was written.

GEOGRAPHICAL AREA

For our purposes 'British' refers to the present land area of the British Isles (including England, Wales, Scotland, the Orkneys, the Shetland Isles, Isle of Man, Northern Ireland, Ireland, Scilly Isles and the Channel Islands), plus the sea areas defined by the northern section of the Shipping Forecast that *directly abut* onto the British coastline (thus the story of Dunkirk is included, but Trafalgar is not). We have allowed only a handful of exceptions – all stories, to our minds, with sufficient resonance for this rule to be broken.

THE DATING GAME

By the sixteenth century, the Julian Calendar was ten days out of step with the passage of the seasons (365.25 days is slightly longer than the true length of a year). Pope Gregory introduced the Gregorian Calendar, in 1582, to correct this. Initially adopted by Roman Catholic countries only, it was enacted by Parliament in Britain in 1752. Thus, the day after 2 September 1752 was, magically, 14 September – causing riots as people believed their lives had been shortened. The same Act also moved the official start of the year from 25 March to 1 January.

For the book to present a consistent portrait of daily weather, through history, all dates pre-2 September 1752 obviously had to be changed to the 'new style' Gregorian calendar. This causes all kinds of problems. First, as the difference between the calendars grows by three days every four centuries, the further back you go, the smaller the margin of correction required. For example, a date in 1600 is corrected to the new style by adding ten days; a date in 1200, by adding seven days, and so on, back to the third century when they correspond. Secondly, historians and authors are often unclear about which calendar they are using, especially around 1752, making many references suspect. We have done our best to tie down and correct all dates to the Gregorian Calendar, but, inevitably, there will be some mistakes. Because of these date corrections, pre-1752 quotes (Samuel Pepys, for example, or Jonathan Swift) do not appear on the date where they appear in other published forms.

KEY TO 'BACKCAST' WEATHER SYMBOLS

 Mainly sunny, winter.

 Mainly sunny, summer.

 Scattered cloud, dry, winter.

 Scattered cloud, dry, summer.

 Broken cloud, dry, winter.

 Thundery showers, summer.

 Frost.

 Overcast, dry.

 Broken cloud, light rain, winter.

 Broken cloud, light rain, summer.

 Broken cloud, heavy rain, winter.

Broken cloud, heavy rain, summer.

 Overcast, light rain.

Overcast, heavy rain.

Fog

For our purposes, 'dry' means less than 0.2 mm rain; 'light rain' means greater than or equal to 0.2 mm rain, but less than 1 mm rain; 'heavy rain' means greater than or equal to 1 mm rain. In the central England area defined by our data sets, and according to our thresholds, there are, annually, approximately 115 days of 'heavy rain' (31.5%), 45 days of 'light rain' (12.3%) and 205 'dry' days (56.2%). To give the year meteorological shape we have divided it in half, from the spring to the autumn equinox. Weather symbols for the 'winter half' (21 September-20 March) feature a small sun, those for the 'summer half' (21 March-20 September) feature a large sun. The 'frost' symbol represents the ten coldest days of the year (from the Central England Temperature Series) over the same period as our sunshine and rainfall data. We have used 'fog' and 'thundery showers' symbols only where our sunshine/rain thresholds coincide with recognised 'singularities'.

January brings the snow
Makes your feet and fingers glow
February's ice and sleet
Freeze the toes right off your feet
Welcome March with wintry wind
Would thou wer't not so unkind
April brings the sweet spring showers
On and on for hours and hours
Farmers fear unkindly May
Frost by night and hail by day
June just rains and never stops
Thirty days and spoils the crops
In July the sun is hot
Is it shining? No, it's not
August cold, and dank, and wet
Brings more rain than any yet
Bleak September's mist and mud
Is enough to chill the blood
Then October adds a gale
Wind and slush and rain and hail
Dark November brings the fog
Should not do it to a dog
Freezing wet December then:
Bloody January again!

Michael Flanders

JANUARY

1 JANUARY

FIRST 'HILLWALKERS' DIE

New Year's Day brings the start of, statistically, the coldest month (40 per cent of the time) and the windiest (44 per cent of all major windstorms). The sun has little power now, which is why days of bright sunshine are often very cold. Ten days past the shortest day (*21 December*), mornings are still getting darker: this is due to the slightly elliptical orbit of the earth round the sun. A North Sea storm in 1999 reveals a circle of wooden timbers in an ancient saltmarsh on the Norfolk coast (♀): dated to 2050 BC, 'Sea Henge' is heralded as the greatest-ever Bronze Age discovery.

> 'Through a chink in the bedroom curtains my unenthusiastic eye caught an early-morning glimpse of the New Year: it looked battleship grey ... I never liked New Year's Day anyway.'
> *Alec Guinness, actor, 1995*

> 'Cold, & Foggy, & dismal the New Year opens. I pray most earnestly that I may never see another.'
> *Edward Leeves, diarist, 1850; he lives until 1871*

Caught in a 'hellish' blizzard on Cairngorm (♂) today in 1928, two Glasgow students, Thomas Baird and Hugh Barrie, share the unhappy distinction of being the first official British hillwalking deaths. Desperate to get off the highest, bleakest plateau in Britain, they either fall or are buried by an avalanche. Barrie's body is not discovered until spring.

In 1928, hillwalking is still an eccentric pastime. Although climbing clubs have existed since the 1850s, the first rudimentary mountain rescue service, established by the Rucksack Club, is five years away – the Ramblers' Association won't form for another seven years. Our hills are still remote, empty and inaccessible. There are few phone boxes. Walkers have no long-range forecasts, no breathable 'windcheaters', no fibre pile, isotonic drinks or survival bags. Anyone in the hills running into bad weather is on their own, save for the odd rudimentary stone shelter. For the missing, ad hoc search parties of locals are convened from police, shepherds and keepers to recover the bodies.

155.5mm Achnagart 1992

Today's accident sets a grim precedent for New Year's Day in the Highlands. In 1933, at Glen More, an 'elemental storm' prompts the largest-ever pre-Mountain Rescue search – involving more than three hundred – for two walkers, both eventually found dead. In 1959, five walkers die in a blizzard on Glen Doll, the first major civilian tragedy to involve helicopters and RAF Mountain Rescue (*14 March, 30 May*).

15C Colwyn Bay 1922
15C Welshpool 1922
-17.2C Wallingford 1962

2 JANUARY

1.5C min / 6.7C max

IT'S A FROSTY MORNING, SOD WORK

'It's a frosty morning with a red, suspended ball of sun. I think I shall go out for a walk, burbling and muttering to the amazement of all I pass. I have a lot of work to do but sod all that. My sister has a bastard friend staying the weekend and I want to get out of the house. Do you know the kind of morning – cold, with a pale diffused light over everything, with frost on the grass and hedges, and ice in the puddles and cartruts? I went round Tachbrook Malory, and Bishop's Tachbrook, through Moreton Morrell and Wellesbourne Hastings, and home via Barford. The sky was half ice-blue, and half misty and dove-coloured. Occasionally an aeroplane swam across. And the land was so richly brown and green, with occasional flocks of grey and golden sheep; and red brick farms rising up ... The sun flashed blindingly from frozen puddles and there wasn't a breath of wind.' *Philip Larkin, poet, Warwickshire, 1943*

'Cold weather brings out upon the faces of the people the written marks of their habits, vices, passions, and memories, as warmth brings out on a paper a writing in sympathetic ink.' *Thomas Hardy, poet and novelist, 1886*

'The roads are dangerous – the horses soon knock'd up – The outside to a Man who like me has no great Coat, is cold and rheumatismferous – the Inside of a Coach to a man, who like me has very little money, is apt to produce a sickness on the Stomach – Shall I walk? I have a sore throat – and am not well.' *Samuel Taylor Coleridge, in a letter to fellow poet Robert Southey, 1795. Coleridge, still unpublished, is struggling to find his way in life*

'I no longer wonder why these people talk so much of the weather; they live in the most inconstant of all climates, against which it is so difficult to take any effectual precaution, that they have given the matter up in despair, and take no precautions at all. Their great poet, Milton, describes the souls of the condemned as being hurried from fiery into frozen regions: perhaps he took the idea from his own feelings on such a day as this, when, like me, he was scorched on one side and frost-bitten on the other; and, not knowing which of the two torments was the worst, assigned them to the wicked both in turn.' *Robert Southey, writing under the pseudonym of a fictitious foreigner, Don Manuel Alvarez Espriella, Letters from England, 1803*

15.6C Llandudno 1933
-21.1C Corwen 1962
15.6C Wistanstow 1925

3 JANUARY

THE MOST FAMOUS LIGHTHOUSE IN THE WORLD

When a gale drives Henry Winstanley's ship, *Constant*, on to the Eddystone Rock today, in 1695, it's too much (☹). It's the second ship he's lost to the rock in a year and he decides, at his own expense, to erect a lighthouse. And so begins the story of an engineering wonder of the world.

The Eddystone lies 15 miles (24 km) out to sea. At high tide, just one steep-sided, jagged rock is visible, ten paces wide. Building anything in such a place is as ambitious, in the seventeenth century, as putting a man on the moon. Prior to state control, no guiding precedents exist on how a lighthouse should look. Yet, against all odds, in 1698 a dim candle shines out from a sort of nautical gazebo.

> 'An east wind, not knowing, I suppose, who I was, has given me a cold.'
> *Oscar Wilde, poet and dramatist, 1893*

The following year Winstanley doubles its height, to over 100 feet. It's an immediate, triumphant success. For five years no vessel wrecks on the Eddystone. Delighted, Winstanley boasts that his dearest wish is to witness, from his masterpiece, 'the greatest storm there ever was'. Fate brutally complies, and he is there on the night of the Great Storm of 1703 (*7 December*). All that remains – of the tower and Winstanley – are a few twisted lumps of iron. Two days later, the Eddystone claims its next ship.

John Rudyerd's lighthouse, in 1709, dispenses with Winstanley's artistic fripperies. Narrow and smooth, offering the least possible resistance to wind and sea, his tower lasts forty-seven years. Its only drawback is that, clad in timber, it isn't fireproof. When the lanterns ignite the roof in 1756, it burns to a crisp. With the next tower, by John Smeaton, the lighthouse finally finds its form in stone throughout. Smeaton's genius is to cut granite blocks that interlock for impregnable strength. From a colossal base, its tapering form lowers the centre of gravity and provides maximum buttressing – 'Like an oak,' he says. It's completed in 1759, and every future offshore lighthouse will copy it. When finally taken down, 127 years later (to be re-erected, by popular subscription, on Plymouth Hoe), the reason is not its own weakness but that of the fissuring rock on which it stands. Its replacement – the fifth, and present, Eddystone light – is almost twice as tall. If Smeaton's design had a fault, it was that, in bad weather, it was frequently so engulfed in spray that the light became invisible – just when it was needed most.

-20C West Linton 1941 ❄

15.6C Cannington 1948 ☀

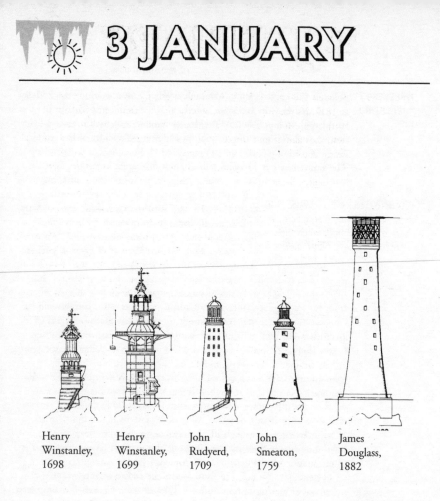

Henry
Winstanley,
1698

Henry
Winstanley,
1699

John
Rudyerd,
1709

John
Smeaton,
1759

James
Douglass,
1882

1.4C min / 6.4C max

WRECKERS /
BLUEBIRD

Between 1700 and 1850, in Cornwall, a gale means a wrecking story. Today, in 1817, it's the brig *Resolution*, approaching the unfinished harbour of Porthleven, Mounts Bay (☒). Deliberately grounding herself to save the crew, she's almost undamaged – until the tide recedes and hundreds of locals swarm aboard, a-plundering: 325 pipes and 25 hogsheads of wine and all private property are snapped up in minutes, the wreckers fighting over the best booty. In two days the vessel is stripped. Even hull timbers are taken.

'A base gloomy day and dispiriting in proportion ... everything gloomy as the back of the chimney when there is no fire in it.' *Walter Scott, author, 1831*

'Such a beautiful day, that one felt quite confused how to make the most of it, and accordingly frittered it away.' *Caroline Fox, diarist, 1848*

Of all the 'bad weather' tradesmen, the wreckers, especially those of Cornwall and the Scillies, are notorious. Ships in trouble are watched greedily from ashore by men bearing sharpened hatchets. In 1753 Cornishman George Borlase, an early weather observer, reports 'great barbaritys' and 'things which shock humanity ... they'll cut a large trading vessel to pieces in one tide and cut down everybody that offers to oppose them.' It's lucrative work. In the 1820s, 131 vessels are lost in two decades between Land's End and Trevose Head alone. Did wreckers really hang out lights in bad weather to lure ships to false harbours and doom? Probably. Certainly the Wolf Rock off the Lizard – so-named because of the wolf-like howl it emits when air is trapped and released by a cavern in high seas – is carefully blocked with stones and silenced.

'Conditions were as perfect as I have seen them,' records Norman Buckley, chief observer for Donald Campbell's world water speed record attempt on Coniston Water (⚲) today in 1967. Arguably, it's this perfection that leads to disaster. Lousy weather messed up trials a few weeks earlier, and the arrival of ideal conditions prompts the impatient Campbell to snatch the window before he is, strictly, ready – he has a bad cold. Is this what affects his decision-making? After a successful first run, for some reason Campbell starts his second without either refuelling or waiting the few minutes for the wake from his first run to die away. Why?

-20C West Linton 1941 ❄

15.6C Llandudno 1921 and 1957 ☀

Despite the glassy water and clear skies, on his second run *Bluebird* somersaults at 320 mph (510 kph), breaking up and killing Campbell. The reason for the crash is thought to be the lighter, unrefuelled boat hitting the remains of his previous wash. The damage a fine day can do ...

5 JANUARY

1.3C min / 6.4C max

AMY JOHNSON / JAMESTOWN

Into thick, freezing fog today, in 1941, Amy Johnson takes off from Blackpool airport, never to be seen again. The mystery of what happens to Britain's most famous aviatrix has never been satisfactorily explained. The ninety-minute flight is a routine delivery of an Airspeed Oxford aeroplane to Kidlington airbase near Oxford – hardly taxing for someone who's made record-breaking solo flights to Australia and Cape Town. Four and a half hours later, the wreckage of her plane is found in the Thames estuary, 100 miles (160 km) off course. Hearsay and conspiracy theories surround the crash. Was she flying a spy out of the country? Was she accidentally shot down by British anti-aircraft guns? Was it an elaborate plan to fake her death? Much the most likely explanation is that, without modern navigational aids, she simply loses her bearings over a landscape buried beneath a blanket of fog. Her body is never found. One grisly theory, from an RAF clerk who says he typed up an official report, explains why: HMS *Haslemere* spots a parachutist, calling out that she is Johnson, in the Thames estuary. In the icy water, she cannot grasp the rope she is thrown. 'Then someone dashed up to the bridge and reversed the ship's engines . . . she was drawn into the propeller and chopped to pieces.' Not a fitting end for one of Britain's greatest heroines – so the incident is officially covered up.

> 'Rose late – dull and drooping – the weather dripping and dense. Snow on the ground, and sirocco above in the sky, like yesterday. Roads up to the horse's belly, so that riding (at least for pleasure) is not very feasible . . . Clock strikes – going out to make love. Somewhat perilous, but not disagreeable.'
> Lord Byron, poet, 1821

Bad weather in the Channel in 1607 forces three America-bound ships that have just set sail from London – *Godspeed, Discovery* and *Susan Constant* – to drop anchor. There they remain, stormbound, for nearly a month. Not that the crew and passengers are deterred by such a trifle. These men, hand-picked for their mettle and self-reliance, captained by the swashbuckling, one-armed adventurer Christopher Newport, have a charter granted to the Virginia Company by James I for land along the mid-Atlantic coast. Four months after the weather breaks and the ships are able to sail out of the Channel they found Jamestown – the first permanent English colony in North America.

Braer oil tanker sinks in a gale, 1993

15C Turnberry 1928

-21.1C Houghall 1941

6 JANUARY

1.3C min / 6.5C max

THE NIGHT OF
THE BIG WIND

Such is the menacing, unremitting drone of the wind that few people in Ireland sleep a wink this fearful night in 1839. Many wonder if 'the gates of heaven are closing'. Generations later, the night remains a milestone by which other events are dated. When the national state pension system is introduced in 1909, people trying to prove their age without documents are simply asked if they remember '*Oiche na Gaoithe Moire*' – 'The Night of the Big Wind.' In the run-up to the storm, the weather is 'unearthly', newspapers report. After a fall of snow, the wind drops completely. An eerie sense of tranquillity settles over the countryside. A massive depression – one of the deepest ever recorded near the British Isles – is roaring across the Atlantic Ocean. The worst storm in five hundred years is about to hit Ireland.

> 'I go out, & spend the afternoon in walking alone over the very desolate Gower cliffs, communing with the cold and the quietness. I call this taking my devils for an airing.'
> *Dylan Thomas, poet, 1934*

The west coast takes the initial impact. Huge quantities of seawater are thrown ashore: in widespread flooding, sea fish are found several miles inland. By midnight, winds up to 150 mph (240 kph) are stripping the midlands – where there are practically no natural windbreaks – of all structures. Because the storm arrives at night, only three hundred are killed – but the damage is unprecedented. In Carlow (♥), the tower of the cathedral is torn off. Many large country houses are de-roofed. Windmills, factories, trees, ancient monuments and barns are blown down. Tombstones are lifted from the ground and found up to a mile away. Tens of thousands are left homeless; many more lose their savings, hidden in the thatched roofs. In spring, birdsong is absent – scarcely any birds survive. In an age before comprehensive relief measures and insurance, the nation is kneecapped. Devout Christians believe it's Judgement Day. But for one man at least, it is simply a colossal storm: the Reverend Romney Robinson, Director of the Armagh Observatory, is inspired to develop his cup-anemometer, a breakthrough in the accurate measurement of wind speed.

-21.1C West Linton 1941

17.2C Rhyl 1916

In 1947 (*4 March*), wintry weather divides the British Isles today. In London skiers practise parallel turns on Hampstead Heath and ice slabs wash on to Southend foreshore in Essex, while in Torquay in Devon 'coatless promenaders' enjoy nearly seven hours of sunshine and primroses are being picked in Pembrokeshire.

7 JANUARY

1.1C min / 6.2C max

LONDON FLOOD / CROSS-CHANNEL BALLOONING

A white Christmas followed by a general thaw, heavy rain, a deep depression moving south over the North Sea, gale-strength north winds and a spring tide: they all combine on the night of 6 January 1928, to bring the catastrophic flood that London's always dreaded. The Thames, in spate, meets the rising sea. Water surges over the banks before dawn this morning, flooding Greenwich, Temple underground station, Battersea and Westminster. The Embankment in front of the Tate Gallery gives way under the strain as fetid, stinking liquid gushes into the ground floor rooms, filling them to the tops of the doors. Turners and Landseers disappear beneath sewage diluted by the North Sea.

'We have had the mildest weather possible. A great part of the vegetable world is deceived and beginning to blossom, not merely foolish young plants without experience, but old plants that have been deceived before by premature springs; and for such, one has no pity. It is as if a Lady were to complain of being seduced and betrayed.'
Sydney Smith, clergyman, 1832

Personal horror stories abound. When 80 feet (25 metres) of embankment burst near Lambeth Bridge, the basements of houses flood so fast that many can't escape. By the time one man is woken by icy water swilling round his bed, the pressure against the door of his children's bedroom is so great he can't open the door. All four drown. Ten others drown in similar incidents. What becomes clear after the tragedy of the 'London Flood' is that the tide is by no means the highest, nor the winds the strongest, of the year. It might have been far worse. Yet nothing is done about the threat until the Thames Barrier is built fifty years later.

In 1784 a light northerly wind assists the attempt of Frenchman Jean-Pierre Blanchard and American doctor John Jeffries (*30 November*) to cross the Channel by hot air balloon. The journey takes two and a half hours and they arrive semi-naked, having discarded most of their clothes to lose weight and stay airborne. A storm in 1839 damages Thomas Telford's masterpiece, the suspension bridge across the Menai Strait to Anglesey – the structure which inspires Samuel Fox's design for the folding umbrella (*6 December*). Meanwhile, in 1982, forty-three hours of continuous snowfall brings an unexpected consequence – the spread of the Ruddy Duck. This invasive North American species, originally escapees from the Slimbridge Wildfowl Trust, Gloucestershire, scatters widely in freezing conditions searching for unfrozen water. Inter-breeding with the European White-headed Duck, they threaten to become the Japanese knotweed of the duck world.

-22.6C Braemar 1982

15C St. Abb's Head 1989

180.4mm Rydal Hall 2005

15C Rhyl 1934

8 JANUARY

1.3C min / 6.4C max

COLDS, FLU
AND THE
FORGOTTEN
WINTER
OF 1982

Colds, flu, respiratory viruses and chronic obstructive pulmonary disease reach an annual peak around today, according to Met. Office research, even though they are in constant circulation all the year round. One explanation is that we mix more during the Christmas holiday, swapping germs liberally under the mistletoe. But the weather contributes too. Cold air makes people more vulnerable to viral infection, as well as making it more difficult to breathe. Also a deficiency in vitamin D (our main source of which is the sun) can weaken the immune system: twenty minutes outside in the sunshine can make all the difference to our health at this time of year.

'A wind from the north, beyond measure violent, caused irreparable damage ... so that the disturbed state of the elements seemed well suited to the state of the human race.'
Matthew Paris, historian,
Chronica Majora, 1241

'Off to rehearse Oberon, which is a good exercise on a chilly morning.'
John Gielgud, actor, 1945

'Intense cold. Mass. Ordered central heating and donned woollen underclothes.'
Evelyn Waugh, author, 1954

'Wales is cut off,' Terry Wogan tells Radio 2 listeners today in 1982. Although rarely cited alongside the 'great' twentieth-century winters of 1947 and 1963, 1982 deserves an honourable mention. The worst blizzard for four decades rages for forty hours and high winds create 20-foot (6-metre) drifts in Wales. Hundreds of cars are buried and roads are impassable. When the M4 is closed, mountain rescue teams are called in to lead people to safety and several hundred drivers spend five nights stranded at a community centre in Bridgend. Plymouth, Okehampton, Tavistock and Torquay are also cut off and in York the River Ouse floods, carrying blocks of ice down the streets. For three weeks, temperatures remain stubbornly below freezing and the snow lies. As food begins to run out in Wales *Operation Snowman*, involving both the regular and territorial armies, sets about clearing the roads with tracked vehicles and delivering essential supplies by helicopter. The most isolated communities are not reached until 19 January.

-26.8C Grantown-on-Spey 1982

In 2005, the police appeal to boat owners in Carlisle to help with the rescue operation when the rivers Eden, Petteril, Caldew and Little Caldew all overtop their banks. Three thousand are evacuated and many more are without electricity. But only days after the Asian tsunami, who cares?

14.9C Aber 1988

102mph gust, Port Talbot 1974

9 JANUARY

1.4C min / 6.5C max

THE BRAVEST MAN WHO EVER LIVED

As a North Sea gale batters the Norfolk coast this morning in 1917 a distress call arrives from the *Pyrin*, just in sight of the coast (↺). The shallow beach at Cromer makes launching the heavy rowing lifeboat *Louisa Heartwell* treacherous in the colossal breakers. With the youngest and strongest of the village gone to the Royal Navy and merchant fleet for the war, the average age of the crew is over fifty – some are nearly seventy. Nevertheless, under the forty-year-old coxswain, Henry Blogg, they launch. Repeatedly it looks as if the boat will be smashed against the pier, or swept back ashore, but eventually, after three hours, she makes it to the *Pyrin* and rescues sixteen.

BIG BEN AND SNOW PLOUGHS STOP
Heavy rain today in 1968 cools so suddenly and dramatically that at 2.58 p.m. Big Ben freezes solid for four hours. The storm goes down in history when three snowploughs themselves get stuck on the Berkshire Downs.

Hardly are the lifeboatmen back in dry clothes when another SOS arrives: the Swedish ship *Fernebo*, 4 miles (6 km) out, has hit a mine and blown in two. Blogg rallies his aged crew. Launching is less straight-forward this time. The mountainous seas drive the lifeboat relentlessly back on shore. Eventually, half of the *Fernebo*'s men are brought to land by human chain. But for the crew trapped on the second stricken half, hope evaporates. It's dark. Rocket-lines launched by searchlight fail. For the third time, Blogg and his crew launch. Five oars smash immediately and three more wash overboard, as the lifeboat is hurled back on to the beach. New oars are brought. Blogg spots that the tide is at a point where an 'outset' (seaward flow) might carry them to the wreck. Watched now by a crowd of five thousand lining the cliffs, the coxswain executes his plan under searchlights. At 1 a.m., after fourteen hours, eleven Greeks and Swedes are plucked to safety.

-25.9C Grantown-on-Spey 1982 ❄

17.2C Aber 1971 ☼

The rescue propels the name of Henry Blogg into the public eye. This shy fisherman is acclaimed, even among the selfless heroes of the Royal National Lifeboat Institution, as 'the greatest of the lifeboatmen'. Like most supremely courageous men, Blogg shuns attention and publicity. He is nevertheless awarded the RNLI's gold medal. He wins it twice more, plus four silver medals – an unprecedented total – along with the George Cross. By the time of his death in 1954, after fifty-three years of service, he has rescued 873 people. The plaque beneath the bronze bust of Cromer's favourite son states simply: 'ONE OF THE BRAVEST MEN WHO EVER LIVED'.

10 JANUARY

1.4C min / 6.7C max

THE WINTER OF DISCONTENT

On a bitingly cold day in 1979, an orange-tanned Jim Callaghan holds an impromptu press conference at Heathrow on his return from the Caribbean. The image of a relaxed, sun-kissed Prime Minister is more than shivering Britain can swallow in the middle of the most miserable winter of the twentieth century. For this is the 'Winter of Discontent'. The freezing weather sets in just before New Year, when snow covers the whole of Great Britain for the first time since 1963. The *average* temperature for January is below freezing. Thick fog, heavy frosts, sleet and more snow alternate until the end of March, punctuated by brief thaws, compounding the anguish with burst pipes and flooding.

As the winter tightens its grip, so does the industrial action. Go-slows and strikes affect industry after industry. The BBC goes off air, oil is undelivered, the criminal justice system is crippled and public transport breaks down. In Northern Ireland, a state of emergency is declared. An entire round of the FA Cup is postponed for the first time in the 108-year history of the competition. With petrol and food shortages, the country descends into an unprecedented state of chaos.

Failed pay talks lead to strikes, strikes and more strikes – by hospital workers, school caretakers, refuse collectors, road gritters and, most famously, grave diggers. As it all spirals out of control, the impact is chronically exacerbated by the weather. By mid-February, accusations of a paralysed government are increasingly hard to refute. Dennis Howell (remember him? Formerly Minister for Drought in the summer of 1976 [*3 July*]) is appointed Minister for Snow. Asked his immediate plans, he replies, 'Well, I've offered my prayers to the Almighty.'

916mb (record low sea-level pressure)

94mph gust, Benbecula 2007

-27.2C Braemar 1982

18.3C Aber 1971

-26.1C Newport 1982

For a Prime Minister usually sensitive to his electorate, today's press conference is a public relations disaster. 'Crisis? What Crisis?' screams the *Sun* next day, and the papers are full of photographs showing 'Sunny Jim' enjoying the swimming pool at his Caribbean hotel. At May's general election, it's curtains for Labour for a generation as Margaret Thatcher is voted in on a wave of anti-union hatred.

11 JANUARY

THE FLANDERS MARE

Fearing a sea voyage in dreadful weather might 'alter her complexion', Anne of Cleves, bride-to-be of Henry VIII, waits for over two weeks in Calais for winds in the Channel to abate. The King has never met Anne, a German noblewoman, but with every stormy day his impatience to set eyes (and, indeed, more) on his fourth wife grows. Henry has agreed to marry Anne on the evidence of ambassadors' reports of her beauty ('she excelleth as the golden sun excelleth the silver moon') and a portrait by Holbein, the court painter. Everything he has seen and heard leads him to expect love at first sight. Alas, at their first meeting at Rochester today in 1540, it is not to be. Perhaps the portrait is too flattering. Perhaps the weather *has* damaged Anne's looks (certainly the journey was treacherous because of the continuing storms). Perhaps the delay has sent the King's imagination – that mistress of self-deception – spiralling out of control. Whatever, Henry finds his bride awkward, unattractive and plump. The betrothed couple exchange twenty words before Henry departs in a huff, later declaring, 'She is no better than a Flanders mare.'

Despite his best efforts, however, the marriage goes ahead. But it is never consummated, and later annulled. 'I have left her as good a maid as I found her,' the King says. It's the beginning of the end for Thomas Cromwell, the Machiavellian Chancellor who suggested the union for politico-religious reasons. Anne is ordered to leave court and live in Richmond Palace where, apparently, 'the climate is much better.' Henry turns his attentions to Catherine Howard.

In 1978, the worst gales and sea floods to strike the East Coast since 1953 (*31 January*) rip down some of the last of our cheerful, seaside Victoriana: 150-year-old piers at Margate, Skegness, Hunstanton and Clacton are all damaged or swept away.

'People went over and along the Thames on ice from London Bridge to Westminster. Some played at the foot-ball as boldly there as if it had been on dry land; and the people, both men and women, went daily on the Thames in greater number than in any street of the city of London.'
Raphael Holinshed, chronicler, 1565

'Greatest snow, & severest weather I ever remember. Only one woman at church.' *Reverend William Cole, antiquary and rector of Bletchley, Buckinghamshire, 1767*

'It is impossible to have health in such desperate weather ... our kingdom is turned to be a *Muscovy*, or worse.' *Jonathan Swift, writer and satirist, 1739, at the start of a memorably cold winter*

16.1C Balmacara 1971

-26.6C Bowhill 1982

166.6mm Honister Pass

BELL ROCK AND
THE LIGHTHOUSE
STEVENSONS

The gale that hurls the 74-gun HMS *York* on to the Bell Rock (✸) today in 1804 claims 600 lives, but establishes one of Scotland's most extraordinary dynasties. Hundreds of vessels have previously foundered on this sandstone reef, Scotland's deadliest shipping hazard (six ships every winter, on average), but this is the Navy, by Jove! There is such a furore in Parliament that the notoriously inert Commissioners of the Northern Lights have to act. Over the centuries, numerous attempts have been made to mark the Bell Rock – it takes its name from the warning bell fixed in the fourteenth century, subsequently removed by wreckers (*4 January*). Following the HMS *York* disaster, a lighthouse is finally built in 1811. But the story of its building – an epic to match that of the Eddystone (*3 January*) – is eclipsed by the story of the man who constructed it.

Yes, the Bell Rock lighthouse, 11 miles (17 km) out to sea on a rock covered by 16 feet (5 metres) of water at high tide, makes the name of Robert Stevenson, Chief Engineer of the Northern Lights. But it is merely the first of ninety-seven manned lighthouses that eight members of the Stevenson family will design and build, between 1790 and 1940. The works of the 'Lighthouse Stevensons', as the family becomes known, dot the 6200 or so miles (10,000 km) of Scottish coastline, gracing windswept headlands with magical names like Muckle Flugga, Dhu Heartach, Cape Wrath and Skerryvore (so remote and bleak, according to Sir Walter Scott, that it makes 'the Bell Rock and Eddystone a joke to it').

Was a single family ever more bound up with the weather? Thomas Stevenson is the first person to measure the force of an ocean wave, with his 'wave dynamometer' (in winter gales, he discovers, the force can be 6000 lb per square foot, or just over 100 kg per square metre). He also invents the Stevenson Screen, the characteristic raised, white-painted, louvred instrument shelter that revolutionizes meteorological data collection. Gradually the Stevensons accumulate, from around the world, their arcane, eclectic expertise: in interlocking granite, argand lamps, mirrored reflectors, giant lenses, coloured glass and clockwork (for turning their lights). As Thomas's one son who takes a different trade – the author Robert Louis Stevenson – later writes: 'Whenever I smell saltwater, I know that I am not far from the works of one of my ancestors.'

-21.8C West Linton

124mph gust, Kilkeel

16.5C Colwyn Bay 1984

ADMIRAL
BEAUFORT'S
WINDSCALE

'Hereafter I shall estimate the force of the wind according to the following scale, as nothing can convey a more uncertain idea of wind and weather than the old expressions of moderate and cloudy, etc etc,' Francis Beaufort notes in his journal today in 1805 aboard HMS *Woolwich*. His idea will change our relationship with the wind for ever.

'Storms, Rain, and all the various Inclemencies of the Sky. The new year finds me in the same Situation the old year left me, a domestic Animal fond of my own House, and loth to quit my Chimney Corner.'
The Earl of Orrery, in a letter to Jonathan Swift, 1739

'Rainy weather. Does weather matter in a journal? Lunched alone; does that matter? (Grilled turbot and apple-pudding if you want the full details.)'
Siegfried Sassoon, anti-war poet, 1921

The 'Wind Force Scale' that still bears his name is one of the most important developments in maritime history and in the sciences of meteorology and hydrography. Without a scale, wind is a subjective thing: a gust that blows a pensioner off his feet may delight a boy flying a kite. The inspiration for his scale to standardize degrees of wind is not, however, entirely Beaufort's own. He borrows ideas accumulated by sailors and engineers over generations, but refines them into a qualitative scale that measures wind force on a ship's sails. Crucially, he gets his scale accepted, thirty years later, by the Royal Navy.

The brilliance of Beaufort's thirteen-point scale (later modified to twelve – see *Appendix*) lies partly in his poetically evocative descriptions, partly in the adaptability of the scale – proven over two centuries. The terms (originally devised with a full-rigged man-o'-war in mind, the principal ship of the Navy) have been revised for steamships and use on dry land, and updated on several occasions. Thus Beaufort's original Force 10 ('Whole Gale: Or that with which she could scarcely bear close-reefed maintopsail & reefed foresail') has become the more practical, if prosaic, 'Very high waves. The sea surface is white and there is considerable tumbling. Visibility is reduced.'

-24.6C Carnwath 1979

On this day in 1205 'Began a frost which continued till the two and twentieth day of March, so that the ground could not be tilled; whereof it came to pass that, in summer following, a quarter of wheat was sold for a mark of silver,' *Stowe's Chronicle* records. It is thought that this great and fatal frost gives rise to the medieval belief that 13 January, St Hilary's Day, is the coldest of the year.

15.6C Torquay 1915

-7C Frost destroys
Abbey Garden, Tresco 1987

14 JANUARY

WRITER IN A STORM

A storm which envelops a coal ship off the Cornish coast today in 1882 would not be exceptional were the second mate not later to become a writer and record the experience in an autobiographical short story (*21 September*). Few masters of the English language witness such events. The story is *Youth* and the writer is Joseph Conrad:

'The world was nothing but an immensity of great foaming waves rushing at us, under a sky low enough to touch with the hand and dirty like a smoked ceiling. In the stormy space surrounding us there was as much flying spray as air. Day after day and night after night there was nothing round the ship but the howl of the wind, the tumult of the sea, the noise of water pouring over her deck. There was no rest for her and no rest for us. She tossed, she pitched, she stood on her head, she sat on her tail, she rolled, she groaned, and we had to hold on while on deck and cling to our bunks when below, in a constant effort of body and worry of mind . . .

And there was no break . . . The sea was white like a . . . caldron of boiling milk; there was not a break in the clouds, no – not the size of a man's hand – no, not for so much as ten seconds . . . We pumped watch and watch, for dear life; and it seemed to last for months, for years, for all eternity, as though we had been dead and gone to a hell for sailors. We forgot the day of the week, the name of the month, what year it was, and whether we had ever been ashore. The sails blew away . . . the ocean poured over her, and we did not care. We turned those handles, and had the eyes of idiots. As soon as we had crawled on deck I used to take a round turn with a rope about the men, the pumps, and the mainmast, and we turned, we turned incessantly, with the water to our waists, to our necks, over our heads. It was all one. We had forgotten how it felt to be dry.

And there was somewhere in me the thought: By Jove! this is the deuce of an adventure – something you read about; and it is my first voyage as second mate – and I am only twenty – and here I am lasting it out as well as any of these men, and keeping my chaps up to the mark. I was pleased. I would not have given up the experience for worlds.'

-23.5C Lagganlia 1979

15.6C Rhyl 1930
Aberystwyth flood 1938

15 JANUARY

WINTER GARDENING

In 1867, during a very frosty winter, thousands are daily lacing up their steel skates (*11 December*) across the British Isles. In fact, there is something of a skating craze – until tragedy strikes today. The ice on the boating lake in Regent's Park, London, cracks under the weight of skaters and forty drown. The depth of the lake is subsequently reduced to 4 feet (a little over a metre). The Clyde valley takes the brunt of a short but very sharp storm (wind speeds reach 100 mph) in 1968 – chimneys litter the streets of the tenements in Glasgow and two thousand are left homeless.

'The weather has shrunk me into a mere librarian and I do little but make patrol activity in my small world.' *Philip Larkin, poet, recently appointed to his first job as a librarian in Wellington, Shropshire, 1946*

'I have been sitting in the Sun whilst I write this till it became quite oppressive, this is very odd for January – The vulcan fire is the true natural heat for Winter: the sun has nothing to do in winter but to give a "little glooming light much like a shade'". *John Keats, poet, letter to Georgiana Keats, 1820*

'We are now in the depths of winter ... my first winter at the cottage ... and the first winter when I went mad.

The average gardener, in the cold dark days of December and January, sits by his fire, turning over the pages of seed catalogues, wondering what he shall sow for the spring. If he goes out in his garden at all it is only for the sake of exercise. He puts on a coat, stamps up and down the frozen paths, hardly deigns to glance at the black empty beds, turns in again. Perhaps, before returning to his fireside, he may go and look into a dark cupboard to see if the hyacinths, in fibre, are beginning to sprout. But that represents the sum total of his activity.

I wrote above that, on this first winter, I went mad. For I suddenly said to myself, "I WILL HAVE FLOWERS IN MY GARDEN IN WINTER" ...

I wonder why. And yet, perhaps I know. For this passion for winter flowers has its roots deep, deep within me. I have a horror of endings, of farewells, of every sort of death ... I believe that my love for winter flowers has its secret in this neurosis.' *Beverley Nichols, writer, 1932*

-20.6C Dalwhinnie 1955

Highest equal mean hourly wind speed,
99mph Great Dun Fell 1968

16C Colwyn Bay 1990

1.3C min / 6.1C max

REBECCA / TILL FLOOD / OPERATION SNOWDROP

In a white wall of fog today in 1930 the *Romanie*, a merchant ship from Antwerp, is wrecked in Polridmouth Cove, Cornwall (⊕). The event would be no more than a footnote in the log of the Fowey lifeboat – the sea is calm, no one drowns – were it not for the fact that one of Cornwall's literary giants is on the clifftop path, watching. Seven years later, Daphne du Maurier re-creates the shipwreck in her most famous novel, *Rebecca*. The day after the masked ball, hosted by Maxim and his second wife at Manderley, the fog – 'stifling, like a blanket, like an anaesthetic' – is 'rent in two by an explosion' when a ship goes ashore on the reef in the cove below the house. The sea is 'so calm that when it broke upon the shingle in the cove it was like a whisper'. The crew are rescued, but when a diver is sent down to check damage to the hull he discovers a small boat on the seabed containing a body – the beautiful Rebecca, the woman whose beauty haunts the book . . . we won't say any more, in case you haven't read it. From the coastal footpath above Polridmouth Cove – du Maurier's favourite view in Cornwall – you can still see the remains of the *Romanie*, at very low tide.

> 'Read – wrote – fired pistols – returned – wrote – visited – heard music – talked nonsense – and went home . . . The weather is still muggy as a London May – mist, mizzle, the air replete with Scotticisms which though fine in the descriptions of Ossian, are somewhat tiresome in real, prosaic perspective.'
>
> *Lord Byron, poet, 1820*

Today in 1841 a fast thaw, exacerbated by heavy rain, fills the River Till in Wiltshire. It pours over its banks, sweeping away seventy-two houses in the village of Shrewton (☔), no mean achievement for an effete English chalk stream.

Heavy snowfalls in 1955 – the coldest winter between 1947 and 1963 – almost halt construction work on the new Dounreay nuclear power station in the far north of Scotland (☔). Caithness becomes a white island unto itself. Large areas are without power for weeks, isolated by 30-foot (10-metre) drifts and icy gales, triggering the first ever coordinated airborne relief operation, nattily christened Snowdrop. Navy helicopters from Wick deliver food and medical supplies to the remotest farms, while RAF Kinloss despatches the latest high-tech planes – Lockheed Neptunes, equipped for Russian submarine surveillance in the North Atlantic – to deliver hay.

-22.2C Kelso 1881 ❄

15.6C Cullompton 1920 ☀

17 JANUARY

1.1C min / 6C max

ARTIFICIAL FOOTBALL PITCHES

A cold snap today in 1987 causes mayhem in the Football League calendar. Only ten matches in England, Wales and Scotland go ahead, and the inevitable debate follows about whether football is, in fact, a game better suited to the summer. Four League clubs see no reason to enter into the argument, however: they have artificial pitches. Artificial pitches are invented in the 1960s, but the first British club to dig up the hallowed turf and roll out the 'sporti-felt' carpet is Queen's Park Rangers, in 1981. Luton Town, Oldham Athletic and Preston follow, in the mid-1980s. The advantages are obvious: games are never cancelled; grass pitches have to be relaid and cost a packet to maintain. Unfortunately, the future has arrived a little too soon for these clubs. The weird bounce, carpet burns and increased risk of a condition called 'turf toe' mean the character of the contest changes. A degree of edge is lost. 'It's football, but not as we know it,' fans complain. The players don't like the pitches either and the clubs are a laughing stock. The FA bans artificial pitches in April 1988.

> 'Snow means such special things to me. It means a fat soft plop, plop, as it is shovelled off the roofs and falls into the courtyard below. It means the strange melancholy halloo by which the deer are called to be fed ... It means these things in an intimate way, like the ticking of the clock in one's own room means something; and is part of one'.
> *Vita Sackville-West, writer, Knole, Kent, 1926*

The problems of creating and keeping a playing surface that remains in good condition throughout a British winter refuse to go away, though. The reason the great brown cow fields of the 1970s have finally disappeared is not that the climate has improved. It is because of undersoil heating, improved drainage and remarkable advances in turf technology. Most leading soccer clubs spend a fortune on their pitches, using either scientifically modified grass or an engineered artificial fibre mixed in with natural grass. The pitches, and consequently the way the game is played (*25 November*), have changed dramatically because of this quiet revolution. Perhaps the best example is Arsenal: having nearly thrown away the 1989 championship because of the famous morass at Highbury, the Gunners re-lay their pitch with excellent drainage and modified grass. On this new surface, Arsenal quickly establish a reputation for beautiful football. Major disruption to the football calendar is now a thing of the past. When Saturday comes, the show goes on, whatever the weather (to mix three metaphors).

238.4mm Loch Sloy 1974

-26.7C Kelso 1881

17.2C Colwyn Bay 1982

18 JANUARY

0.9C min / 6.4C max

EVEREST TRAINING AND COMFORT SHOPPING

On an 'exceptionally sunny day' in 1953, the British Mount Everest Expedition team assembles in North Wales for 'oxygen trials'. They torture themselves hiking up Snowdon (♪) on this unusually hot morning, testing the cumbersome 'closed-circuit' oxygen apparatus with close-fitting masks. The equipment is designed for high-altitude, sub-zero environments. In the exact opposite conditions, Colonel John Hunt – later Lord Hunt – is 'purple and perspiring' after ten minutes: not exactly ideal preparation for an assault on the world's highest mountain, he notes.

TOP TEN COLD WEATHER COMFORT FOODS

Spaghetti Bolognese (cooked twice a week by 6.1 million)
Heinz tomato soup
Shepherd's pie
Toast and Marmite
Roast chicken
Chicken Tikka Masala
Apple crumble
Bangers and mash
Beef stew
Roast lamb

Today in 2007 the fiercest winds for twenty years leave a trail of destruction across southern Britain. People count the cost of uprooted trees, overturned lorries, cancelled flights and damaged chimney pots, but arguably the greatest economic impact is on the high street. Shopper numbers, according to the Retail Footfall Index, are down 17.2 per cent on today in 2006, highlighting just how direct the relationship between weather and consumer spending is. Not surprising, perhaps – you have to be dedicated to shop in gales or heatwaves. The wider impact of weather on consumer trends is bizarre, though. In hot weather, sales of leg wax, shower gel and lager soar; during cold snaps, bags of cat litter (used to grit driveways), lip balm and 'comfort foods' regularly sell out. Fizzy drinks sell well up to about 23C, then bottled water takes over. Sherry is a popular purchase in the cold; white wine and salads in the heat. In fact, it is claimed that 75 per cent of goods traded in the shops are weather-sensitive. To get pre-season planning and stock management right, retailers want weather that is 'average' for the season.

16.1C Gordon Castle 1932

-24.4C Kelso 1881 ❄

99mph gust, Isle of Wight 2007 🎐

In 1881 the Great Victorian Blizzard, one of the worst snowstorms on record, brings the south of England to a standstill. Sixteen-foot (5-metre) drifts are reported in London where the snow arrives on the wings of a violent easterly gale. Road and rail transport is interrupted for a week, a hundred die and the event freezes itself deep into the national consciousness. Fifty years later T. H. White, author of *The Once and Future King*, plays darts in a Wiltshire pub where he learns that the score 81 is called a 'Snowstorm'.

19 JANUARY

THE GREAT FROST FAIR 'I went crosse the Thames on the ice, now become so thick as to beare not onely streetes of booths, in which they roasted meate, and had divers shops of wares, quite acrosse as in a towne, but coaches, carts, and horses, passed over.' So records the diarist John Evelyn of the river's transformation into an icy shopping mall today in 1684.

'We went upon the top of London Bridge, from whence we viewed with a pleasing horror the rude and terrible appearance of the river, partly froze up, partly covered with enormous shoals of floating ice which often crashed against each other.'
James Boswell, diarist, 1763

'The cold increases, the snow is getting deep, and I hear the Thames has frozen over very nearly, which has not happened since 1814.' *Queen Victoria, 1838*

During this, the coldest period of the so-called Little Ice Age (*3 March*), the Thames freezes several feet thick a number of times. This is partly because compared to today the river is wider – there are no embankments yet – shallower, and practically dammed by the closely spaced stone piers of Old London Bridge, making it easier for ice to form. But it's also significantly colder. The first recorded 'Frost Fair', when Londoners set up stalls on the ice, is in 1608 (*24 February*). The river freezes on at least twenty more occasions before the last fair, in 1814 (*4 February*). But for sheer splendour and ribaldry, no event surpasses the Great Frost Fair of 1684. It continues until early February. The Merry Monarch, Charles II, and his court spend a night on the ice in late January. Printers sell souvenir cards to tourists. Oxen are roasted at makeshift eateries ('Walk in, kind sir, this booth is the chief / We'll entertain you with a slice of beef') amidst horse races, a fox hunt and wrestling matches. 'There was likewise, Bull-baiting, Horse and Coach races, Pupet-plays and interludes, Cookes and Tipling, and lewder places,' continues Evelyn. 'So as it seemed to be a bacchanalia, Triumph or Carnoval on the Water.'

In snow, mist and squalls today in 1915, Martha Taylor, a pensioner, and Sam Smith, a shoemaker, become victims of the world's first strategic bombing mission when a German Zeppelin attack loses its way. Kapitänleutnant Hans Fritz, commander of the dirigible from Deutschland, has orders to attack dockyards on the River Humber but the Zeppelin is blown miles off course and bombs Great Yarmouth (⚑). It's bungling and amateurish. But this military mission is also epoch-making. The civilian deaths herald a new era of twentieth-century fighting: Total War.

16.1C Hawarden Bridge 1930

-18.9C Elmstone 1966

MURPHY'S ALMANAC When the temperature sinks to -26C at Walton-on-Thames today in 1838, one man's fortune is made. Patrick Murphy Esq., MNS ('Member of No Society') of Cork, is a scientist and writer of little or no repute – the author of assorted books on gravity, magnetism and the solar system. Murphy's fleeting fame and fortune follow the publication of his *Weather Almanac (on Scientific Principles, showing the State of the Weather for Every Day of the Year)*. In his *magnum opus*, he uses his theories – relating the weather to the phases of the moon, other astronomical criteria and even animal behaviour – to give a forecast for every day of 1838. An analysis of his predictions reveals that he is entirely wrong on 197 days and partly wrong on nearly all of the rest.

On just one day, however, today, 20 January, the coldest day in living memory, he is right. His prediction says simply: 'Fair. Prob. lowest deg. of Winter temp.' The book becomes an immediate bestseller. The offices of his publishers are over-run by people desperate for a copy (the event is dubbed 'Al-maniac Day'), and the Almanac runs to forty-five editions. Murphy becomes a sensation – a weather prophet for his time – and rich. The winter of 1838 is still known as 'Murphy's Winter'.

'Mr Hool's man says that he caught this day in a lane near Hackwood-park many rooks, which attempting to fly fell from the trees with their wings frozen together by the sleet, that froze as it fell'. *Gilbert White, naturalist and writer, Selborne, Hampshire, 1775*

'"How is the King?" is our first question. "The 11.45 bulletin was bad. It said that His Majesty's life was moving peacefully to its close." How strange! That little hotel at Dingwall, the journalists, the heated room, beer, whisky, tobacco, and the snow whirling over the Highlands outside. And the passing of an epoch'. *Harold Nicolson, diplomat and writer, 1936. King George V died at 11.55 p.m.*

Published annually, almanacs traditionally included information on astrology, agriculture, politics and meteorology. Before the science of forecasting arrives (*1 August*), many people turn to almanacs for daily weather predictions based on the movement of stars and planets. By the 1800s, almanacs are big business, outselling everything but the Bible. Sales of *Old Moore's Almanac*, first published in 1697, peak in 1839 with 560,000 copies. Were *Moore's* weather predictions ever right? Occasionally. Or as the verse goes: 'If you would make a Weather Almanac, attend to what I say / Throw frost, wind, rain and snow into a sack, and draw out one per day.'

-23.6C Grantown-on-Spey 1984

Highest equal hourly mean wind speed, 99mph Lowther Hill 1963

15C Stratford Upon Avon 1954

21 JANUARY

1.3C min / 6.4C max

SNOWMELT IN THE LAKES / ST MAURY'S WIND

'What a fantastic transformation within the space of a day! Yesterday morning . . . a white world, the snow inches deep . . . today . . . the worst floods,' records Lakeland diarist Harry Griffin of today's dramatic thaw in 1960. 'The river, one of the fastest-flowing in England, they say, is racing through our little town this afternoon in a relentless brown torrent, bearing with it great trees, henhouse roofs and oil drums . . . People in the riverside houses, anxiously watching the rising waters all day, are now taking up the carpets. Some of them remember the last time, when the kitchen table floated underneath the ceiling. Out of town, the roads are flooded right across . . . and acres of fields are under water. Trees, fences and telegraph poles are growing out of the water, and the sheep and cattle are huddled together, rather sorrowful-looking, on tiny islands.'

> 'I have taken up with a highbrow gentleman who has rubber soles on his boots. Every ten yards or so he falls flat on his back and it is very enjoyable as he never interrupts his conversation and goes on talking about education and culture and psychology from the road.'
> *Evelyn Waugh, author, to his wife, from a snowbound Marines training villa, Kent, 1940*

The phrase 'the greatest storm there has ever been' is much bandied by the medieval chroniclers, but this evening's gale in 1362 seems a genuine contender. 'St Maury's Wind' (after the saint's day) blows from the south and west, 'from evensong til mydnyte' with such strength that it 'threw down high houses, tall buildings, towers, trees, and other strong and durable things', records John of Reading. Downing bell towers as far apart as London and Norwich, it's taken by many as the 'scourging of God'. Not that the devout are spared. 'A certain churchman . . . clothed day and night in sackcloth, . . . abstemious, mortifying himself continually in prayer, whose name was Ralph' seeks shelter in the church of St Pancras. 'A great beam was thrown by the fury of the wind upon an image of the blessed Virgin, and the said cleric, being prostrated before the image, was killed.' The damage in the British Isles is trifling, however, alongside the catastrophe in Europe, where floods drown one hundred thousand – half the populations of southern Denmark and north-west Germany. The event becomes known as 'Grote Mandrenke' – the 'Great Drowning'.

And in 1876 snow choking a signal at Abbots Ripton, near Huntingdon (♥), prevents the arm mechanism moving from 'LINE CLEAR' to 'DANGER'. Following a disastrous triple collision, all signal arms are redesigned to hinge some distance from the post.

-23.3C Rhayder 1940 ❄

15C Torquay 1969 ☼

22 JANUARY

TRENCH COATS / ENGLAND'S ATLANTIS

The first snowfall of winter sweeps across England today in 1915. In the capital 'the streets are choked with slush', *The Times* reports, and it's 'impossible for a pedestrian to keep his feet dry'. The same weather that so inconveniences fashionable Londoners is making life hell in Flanders, though. During this, the wettest winter on record, the British Expeditionary Force are engaged in digging the 'trench system' that will establish the modus operandi of the First World War. A little respite comes, for the officers at least, in the form of a coat made from a recently invented fabric called gabardine. Thomas Burberry's patented process involves waterproofing the yarn before weaving. The weatherproof, durable and breathable cloth is, on commission from the War Office, cut as a military-style raincoat with epaulettes, buckled cuff straps, D-rings, a storm flap and storm pockets. The advantages of the 'Burberry Safeguard, The Trench Warm Service Weatherproof Coat', are recognized early on in this filthy winter. And the trenchcoat, a style icon of the twentieth century, is born.

> 'When so strange a Clowd of darknesse came over, & especially the Citty of London, that they were faine to give-over the publique service for some time, being about 11 in the forenoone, which affrited many, who consider'd not the cause, (it being a greate Snow, & very sharp weather), which was an huge cloud of Snow, supposed to be frozen together, & descending lower than ordinary, the Eastern wind, driving it forwards.' *John Evelyn, diarist, on 'Black Sunday', when it grew so dark in London that ministers could not read their notes in their pulpits, 1680*

A great North Sea storm hammering England today in 1328 gives rise to one of the legends of the Suffolk coast – the fifty drowned churches of Dunwich (♦). The storm alters the shape of the shingle bank just off the coast, blocking Dunwich harbour and, at a stroke, removing the key to the medieval port's prosperity. There have never been anything like fifty churches in Dunwich, although it is an important town, with a Royal Charter from 1199, three monastic establishments, an eighty-strong fleet, sea defences and two MPs by 1279. But the myth of England's Atlantis, and the church bells which allegedly ring from beneath the North Sea in a gale, has accrued over centuries, just as the cliffs of sand and gravel, the homes, warehouses and churches of so much of the Suffolk coastline, have slowly been eroded, storm by storm.

15C Llandudno 1960

-21.1C Rhayader 1940

Looting frenzy follows grounding of container ship MSC *Napoli*, Branscombe, 2007

23 JANUARY

1.2C min / 6.1C max

THE GONIAL BLAST The ferocious snowstorm that tears up southern Scotland today in 1794 becomes known as the Gonial Blast. *Gonial* or *goniel* is an old Scots word for a sheep found dead and partly decayed. And that's what this storm does. James Hogg, poet, novelist and shepherd in the Borders, later writes: 'When the flood after the storm subsided, there was found on that place [the mouth of the river Esk (🔴)] and shores adjacent, one thousand eight hundred and forty sheep, nine black cattle, three horses, two men, one woman, forty-five dogs and one hundred and eighty hares, beside a number of meaner animals.'

'Heavy snow, very cold. Another rail strike by ASLEF. *Tony Benn, politician and diarist, 1979 (10 January)*

The loss of so much livestock causes great hardship. But this is merely the curtain-raiser. The winter of 1794 continues with poor weather and periodic storms. In the autumn, rains and frost as early as 15 September kill the potatoes and ruin the harvest. By the following winter the rural poor are desperate, but again the weather is exceptionally severe: snow lies in Sussex for three months. In 1795, as the French are defeating the allies on every front in the Napoleonic Wars, the food shortage becomes a crisis.

Commodities reach the highest prices in living memory and these, combined with the shortage of bread as well as various political and military grievances, lead to anarchy. Hundreds of 'bread riots' break out. Crowds of peasants either force vendors to lower prices or simply seize food shipments. The Riot Act is often read. While, in hindsight, it is easy to see the link between the weather, poor harvests and the price of bread, many contemporaries believe their empty stomachs are the result of human wickedness. The chain from seed to loaf includes farmers, corn dealers, millers and bakers – and everyone is suspected of price manipulation and fixing monopolies. Blaming the weather is not an option.

In 1963, a young band called the Beatles have a song in the charts and today they are learning about the glamour of show business. After recording in London for the BBC's *Saturday Club*, they drive back to Liverpool, in freezing temperatures, in a van with no windscreen. Huddled in the back, wearing paper bags on their heads, they slug whisky.

-20.6C Stanstead Abbotts 1963
15C Cardiff 1938

24 JANUARY

1.2C min / 6.2C max

WORST DAY OF THE YEAR

Today is, officially, 'Gloomsday'. According to Dr Cliff Arnall, a Cardiff University psychologist specializing in seasonal disorders, today is when winter misery troughs; when the combination of grim weather, debt, fading Christmas memories, broken New Year resolutions, drained self-esteem and a general lack of motivation conspire to produce a vortex of wintry despair. Dr Arnall's 'January Blues Day Formula', drawn up in 2005, attempts to take into account the feelings that most reliably deliver this winter misery. The formula is originally devised to help a travel company analyse when and why people book holidays. Research indicates that when people reach the slough of despond, when they have the least to look forward to, that's the moment they are most likely to buy a ticket to paradise – or, at least, Tenerife.

'The car taking me to Moorfields wriggled its way through tiny, twisted City streets which were almost deserted; a few thin clerks with blue noses hunched themselves against the bitter wind, walking stiffly and alone, like the black matchstick figures in a Lowry industrial townscape. The women to be seen were, for the most part, dressed as Paddington Bear ... The car slid past St Paul's Cathedral which somehow looked smaller than usual and rather drab'.
Alec Guinness, actor, 1996

JANUARY BLUES DAY FORMULA

$[W + (D-d)] \times TQ\, M \times NA$

Where: W = Weather

D = Debt

d = Money due in January pay

T = Time since Christmas

Q = Time since failure to quit a bad habit (usually smoking or drinking)

M = General motivation

NA = Need to take action to do something about it

And if the credit card's already cut up? Well, take heart: the days are getting longer, the mornings are now getting lighter, and the chances of clear winter days are supposedly higher now than in early January.

What other news to get you through this gloomy day? In 1907, Arctic weather grips Europe, with temperatures down to -34C; in 1915, 30 million people miss the total eclipse of the sun because of cloudy skies; in 1957, during a hailstorm over Northwood in Middlesex a hailstone 5½ x 4 ins (14 x 10 cm) weighing up to half a kilo (almost 1lb), drops in front of a greengrocer's shop – it's possibly wing ice falling off a Heathrow-bound aircraft. So cheer up ... it could be worse.

-20C Lauder 1881

14.4C Bramham 1962

14.4C Wisley 1937

25 JANUARY

1.1C min / 6.3C max

WOMEN AND CHILDREN FIRST / SCARILY MILD 1662

Off the island of Anglesey (✠), a gale today in 1794 that wrecks the *St Patrick*, a packet sailing from Liverpool for Ireland, raises questions about lifesaving drill at sea. As is routine, the *St Patrick* carries only enough boats to accommodate the crew, and while they are busy saving themselves all twenty-eight passengers drown – a not uncommon situation. Astonishing as it may seem, the novel idea of carrying enough lifeboats for everybody doesn't strike anyone until after the *Titanic* disaster in 1912 (*14 April*). And only when HMS *Birkenhead* goes down off Cape Town in February 1852 does Lieutenant-Colonel Seton of the 74th Royal Highland Fusiliers establish – at sword-point – the noble precedent of 'Women and Children First'. As Kipling later writes: 'To stand and be still, to the Birken'ead Drill, is a damn tough bullet to chew.'

> 'A fine frosty day, everything brisk and cheerful ... I don't know when I've enjoyed a birthday so much – not since I was a child anyhow.'
> *Virginia Woolf, author, aged thirty-three, 1915*

In the light of recent climate change fears, it's instructive to hear the diarists Samuel Pepys and John Evelyn noting the holding of a general fast today in 1662 'to avert God's heavy judgement ... there having falln so greate raine without any frost or seasonable cold ... it being neere as warme as at Midsomer' (Evelyn); 'It is, both as to warmth and every other thing, just as if it were the middle of May or June, which doth threaten a plague (as all men think)' (Pepys). Throughout the Middle Ages, today, St Paul's Day, is of paramount importance as a 'prognosticator' of the character of the coming year. Whether the year is to be prosperous, expensive or wartorn, all depends on today's weather.

-21.7C Grantown-on-Spey 1958 ❄

13.9C Nantmor 1997 ☀

In 1990, the Burns' Day Storm (today is the poet's birthday) is the worst in recent history. In sustained winds of 75mph (90 kph), equivalent to a weak Category 1 hurricane, up to a hundred people die, half a million homes lose their power and 3 million trees are downed. Compared to its lack of warnings before the '87 'hurricane', however, the Met Office is praised for the accuracy of its forecasts. Not that these reduce casualties – mainly due to collapsing buildings or falling debris – as the storm hits during the daytime. *'Allo 'Allo!* actor Gordon Kaye receives severe head injuries when part of a billboard blows through the windscreen of his car.

26 JANUARY

1.1C min / 6.3C max

THE POOLS PANEL
A month after the great freeze of 1962–3 begins (*6 February*), with 322 Football League and Cup fixtures postponed and the pools companies haemorrhaging profits, a panel of experts gathers today, like a papal conclave, in a secret London hotel room. Their mission: to predict the results of all the postponed matches in the League programme, so coupons are not declared void. The Pools Results Panel – fantasy football in its earliest form – is born.

> 'Sitting in a bus in London last week, it being a raw day I took out of my pocket my white lip salve and applied it to my chapped lips. An elderly woman sitting opposite put on a strongly disapproving face, and said, "Well!" in a long-drawn out tone. I paid not the slightest notice.'
> *James Lees-Milne, writer, 1977*

Lord Brabazon of Tara, a former government minister, is chairman. Tom Finney, Ted Drake, Tommy Lawton and George Young, all ex-players, and Arthur Ellis, a retired referee, comprise the first panel. Scrutinizing current form, previous meetings between the teams, injuries, tactics, set-piece plays and the disciplinary record of the referee, the panel forecasts 23 home wins, 8 away wins and 7 draws. This is new territory for everyone associated with the 'beautiful game'. With the devised results to hand, several newspapers delight in publishing match reports for games that never took place. One player, it is reported, hears the forecast result for his team and asks the manager for his win bonus. He is told he was dropped.

The Panel is criticized in the media, but survives. In fact it still sits today, even though betting on the pools has so declined that hardly anyone sits around waiting for the magical eight score draws on a Saturday evening. The 1963 season is eventually extended by a fortnight and the FA Cup Final (the third round alone takes 66 days, with 261 postponements) is scheduled three weeks late. The government becomes so concerned about disruption to the sporting calendar that the Met Office starts issuing thirty-day forecasts.

18.3C Aboyne 2003

-21.1C Haydon Bridge 1881

Back in 1884, an exceptionally stormy week ends today with the lowest unchallenged pressure reading ever recorded in the British Isles – 925.6 mbar – at Ochtertyre, near Crieff (☞). A violent gale ensues, blowing down a million trees on one Scottish estate alone. And in 1912, a severe frost lasting a fortnight begins today; *The Times* says, 'According to the map this is Hampshire, but the weather swears a mighty oath it is Caithness.'

27 JANUARY

0.9C min / 6.4C max

GILBERT
WHITE'S
'REMARKABLE
FROST'

A 'singular and striking . . . remarkable frost' in 1776 peaks in intensity today. One of the coldest Januarys since records begin (in 1659), the month is almost as cold as the record-holding 1963 (*6 February*). One of the most vivid accounts of it is by the amateur naturalist the Reverend Gilbert White, in *The Natural History of Selborne* (🖤):

JANUARY
First week: 'Drowned with vast rains.'
7: 'Snow driving all the day . . . followed by frost, sleet and some snow . . .'
12: 'A prodigious mass [of snow] . . . drifting over the tops of the gates and filling the hollow lanes.'
14: 'Cocks and hens are so dazzled and confounded by the glare of snow that they would soon perish without assistance. The hares also lay . . . and would not move compelled by hunger, being conscious poor animals that the drifts and heaps treacherously betray their footsteps, and prove fatal to numbers of them.'
20: 'The sun shone out for the first time since the frost began.'
21: 'The birds in a very pitiable and starving condition . . . crows watched horses as they passed, and greedily devoured what dropped from them.'
22: 'To London, through a sort of Laplandian scene . . . The metropolis being bedded deep in snow, the pavement of the streets could not be touched by the wheels or the horses' feet, so the carriages ran about without the least noise.'
27: 'Much snow . . . in the evening the frost became very intense . . . rime on the trees and on the tube of the glass, the quicksilver sank exactly to zero [-18C] . . . During these four nights the cold was so penetrating that it occasioned ice in warm chambers and under beds.'

FEBRUARY
1: 'Behold, without any apparent cause . . . a thaw took place, and some rain followed . . . frosts often go off as it were at once, without any gradual declension of cold.'
2: 'Thaw persisted.'
3: 'Swarms of little insects were frisking and sporting in the courtyard at South Lambeth, as if they had felt no frost.'

-23.4C Lagganlia 1985 ❄

112mph gust
Quilty 1920

18.3C Aber 1958 ☀

At the other end of the thermometer, today in 1958 logs 18.3C at Aber on the north coast of Wales. Just eight January days in the twentieth century reach or exceed 17C and astoundingly, seven occur here – the 'Föhn effect' at work (*12 March*).

28 JANUARY

1940 ICE
STORM /
FALKIRK 1746

'Everything glass glazed. Each blade is coated, has a rim of pure glass. Walking is like treading on stubble. The stiles and gates have a shiny, green varnish of ice.' So writes Virginia Woolf today in 1940, as wartime Britain is stunned by one of the most extreme weather events of the twentieth century. When a warm Atlantic front confronts continental high pressure over England, rain falls on a white, concrete-hard landscape engulfed by freezing air. On impact, the rain turns instantly to ice: plants turn to glass rods, machines become ice sculptures, trees are split in two, wild ponies on the mountains in Wales are entombed in ice. In Kent, birds die in flight when their wings lock solid. Roads are like skating rinks, railway points cannot be changed, thousands of telegraph poles collapse. The country is paralysed.

'As the day lengthens, the cold strengthens. This is indeed the coldest day I have ever known since the hard frost 1739–40; everything freezes. When I sat down on my chair (indeed a wooden bottom) I started up surprised, thinking I had sat down on a pool of cold water, when it was only the coldness of the weather.' *John Baker, barrister and amateur meteorologist, Horsham, Sussex, 1776*

The ice storm lasts five days, before burst pipes gush back to life. This salutary event, and the winter of 1939–40 as a whole, leaves a deep impression on Britain. Unlagged, exterior water pipes become a thing of the past. Central heating, almost forgotten since the Romans left Britain, grows in popularity.

'The bitter cold, which has broken out here, and the very real lack of coals in our house, forces me to press you once again – even though I find that the most painful of all things.' *Karl Marx, in penury, to his fellow revolutionary thinker Friedrich Engels, London, 1858*

In pelting rain on Falkirk moor (✗) today, in 1746, the Stuart claimant to the throne, Bonnie Prince Charlie, is fighting English troops commanded by General 'Hangman' Hawley. As the Jacobite army advances off the top of the moor, the royal Hanoverian artillery is stuck in the mud at the bottom of the hill. In the soaking mayhem, several royal regiments flee. The commanding artillery officer is court-martialled, and two weeks later the Duke of Cumberland takes command of the campaign.

-23.3C Logie Coldstone 1910 ❄

156.2mm Seathwaite 1906

15.4C Burghill 1887

AGONIZING
AROMA OF
CHOCOLATE
/ FOG CHAOS

Following snow and a freezing gale in Roald Dahl's classic children's story, *Charlie and the Chocolate Factory*, today is the moment when Charlie and his family reach their nadir of despair:

'There is something about very cold weather that gives one an enormous appetite. Most of us find ourselves beginning to crave rich steaming stews and hot apple pies and all kinds of delicious warming dishes; and because we are all a great deal luckier than we realise, we usually get what we want – or near enough. But Charlie Bucket never got what he wanted because the family couldn't afford it, and as the cold weather went on and on, he became ravenously and desperately hungry. Both bars of chocolate, the birthday one and the one Grandpa Joe had brought, had long since been nibbled away, and all he got now were those thin, cabbagy meals three times a day.

. . . Slowly, but surely, everybody in the house began to starve.

And every day, little Charlie Bucket, trudging through the snow on his way to school, would have to pass Mr Willy Wonka's giant chocolate factory. And every day, as he came near to it, he would lift his small pointed nose high in the air and sniff the wonderful sweet smell of melting chocolate.'

Two days later, 'walking back home with the icy wind in his face (and incidentally feeling hungrier than he had ever felt before)', Charlie spots a fifty-pence piece buried in the snow, with which he buys a bar of chocolate. Then he buys another one, in which he finds the golden ticket that is his salvation and sets him off on his whole adventure.

15.5C Lossiemouth 1961

-20.6C Writtle 1947

In 1959, dense fog blankets most of Britain, bringing transport mayhem. 'We had a blind piano-tuner who was tuning the piano that night,' recalls David Whale in London. 'I remember my grandmother saying to my uncle that he should see the piano-tuner home as the weather was so bad. The blind tuner pointed out that he'd have to see *him* (my uncle) back, as he could get along far better.' In the Scottish Highland blizzard, 1978, a travelling salesman survives for eighty hours by wrapping himself in his wares – women's tights. Also in Scotland, in 1644, frost helps the Parliamentarians in the Civil War. When the supporting Scottish army reaches the Tweed (✕), the ice is so thick that the entire cavalcade of horses and guns clatters straight across.

30 JANUARY

1.5C min / 6.5C max

SEVERN TSUNAMI 'Mighty hills of water tumbling over one another' travelling 'with a swiftness so incredible, as that no gray-hounde could have escaped.' Usually it is the East Coast of England that suffers from sea surges (31 January, 29 December), but today at 9 a.m. in 1607 a flood on the River Severn inundates the area from Barnstaple in Devon to Gloucester, and north of the estuary as far as Cardiganshire – some 350 miles (570 km) of coastline and riverbank. In places, the water rises 10 feet (3 metres) above the high tide mark. Bristol is flooded. The Gwent Levels are inundated. Swathes of farmland are washed away. A 60-tonne ship is driven across the marshes, and dozens of Somerset villages are submerged. One contemporary source estimates the death toll at two thousand, another says the flood reaches the foot of Glastonbury Tor, 14 miles (22 km) inland. 'Unspeakable was the spoyle and losse,' the Vicar of Almondsbury records. It's one of the worst natural disasters in British history.

'A very wet day. Did nothing but eat and drink and sit by the fire all day and hard work I found it.'
Thomas Marchant, yeoman farmer, Sussex, 1728

'Indescribable mixture of ice and slush. I fell off three times, and was, of course, hustled into the gutter and drenched in fountains of filthy squelch by those amiable people who drive "private cars".'
J.R.R. Tolkien, academic and author, bicycling around Oxford, 1945

'The wrath of God is upon the established church,' is the Puritan preachers' explanation. More probably, it's a freak combination of powerful south-westerly winds, tidal peaks and an extreme low-pressure system. The ambiguity of some contemporary accounts has led to recent theories that the flood could have been caused by a tsunami. That debate continues. Plaques marking the height of the floodwaters still hang in many of the churches beside the river.

In 'a gale of wind, a heavy running sea, and the atmosphere so thick and hazy', three naval transport brigs – *Boadicea*, *Lord Melville* and *Seahorse* – are lost today in 1816, off the Old Head of Kinsale (✪) on the Irish coast, bound for Cork. Nearly a thousand soldiers returning from sharing in 'the glories of Waterloo', along with their wives and children, drown. On a gloomy and bitterly cold day in 1965, several hundred thousand silently line the streets of London to watch the gun carriage bearing Sir Winston Churchill's coffin travel from Westminster Hall to St Paul's. If ever the weather reflected the mood of a nation, it is today. And in 2003, under an inch (2 cm) of snow brings south-east Britain, including the M11, M4 and M25, to a standstill – to howls of derision from the media.

17.2C Aber 1929

✪

-21.3C Elmstone 1947 ❄

31 JANUARY

1.7C min / 6.6C max

BRITAIN'S WORST NATURAL DISASTER

The late news this Saturday evening in 1953 reports that, in a gale in the Irish Sea, the Stranraer car ferry *Princess Victoria* has sunk with 133 drowned. After the news, the weather report goes on to mention storms and gales battering Scotland. Hundreds of miles away, to the inhabitants of England's east coast, preparing for bed, these two facts do not sound an ominous warning. Yet, by dawn, more than three hundred along the coasts of Norfolk, Suffolk, Essex and Kent will be dead – mainly drowned in their beds. Thirty thousand more will face evacuation: victims of a catastrophic storm surge that is Britain's worst peacetime disaster.

In retrospect, there is a chilling predictability to the way that various apparently unconnected events coincide to create the East Coast Flood. High spring tides, falling pressure over the North Sea, rivers flooded by heavy rain and, finally, northerly hurricane-force winds – all combine to generate a massive sea surge just where the gap is narrowest, the sea shallowest and the land lowest: between Britain and Holland. While the East Coast is no stranger to such 'sea bursts' (*27 December*), this one is of exceptional magnitude: 24,000 houses, 160,000 acres of farmland, twelve gasworks and two power stations are inundated. On Canvey Island, one witness reports waking to 'the smell of death' – and this before the surge reaches its disastrous climax in Holland, where over two thousand are drowned.

Worst of all, the catastrophe could largely have been avoided. Information simply fails to get through. This is due partly to poor collation of weather data, partly to the fact that the countryside, still a set of disparate regions, is linked only by phone lines vulnerable to high winds. The only warning of impending disaster comes from rural policemen on bicycles. A vast rescue operation involves every helicopter in Europe (so beginning the association of that machine with mercy missions). In the immediate aftermath, a Storm Tide Warning Service is instigated jointly by the Hydrographic and Met Offices. The disaster finally jolts funding for the Thames Flood Barrier (*29 September*) – which, fifty years later today, closes a record fourteen times. On a lighter note, on this day the first snowdrops, on average, appear.

-21.4C Kingussie 1895

139.7mm Watendlath Farm 1933

15.2C Teignmouth 1880

FEBRUARY

1 FEBRUARY

1.5C min / 6.7C max

FROZEN BEER

February is, on average, almost exactly as cold as January. Although the days are getting longer, the sea surrounding us continues to get colder for another month, which considerably affects the temperature. ('As the days grow longer, the cold grows stronger' – a meteorological consequence of being an island nation.) The last exceptionally cold month since January 1963 is February 1986 (though, because it breaks no records and the rest of the winter is unexceptional, it is often a forgotten month). February is the driest month of the year, and, with snow in recent years (however much we like to pretend otherwise) a rare anomaly, February really needs reclassifying from a winter month to an early spring month. Two traditional harbingers of spring are, on average, first heard around today: the song thrush and the blackbird, singing in the hour before dawn or at dusk.

'We are now here in high frost and snow, the largest fire can hardly keep us warm. It is very ugly walking; a baker's boy broke his thigh yesterday. I walk slow, make short steps, and never tread on my heel. 'Tis a good proverb the Devonshire people have: 'Walk fast in snow, In frost walk slow, And still as you go, tread on your toe: When frost and snow are both together, Sit by the fire and spare shoe leather.'
Jonathan Swift, satirist and writer, 1711

'Yesterday and today were so warm and lively that it had more appearance of Summer than Winter . . .'
Reverend James Woodforde, diarist, 1787

The great freeze of 1740 is at its most intense today, when the *Caledonian Mercury* reports a boy tumbling over in Edinburgh and breaking the 20-pint (11-litre) cask of beer he's carrying. The liquid freezes so quickly that he's able to rope it up and drag it along. In 1947 Lieutenant Ken Reed braves a snowstorm in one of the Royal Navy's first helicopters, a Hoverfly. Flying through the storm by dead reckoning – blind-flying instruments are not invented for another nineteen years – he lands on the quarter-deck of HMS *Vanguard*, in the Channel, to deliver mail to the Royal Family. In 1956 the temperature in London swings from 8C yesterday to -4C today.

-18.3C Corbridge 1972

16.1C Hodsock Priory 1923

16.1C Geldeston 1898

'It blows a hurricane; tiles fly across the streets, and tops of chimneys fall on the pavement, to the great annoyance of passengers, and danger of their lives. The house we inhabit, built of stone, is sensibly shaken by the wind. There is at the end of our street, on the mound, an itinerant menagerie built of boards; if it should be blown down, the people of Edinburgh might see at large in their streets two lions, two royal tigers, a panther, and an elephant, besides monkies, and other underlings of the savage tribe.'
Louis Simond, An American in Regency England, 1811

2 FEBRUARY

1.2C min / 6.7C max

QUEEN VICTORIA'S FUNERAL

Today marks the halfway point between the winter solstice (*21 December*) and the spring equinox. It's midwinter, a key day in the cycle of the year – known to Christians as Candlemas (church candles are blessed) and to pagans as the Festival of Lights – and a time to drive the darkness out. Daylight is steadily increasing by three to four minutes a day. It is also a traditional time to take stock of the weather.

'Fine warm day, like Summer. The Knats were out in my Bedchamber & my Blackbird sung out so as to be heard up Stairs.' *Reverend William Cole, antiquary and Rector of Bletchley, 1767*

'All has been such Doom. I can't tell you what the weather has been like, absolutely throat-cutting! Fearful gales and storms and teeming rain, day after day non-stop, and I have never known anything like it in all my years in Cornwall. One feels it terribly up at Kilmarth, which is so exposed. I have barely been able to go out, far less on Thrombosis Hill, and even on flat Par Beach one can barely stand up, with all the sand blowing in one's eyes … Only longing for the gales to stop, and the house to cease from shaking!' *Daphne du Maurier, novelist, 1974*

'If Candlemas Day be sunny and bright, winter will have another flight; if Candlemas day be cloudy with rain, winter is gone and won't come again,' is one of numerous weather adages associated with today. British emigrants to America adapted these to create Groundhog Day, made famous by the eponymous film. The theory (which may have come from medieval Europeans who studied hedgehogs at this time of year) is: if an intelligent rodent emerges from his burrow and sees his shadow today, he returns underground forthwith, as winter is set to continue. Obviously, in fact, neither today's weather nor hedge-hogs foretell anything. These theories and sayings are merely warnings: though you may wish it, winter is not over.

Certainly, it's bitterly cold and gloomy in London today in 1901 – fitting weather for the State Funeral of Queen Victoria. It's the passing of an age. As John Galsworthy writes in *The Forsyte Saga*, 'The Queen was dead, and the air of the greatest city upon earth grey with unshed tears … Under grey heavens, whose drizzle just kept off, the dark concourse gathered to see the show.'

16.1C Rhyl 1944

-20C Welshpool 1954

3 FEBRUARY

1.6C min / 7C max

KEATS
CATCHES HIS
DEATH

A thaw today in 1820, after bitterly cold weather, tempts the poet John Keats to travel from central London to Hampstead on the late evening stagecoach. Like any self-respecting poet, he hasn't a coat on his back to keep warm nor a bean in his pocket to pay for a seat inside. So he rides outside. The chill air and strong winds bite him to the bone. He staggers to Wentworth Place, where his friend Charles Brown sends him straight to bed. As Keats pulls the bedclothes around him, he coughs and a drop of blood appears on the sheet. 'That drop of blood is my death warrant,' says the poet. 'I must die.' He's prophetic. Despite undergoing numerous treatments for his consumptive sickness, in a year, aged twenty-four, he's dead (*19 September*).

'My Lord [Bruncker] and I, the wind being again very furious, so as we durst not go by water, walked to London quite round the bridge, no boat being able to stirre; and, Lord! what a dirty walk we had, and so strong the wind, that in the fields we many times could not carry our bodies against it, but were driven backwards ... it was dangerous to walk the streets, the bricks and tiles falling from the houses that the whole streets were covered with them; and whole chimneys, nay, whole houses in two or three places, blowed down. But, above all, the pales on London-bridge on both sides were blown away, so that we were fain to stoop very low for fear of blowing off of the bridge. We could see no boats in the Thames afloat, but what were broke loose, and carried through the bridge, it being ebbing water. And the greatest sight of all was, among other parcels of ships driven here and there in clusters together, one was quite overset and lay with her masts all along in the water, and keel above water.' *Samuel Pepys, diarist, 1666*

-18.9C Braemar 1897 ❄
-18.9C Perth 1956 ❄

16.1C Llandudno 1881 ☼
16.1C Shrewsbury 1914 ☼

Freak storm-surge devastates
Sandgate and St. Leonards 1904

4 FEBRUARY

RUSKIN'S STORM CLOUD

Under a funereal, grey sky today in 1884, Professor John Ruskin makes his way to the London Institution to deliver a grand rant entitled *The Storm-Cloud of the Nineteenth Century*. Now 65, and increasingly potty, the critic and painter is obsessed with a 'plague cloud' that, he believes, is hovering over the world. He hopes to convince scientists and the general public that the sky has been getting darker, the wind more malignant and the sun weaker for at least a decade. In short, he believes the climate is changing. His theory is borne out by his daily weather observations. His diaries are full of bleak, apocalyptic entries: 'grand artillery-peals of thunder . . . settling down again into Manchester devil's darkness'; 'the sulphurous chimney-pot vomit of blackguardly cloud'.

Ruskin believes it's all down to the blasphemy, greed and pollution of a century that, in its godlessness, has started a process of 'uncreation' when 'the earth becomes void again; the word goes forth: "let there be no light."' Reviewing his lecture, the press call his ideas 'fantastical, insane' and wonder why he blames air pollution on the Devil. In fact, we now know, he's dead right. His observations coincide with soaring coal consumption across Britain, especially in London and Manchester, as industrialization peaks. Sulphur dioxide levels have never been higher than in the 1870s and 1880s. The skies do darken, the air does become thicker and more noxious, the weather is indeed damper and colder. Ruskin has identified environmental pollution – he's just a century ahead of his time.

Tonight in 1814, riverside strollers in London witness an unlikely 'fire and ice' spectacle, when a large piece of the frozen, silvery Thames – bearing a temporary ale house, the publican, Mr Lawrence of Queenhithe, and nine merry drinkers – breaks free from the now thawing river at Brooks Wharf and heads for Blackfriars Bridge, all the while in flames. The Frost Fair on the river is ending, after only four days. The skittle alleys, swings, liquor dens and barbecues selling 'Lapland mutton' in shilling slices are being packed up by panicking traders and carried off the ice in haste. In twenty-four hours, there will be nothing left (*24 February*). The drama is fitting, as this is London's last ever Frost Fair. The climate of the British Isles is very gradually getting milder. The Little Ice Age (*3 March*) is almost over.

-21.7C Braemar 1897

15C Colwyn Bay 1933

5 FEBRUARY

OUT OF THE
STORM ...
WHISKY!

The south-easterly gale that wakes the inhabitants of the tiny Hebridean isle of Eriskay this morning, in 1941, brings treasure in the form of a shipwreck (✪). For, as word travels swiftly round the little community, this ship carries no ordinary cargo. The SS *Politician*, bound for America, contains 25,000 cases of ... Scotch whisky. Ring any bells? It is, of course, the true story on which, in 1946, Compton Mackenzie bases his bestseller *Whisky Galore!* – in which the SS *Politician* becomes the SS *Cabinet Minister* and Eriskay becomes Great Toddy. The film, directed in 1948 by Alexander MacKendrick, joins the pantheon of Ealing greats. According to recently released Home Office files, the sum of £290,000 was also aboard the *Politician*, in ten-shilling notes – but who remembers that?

The most weather-affected fixture in the history of the FA Cup concludes today in 1972. After mighty Newcastle United draw non-League minnows Hereford United in the third round, at St James's Park, the fixture is postponed (a waterlogged pitch) and postponed again (snow) before Hereford achieve an unexpected replay with a 2–2 draw. Back in Hereford (☂), the replay is postponed three times (frost) and Newcastle, with several international stars, eventually camp out in a local hotel until it thaws. By the time it does, the pitch is a field where cows would fear to tread. After 82 minutes, the great Malcolm Macdonald puts Newcastle ahead and it looks as if Hereford's dream is over. But then, with four minutes to go, Hereford's Ronnie Radford, a part-time joiner, lets rip from 30 yards, striking the ball off a mud divot – into the Newcastle goal. 'No goalkeeper in the world would have stopped that!' cries BBC commentator John Motson. In injury time Hereford score again, causing one of the greatest-ever shocks in the FA Cup. The game is shown on *Match of the Day*, and the subsequent invasion of the quagmire by Hereford fans, led by a uniformed police officer, remains one of football's greatest moments.

-20C Braemar 1912
✪   ❄

16.5C Aber 1990 ☼
    🌩

In 1805 a storm drives the *Earl of Abergavenny*, a merchant ship, on to the Shambles Bank (✪). Sinking within sight of the Dorset coast, she takes almost three hundred with her, including her captain, John Wordsworth, brother of the poet and friend of John Constable. The painter later gives his seascape of Weymouth Bay a forbidding sky, in memory of his friend.

✪

6 FEBRUARY

1.7C min / 6.9C max

THE COLDEST
WINTER
IN MEMORY,
1962-3

The phenomenal blizzard that begins today in 1963 continues for thirty-two hours. It buries tracts of the West Country, Wales and Ireland beneath 5 feet (1.5 metres) of snow. It leaves the deepest snow ever recorded in central England's towns and cities. It's the snowiest day, of the second snowiest winter of the twentieth century. The American site manager at Fylingdales Early Warning Station, North Yorkshire, newly arrived from Alaska, describes the conditions (80 mph/130 kph winds and snow thick enough to bury a double-decker bus in an hour) as worse than anything he's ever experienced.

The 1962–3 winter is one of only four landmark weather events (*4 March, 3 July, 15 October*) within living memory. Cold weather arrives in London in early December with the last of the old-fashioned, pea-souper smogs (*6 December*). But the real freeze, and the snow, arrives on east winds on Boxing Night. After that, for the next sixty days Britain becomes Siberia. The Thames freezes. Millions of gallons of uncollected milk have to be poured down the drain while livestock die of thirst because water sources are frozen. The remotest farms are cut off for more than two months. Diesel freezes in buses and lorries. As conditions persist, incidents become more and more extreme. In Leicestershire, a woman carrying bread rolls is attacked and knocked down by ravenous pigeons. Army engineers dynamite sea ice to get ships out of east coast ports. Starving sheep eat the wool off each other's backs.

The pools companies lose so much from cancelled fixtures that, to cut losses, the first ever Pools Panel is convened (*26 January*). Industry halts. Sheet ice alternates with freezing fog, black ice and gales, blowing further snowfalls into drifts up to 43 feet (13 metres) deep. In the great scheme of worst ever British winters, how bad is it? Well, it's longer and colder, though not actually snowier, than 1947, and trounces 1929, 1917 and 1895. Indeed, only the winters of 1683–4 (*19 January, 18 December*) and 1739–40 (*2 May*) are colder. Its effect on wildlife, especially song-birds, is catastrophic (the wren population is decimated). Local authorities face vast bills for road repairs following frost-heave and snowplough damage. However, when the thaw finally comes, after 2 March, it comes with sunshine rather than rain, so flooding is much less severe than in 1947 (*16 March*).

196.6mm Ben Nevis 1894

16.1C Aber 1990

-20C Benson 1917

6 FEBRUARY

CASUALTY OF THE COLDEST WINTER – SYLVIA PLATH

'In truth, the weather had trapped her', writes Linda Wagner-Martin in *Sylvia Plath: A Biography*, of the poet's suicide. Separated from the love of her life, poet Ted Hughes, in autumn 1962, Plath moves with their two children to London from Devon. Their arrival coincides with the smog and the plummet in temperature. They all get colds. The snow follows. Because of difficulty starting the car, Sylvia and the children are virtually housebound. Going outside, going shopping, any outing, means getting everyone dressed, worrying about slipping on the snow and ice. She thought she could cope, but conditions then worsen dramatically. 'The English, of course, have no snow plows, because this only happens once every five years, or ten,' she writes to her mother. 'So the streets are great mills of sludge which freezes and melts and freezes.' Thaws cause ice to layer over the snow. Stores sell out of candles because of power cuts. Despite the arrival in bookshops of *The Bell Jar* (a semi-autobiographical novel detailing her battle with depression), she feels ill, exhausted and overwhelmed. In five days' time, early in the morning, Sylvia Plath puts cups of milk beside the children's beds. She tapes round the doors and writes a note asking that her doctor be called. Then she kneels beside the open oven of the kitchen of her second-floor flat and turns on the gas.

7 FEBRUARY

1.6C min / 6.9C max

THE ROARING GAME

At 11.30 this morning in 1979, a cannon fires across the frozen Lake of Menteith, near Stirling (🖈), and the hills begin to echo with the growling of granite scudding over ice. The Grand Match or 'Bonspiel', an outdoor curling contest between the North and the South of Scotland, has begun. The Scots invent curling – 'the roaring game' – in the 1500s, and until the early twentieth century it's played entirely outdoors as and when the weather allows. Most years 'rinks' for local games, often on purpose-dug, shallow ponds, can bear curlers after just three or four days of freezing weather.

A Grand Match, however, involves six hundred teams of four players on three hundred or more rinks. Nothing less than a large loch will do, frozen with several inches of black ice capable of supporting players, thousands of spectators and innumerable cases of whisky. A journalist estimates the weight on the 63 acres (25 hectares) of ice at Carsebreck, Perthshire during the 1929 Bonspiel as 200 tons. Not surprisingly, suitable conditions have been rare since the first Grand Match at Penicuik in 1847, and they're getting rarer. There have been thirty-three contests in all, only three since 1945, the last today. With 'Health and Safety' specifying that the ice now has to be a regulation 8 inches (20 cm) thick, has the roaring game roared its last outside?

'Bright, new, wide-opened sunshine, and lovely new scents in the fresh air, as if the new blood were rising. And the sea came in great long waves thundering splendidly from the unknown. It is perfect, with a strong, pure wind blowing. What does it matter about that seething scrimmage of mankind in Europe?' *D. H. Lawrence, writer, Cornwall, 1916*

'I am so longing for spring. I miss the American snow, which at least makes a new, clean, exciting season out of winter, instead of this six months' cooping-up of damp and rain and blackness we get here.' *Sylvia Plath, poet, 1962*

Forty-eight hours of heavy rain on the west coast of Scotland ends today in 1989, setting a new two-day record for Scotland – 306.1 mm at Kinloch Hourn (🖈). Water gushes eastwards off the mountains, flooding the rivers Ness, Spey, Conon and Oykel on the opposite side of the country – where not a drop of rain has fallen.

-21.7C Aviemore 1895 ❄

16.8C Durham 1993 ☼

8 FEBRUARY

1.6C min / 7C max

FEBRUARY
DREAMS OF
ENCHANTED
APRIL

'It began in a woman's club in London on a February afternoon, – an uncomfortable club, and a miserable afternoon.' So opens one of the great hymns to the awfulness of an English February. Elizabeth von Arnim's novel *The Enchanted April* (1922) is almost forgotten until Mike (*Four Weddings and a Funeral*) Newell resurrects it in 1992 with his Oscar-nominated film. Two dowdy, childless women become soulmates when they catch each other reading the same advertisement in *The Times*.

'To Those who Appreciate Wistaria and Sunshine. Small mediaeval Italian Castle on the shores of the Mediterranean to be Let Furnished for the month of April. Necessary servants remain. Z, Box 1000, *The Times*.'

Mrs Wilkins wishes to escape a thrifty solicitor husband. Mrs Arbuthnot, a woman dried up with her selfless good works, so seldom sees *her* husband she suspects he won't notice if she goes away. Appalled at their extravagance, flinging caution to the winds, they rent San Salvatore for April for £60. To defray this astronomical sum, they advertise for other like-minded holiday-seeking women and secure two suitably incompatible co-lessees: a twenty-eight-year-old temptress trying to escape the grabbing hands of men, Lady Caroline Dester; and an elderly widow living in the past, Mrs Fisher.

The grey, wet February weather is the perfect metaphor for English middle-class repression of the 1920s. Lulled by the Mediterranean spring, the defensive public personas of the four women gradually melt. Von Arnim's story achieves that rare thing for a novel: it puts a place – Portofino – on the map. But the spell she casts depends entirely for its power on the curse of Britain in February: 'Looking out of the club window into Shaftesbury Avenue, Mrs Wilkins, her mind's eye on the Mediterranean in April, and the wistaria, and the enviable opportunities of the rich, while her bodily eye watched the really extremely horrible sooty rain falling steadily on the hurrying umbrellas and splashing omnibuses, suddenly wondered whether perhaps this was not the rainy day Mellersh – Mellersh was Mr Wilkins – had so often encouraged her to prepare for.'

-25C Aviemore 1895

15.9C Hodsock Priory 1903

9 FEBRUARY

1.3C min / 6.7C max

MUNRO ALMOST BAGGED

'The mist was rolling over the top of Braeriach, and creeping up the valleys ... Prudence certainly dictated a return to Speyside, but I was well accustomed to finding my way among the mountains alone, in winter and in all weathers,' writes the mountaineer Sir Hugh Munro of an expedition up Cairn Gorm (4081 feet/1244 metres (♀)) today in 1892. And it's no understatement: in 1891, sanctioned by the Scottish Mountaineering Club, he publishes the first Munro Table, classifying some 283 mountains over 3000 feet (914.4 metres) by name, height and grid reference. So he sets off across the Cairngorm plateau for Ben Macdui (4296 feet/1309 metres) without misgivings. But the weather worsens. 'I was soon in dense mist, which froze to one's hair, clothes, and beard ... I dashed off down the snow, in what I believed to be the direction of Loch Etchachan ... [but] found myself at five P.M. ... on the top of the precipices over-hanging Loch Avon. Here was a balmy place to be in! Dangerous cliffs all round, the cold so intense that one could scarcely have lived an hour without moving. It was long after sunset, and the chances of getting out of difficulties before it became quite dark seemed slight.' In fact, Munro does make it down safely, perhaps guided by the 'heavenly music' that he, some-what idiosyncratically for the age, often claims to hear in the mountains.

'The first snowdrops appeared in the Churchyard.' *Reverend Francis Kilvert, diarist, Clyro, Wales, 1879*

'The cold more violent than in any one night of the great frost of 1683.' *Thomas Hearne, Assistant Keeper of the Bodleian Library, Oxford, 1718*

Today in 1649 it's so bitterly cold that the Thames is frozen as Charles I is led to the scaffold in Whitehall to be beheaded. The King has taken the precaution of wearing two shirts so that he will not shiver with cold in front of the huge crowd, thereby giving the impression he's afraid. 'I would have no imputation of fear,' the King says. 'I do not fear death.'

-23.9C Braemar 1895

In 1999 a gale tears the marquee down and it snows as the Eden Project prepares for the Secretary of State to turn the first sod. The venture has been dogged, all winter, by rain 'to break Noah's heart, the worst weather in Cornwall's recorded history,' writes Tim Smit, the creator. And in 1919 the first air service from London to Paris takes off, battered by strong winds. Passengers are reputedly 'too busy smoking and playing cards to notice'.

16.3C Cambridge 1903

THE REMARKABLE STORY OF ELIZABETH WOODCOCK

Returning on horseback from a shopping trip to Cambridge across snow-covered countryside on a winter evening in 1799, Elizabeth Woodcock is startled and bucked off. An hour later, the hapless, horseless housewife is still wandering across open land in a blizzard, exhausted. Eventually, she falls into an icy cocoon created by drifting snow. Quickly, the snow piles up several feet above her head. And here she remains for eight days, with no sustenance save for a pinch of snuff, while her husband and friends – she is only half a mile from home – search the neighbourhood. On the fifth day a thaw begins, but Elizabeth has no strength to extricate herself from her snow-hole. Her family abandon hope.

In fact, a snow-hole or snow-cave is the safest place to be in a blizzard. Snow traps large amounts of air between crystals, providing insulation that more than compensates for the heat it absorbs. The walls of a hole also provide protection from the wind. Learning to dig a snow-hole is one of the first lessons of mountain survival. Elizabeth remains conscious throughout: she hears her local church bells ringing and carts passing on the road. Today, after eight days, a passer-by spots the handkerchief she has managed to thrust above the snow and she is rescued.

'This was a day made absolute, the sun unflawed, the blue sky pure. Slate roofs and crows' wings burned white like magnesium. The shining mauve and silver woods, snow-rooted, bit sharply black into the solid blueness of the sky. The air was cold. The wind rose from the north, like cold fire. All was revealed, the moment of creation, a rainbow poured upon rock and shaped into woods and rivers.' *J. A. Baker*, The Peregrine, *1967*

'A very brief, violent hailstorm this afternoon turned the drive crumbly white, scared the dogs stiff, amazed the cat and made me wonder if the windows might buckle. The suddenness and viciousness of it brought to mind the ghastly attack yesterday evening of the IRA in the Canary Wharf area.' *Alec Guinness, actor, Hampshire, 1996*

-25.6C Braemar 1895 ❄

'9st 2 (extra fat presumably caused by winter whale blubber), alcohol units 4, cigarettes 12 (v.g.), calories 2845 (v. cold). 9 p.m. V. much enjoying the Winter Wonderland and reminder that we are at the mercy of the elements, and should not concentrate so hard on being sophisticated or hardworking but on staying warm and watching the telly.' *Helen Fielding*, Bridget Jones's Diary, *1996*

18.3C Regent's Park London 1899

11 FEBRUARY

1.4C min / 6.5C max

THE WRONG KIND OF SNOW

'BRITISH RAIL BLAMES THE WRONG TYPE OF SNOW' is the headline in today's *Evening Standard* in 1991: the company's fancy new Norwegian snow-clearing kit is failing to prevent massive rail delays through a fleeting cold snap. Ironically, the catchphrase that will become shorthand for bureaucratic, nationalized-industry buck-passing is never actually uttered. It is in fact coined by a journalist interpreting a spokesman's excuses – but it sticks to the beleaguered BR like flesh to ice.

What *is* the wrong kind of snow? Snow forms when ice crystals collide and freeze together to become flakes. At temperatures around 0C, snowflakes fall as wet snow – snowball material. At lower temperatures, the crystals won't stick – and the snow remains powdery. This is the 'wrong kind'. It's fine enough to infiltrate electrical systems, causing short circuits, forcing British Rail to substitute electrical services with diesel trains during the 1991 freeze. As the snow isn't even deep enough to warrant a snowplough, however, BR gets no sympathy.

'Dazzle mine eyes, or do I see three suns?' Yorkist soldiers exclaim at dawn today in 1461 as they prepare for the Battle of Mortimer's Cross (⚔). It's the Wars of the Roses, and the nineteen-year-old Prince Edward is quick to interpret the strange phenomenon low in the sky as a symbol of the Trinity and therefore a good omen. As Shakespeare has Edward say in *Henry VI, Part III*: 'It cites us, brother, to the field.' It is, in fact, an optical phenomenon called a parhelion or sundog: the two mock suns are coloured spots caused by ice crystals in the atmosphere refracting light on a cold morning. Whatever it is, it works: the Lancastrian army is routed and a bright sun is added to the badges of the House of York. In two months, Edward is crowned king.

In 1895, Britain's coldest-ever temperature (-27.2C) is recorded at Braemar (it's been equalled twice since – *10 January, 30 December*). And in the middle of winter 1963 (*6 February*), the Beatles record their first LP *Please, Please Me* today. Abbey Road studios (*8 August*) are freezing. John Lennon, sucking cough tablets, delivers a ragged, throat-shredding rendition of 'Twist and Shout' in the final take of the day: it's an instant classic, a turning point for the Beatles, and rock and roll enters a new era.

16.1C Dyce 1998
178.8mm Keil 1998
-27.2C Braemar 1895
16.1C Hawarden Bridge 1939

12 FEBRUARY

1C min / 6.5C max

THE CHANNEL DASH

Fog, snow and good, old-fashioned, cock-up conspire today, in 1942, as the Allies try to prevent three German warships sneaking up the Channel in broad daylight from Brest on the French coast to Germany. For eighteen hours, in what's become known as 'The Channel Dash', three of Hitler's prize warships present a giant, slow-moving, unmissable sitting duck. Miss them we do, however. Even though we know almost exactly when they depart, yesterday evening two patrols of Coastal Command bombers fail to spot them before, at daybreak, fog banks form. Bombers despatched to intercept them have to be diverted when their destination airfield is snowbound. More bombers, mistakenly armed with armour-piercing bombs which must be dropped from at least 7,000 feet (2130 metres), can see nothing from that height. By the time the Dover gun batteries finally fire the first shot, the squadron is invisible in murk.

And so the operation continues – a saga of missed rendezvous, unsynchronized radio frequencies and unspotted targets. Bomber Command finally arrives at 14.20 p.m., correctly armed and with the enemy in sight. Yet despite 242 sorties before nightfall not a single hit registers, though 15 bombers are shot down – bringing the total of aircraft lost during the operation to 127. As night falls on the North Sea, the Brest squadron slips quietly into the darkness.

> 'The slate slabs of the urinals even are frosted in graceful sprays.'
> *Gerard Manley Hopkins, 1870*

'FOG IN CHANNEL – CONTINENT CUT OFF'
Did this island nation's favourite newspaper headline ever appear? Sadly, every indication is that it didn't. As with 'PHEW! WHAT A SCORCHER!' (*6 July*), fact and legend merge: the anecdote has been around since before 1948, when Russell Brockbank's cartoon (below) appears in *Punch*. The *New York Times* mentions the headline in 1936, and a letter to *The Times* in 1939 ascribes the 'venerable chestnut' to *The Times* in 'the eighties of the last century'. The archives, however, reveal no such headline, in the 1880s or at any other time.

-20.6C Pinmore 1895

17.2C Ashburton 1896

13 FEBRUARY

0.9C min / 6.4C max

THE FORGOTTEN FAMINE

'I hope we have done with this cursed weather, yet still my garden is all in white,' Jonathan Swift writes in Dublin today in 1740. The great Anglo-Irish satirist may now be a mad old crank, but he's right: it's one of the severest winters in Irish history. The 'hard shower of frost which was driven out of Styx' begins with an easterly storm six weeks earlier, freezing rivers and loughs (lakes). The hint of a thaw that Swift refers to doesn't materialize. The country remains iron-shod until the end of the month and the potato crop, the principal winter foodstuff, is ruined.

'The weather fearful, violent deadly E. wind and the hardest frost we have had yet. Went to Bettws in the afternoon wrapped in two waistcoats, two coats, a muffler and a mackintosh, and was not at all too warm. When I got to the Chapel my beard moustaches and whiskers were so stiff with ice that I could hardly open my mouth and my beard was frozen to my mackintosh. The baby was baptized in ice which was broken and swimming about in the font.'
Reverend Francis Kilvert, diarist, 1870

The winter gives way to an abnormally dry spring. By April, the land is rubbed raw by northerly winds and the usually emerald fields of Ireland are parched. Several towns – Carrick-on-Suir, Thurles and Wexford – are destroyed by fire before the snow returns in early May to spoil the wheat and barley crops. By midsummer the rural poor are eating 'nettles and charnock', and there are food riots in the cities. It's not over yet, though: there are storms in August, blizzards in October, frosts and floods in December. By the beginning of 1741, a social crisis is imminent.

'The dreadfullest civil war, or most raging plague never destroyed so many as this season,' a man reports from Cork. The year 1740 becomes known as *bliain an áir*, the year of slaughter. Hunger, poor nourishment, deteriorating sanitation, increased itinerancy and another freezing winter lead to epidemics of dysentery and typhus. The aloof Anglo-Irish administration in Dublin fails to coordinate a national response. Three hundred thousand – an eighth of the population – perish. Emigration from Ulster resumes, as fiercely self-sufficient Scots-Irish Presbyterians head west to America. As normality returns, and new European wars distract, the horror of these twenty-one months of extraordinary weather fades and the crisis becomes known as 'The Forgotten Famine'.

142mph Fraserburgh 1989
(record low level gust)
-21.9C Braemar 1895

19.7C Greenwich 1998

14 FEBRUARY

SAD On average, St Valentine's Day is one of the coldest days of the year, but seasonal affective disorder (SAD) sufferers typically report that they start to feel better today. SAD is related to quantity and quality of daylight, not temperature, and each day is now getting longer by five minutes if you live in Aberdeen, four in London. SAD or 'winter depression' is first documented in a sixth-century book about Scandinavia, but the scientific community doesn't accept the disorder, which only occurs at latitudes higher than 30°N or S, until 1995. What causes SAD is still not clear. Almost certainly it's related to a biochemical imbalance affecting how light hitting the retina at the back of the eye triggers messages to the hypothalamus, the part of the brain which governs sleep, sex drive, appetite and activity. The likely causes are low serotonin (the feel-good body chemical) or melatonin (which makes us sleep).

'This frost so severe, that the harbours of several places were frozen up that no ship could goe out or come in: no packet boats went out; the sea was frozen some miles out from the shore; vast flakes of ice of severall miles were seen floating in the sea; nay, divers ships were so besett with ice that they could not sail backward or forward, but driven to great distresse.' *Narcissus Luttrell, diarist and bibliographer, London, 1683*

Though severely disabling for only a few, a less severe form – sub-syndromal SAD or 'winter blues' – may affect as much as a third of the population, according to Mind, the National Association for Mental Health. Symptoms include depression, lethargy, sleep problems, anxiety, irritability, loss of libido, and alcohol and drug abuse. Sounds like the archetypal student? Interestingly, more people develop SAD before the age of twenty-one than after. Much the most effective treatment is phototherapy – regular exposure to bright light (many times brighter than a normal bulb). Exercise, cognitive therapy, St John's wort and antidepressant drugs can also work. The only permanent cure is wintering in the tropics.

Today in 1809 the new British warship HMS *Warren Hastings* is struck by three 'distinct balls of fire' during a thunderstorm (*12 July*). And in 1579, according to the chronicler Raphael Holinshed, a north-east wind brings a four-day snowfall. When it thaws, 'the water of the Thames rose so high into Westminster Hall, some fishes were found to remain'.

-21.7C Drumlanrig 1895

109.2mm Llydaw Intake 1926

19.1C Tivington 1998

15 FEBRUARY

0.9C min / 6.7C max

SHROVETIDE FOOTBALL / PLUMBER'S DELIGHT

In 4 inches (10 cm) of snow, a hand-painted, cork-filled ball is 'turned up' in a supermarket car park in Ashbourne (♠), Derbyshire today in 1995 to begin the annual Shrovetide football match. Played in this town on Shrove Tuesday and Ash Wednesday since at least 1667, Shrovetide football goes ahead in any weather. Today it's snow, but in recent years the medieval contest (the game dates from the twelfth century) has proceeded in torrential rain (2003), heavy frost (1986) and a gale (1990). Also known as 'hugball' and involving several thousand players in two teams (Up'ards and Down'ards), on a 3-mile-long pitch that covers village streets, fields of mud and a river, Shrovetide football is the mob football from which Association Football emerges in the nineteenth century. Certainly it's the game that's best adapted to our capricious climate, as the royal Shrovetide anthem, sung before each game, notes: 'And they play the game right manfully,/ In snow, sunshine or rain.'

Two supremely hard frosts, peaking mid-month in 1895 and 1929, stand out for plumbers. Following the 1895 frost, which began five days ago, lasts ten days and affects the entire country, so much damage is done to water mains that new bye-laws are introduced requiring mains to be sunk deeper. After the 1929 frost, more than 5 million burst water pipes have to be repaired, taking until mid-May. Ideal conditions for such a 'harvest', as plumbers call this kind of freeze, no longer exist: unlagged external pipes of lead or copper, a sudden sharp frost so that water in the pipes freezes, expands and splits the pipe, and an equally rapid thaw so that everyone needs a plumber. Before the introduction of plastic external water pipes (post-1980s) burst water pipes, in even mild frosts, are a seasonal fixture following cold weather. The most recent periods to inspire a wistful, misty gleam in plumbers' eyes are January 1982 and December 1995 – though the combination of plastic piping with more 'give,' milder winters and more exacting building regulations means that harvests are now rare.

-21.5C Lagganlia 1978 ❄

156.5mm New Dungeon Ghyll 1935 💧

18.1C Prestatyn 1998 ☀

Otherwise, today in 1760 HMS *Ramillies*, an 82-gun second-rate ship-of-the-line, returning to Plymouth before an approaching gale, is driven aground off Bolt Head (⊕) and 699 drown. And in 1925, London fogs are so bad that Regent's Park Zoo announces lights will be installed – so that the animals can see each other (*6 December, 16 December*).

⊕

16 FEBRUARY

1C min / 6.6C max

DISASTER IN SHEFFIELD

In 1962, Sheffield wakes – at least, anyone who's managed to sleep – to an 80 mph (130 kph) wind-storm and a blitzed city (☛). Two-thirds of the houses – a hundred thousand buildings – are damaged or destroyed. A 130-foot 40-metre tower crane lies crumpled across a construction site. Pianos and furniture scatter the streets. Since 4 a.m., the Fire Brigade has been receiving six emergency calls a minute.

What's happened? The answer is that Sheffield, in the lee of the Pennines, has fallen prey to a freak set of conditions. A north-west Atlantic gale becomes trapped, passing over the Pennines, beneath warmer air above it (a 'temperature inversion' – see *Definitions*). Unable to rise and so dissipate, it bounces down, then back up, then down again. The resulting vertical wave motion compresses the air, doubling its speed. The effect is devastating.

The city is declared an official disaster area and 'Dunkirk spirit' kicks in. The building trade withdraws union rules and teams of builders arrive from towns and cities across Britain to help clear up. The Minister of Housing promises – eventually – government help with costs. But it's months before Sheffield is normal again.

'There is a saying: snow is the blind person's fog ... I become unsure about where I am going ... All my familiar points and markings, the different ... textures of grass, gravel, asphalt and concrete are obliterated. If the snow is deep enough to cover the kerbs, then I really have a problem, because I cannot tell ... the road from the footpath. I explain to people that I have ... never slipped yet ... The problem, I explain, is that I cannot tell my route. Sighted people find it difficult to realise that, for a blind person, the body itself has become the organ of sense. Apart from the white cane, and sounds ... the body's knowledge of its surroundings does not exceed its own dimensions. This is such a strange kind of reality that the sighted can hardly grasp it.'
Professor John Hull, Touching the Rock: An Experience of Blindness, *1985*

Today in 1929 has the distinction of receiving the deepest fall of snow ever measured in a single day in Britain below 1000 feet (305 metres). Over 6 feet (2 metres) falls on the south-east fringes of Dartmoor (☛) in fifteen hours, without drifting (*6 February*). Eyewitnesses describe the snow falling as if 'shovelled'.

-20.5C Braemar 1978 ❄

16.1C Wakefield 1927 ☼

WINTER
HOLIDAY

The *Kendal and County News* estimates that ten thousand people are skating on Lake Windermere (♥) today, during the six-week freeze of 1895. Special trains and charabancs bring them from Manchester, Liverpool and Lancaster for the ice carnival on England's largest natural lake.

'The little yacht, with Dick and Dorothea clinging to it, as flat as they could lie on top of their sheepskins and knapsacks, flew on in the snowstorm. The larch-pole mast bent and creaked. The sail, full like a balloon, swayed from side to side and the sledge swayed with it. The ropes thrummed. The iron runners roared over the ice ... Dick clung on, blind but happy. Nobody could say this was not sailing ... "Where are we now?" shouted Dorothea. "Arctic!" shouted Dick. And suddenly Dorothea knew she was afraid.'
Arthur Ransome, Winter Holiday

One of the skaters is an eleven-year-old pupil at Old College in Bowness called Arthur Ransome. Ransome can skate, and he later remembers this as a happy month in an otherwise miserable schooling: 'Our headmaster liked skating ... lessons became perfunctory ... We spent the whole day on the ice, leaving the steely lake only at dusk when the fires were already burning ... Those weeks of clear ice with that background of snow-covered, sunlit, blue-shadowed hills were, 40 years later, to give me a book called *Winter Holiday*.'

The fourth novel in Ransome's *Swallows and Amazons* series begins with a prolonged period of freezing weather, when 'Softly, at first, as if it hardly meant it, the snow began to fall.' Soon enough, conditions are right for an expedition down the lake to the 'North Pole' on an ice-yacht:

Fog, rather than ice, is the reason the Turkish Prime Minister's plane is diverted from Gatwick (♥) today in 1959. Adnan Menderes' Turkish Airlines Vickers Viscount crashes in nearby woods, killing 12 of the 22 on board. Miraculously, Menderes survives. The accident highlights the great flaw of Britain's second airport: Gatwick is situated in a notorious fog and frost hollow. On calm nights, cold air slips down the surrounding slopes and collects in an invisible pool over the airport, where temperatures can be several degrees colder than the surrounding low hills (*10 December*). In its former life in the late nineteenth century, as a racecourse, Gatwick meetings continually have to be cancelled because of fog. In the late 1960s, the huge increase in air travel means planes have to be incessantly diverted (*27 December*). Development at the airport in the last two decades has warmed the site up, alleviating the fog problem.

-23.9C Aviemore 1895

17.4C Llandudno 1878

18 FEBRUARY

0.8C min / 6.8C max

WHAT IT'S LIKE
TO EXPERIENCE
A BLIZZARD

When weather forecasts predict heavy snow with gales over the south-west today in 1978, a friend of G. A. Southern dares him to experience it for real. 'I thought it foolhardy, but the once-in-a-lifetime chance of high drama on Dartmoor could not be lightly dismissed.' Parking their car on the B3212 (♥) (the Tavistock–Moretonhampstead road), by 12.30 p.m., 'booted and muffled up against a force 8 easterly gale', they are already halfway along their planned 3-mile (5-km) walk. 'It was not daunting, indeed it was exhilarating.' They can even see a watery sun and, when the drifting snow allows, a horizon line. But then . . .

'In the afternoon we had battles with the Tugs with snow-balls . . . in the school yard. I received several cuts but gave as good as I got. Tarve Major . . . had his eyes so knocked about that he is perfectly blind . . . Several others, collegers and oppidans, have black eyes and swelled noses.' *Melville Lawford, Eton schoolboy, 1843*

'Another arctic day . . . I went to shop in Harrods, knowing that they generate their own electricity. At the centre . . . is a large hall with rows of armchairs, in which a posse of weary elderly people had come to roost.' *Cynthia Gladwyn, diarist and diplomatic wife, 1947*

'I looked back. "Here it comes," I bawled. To the west the horizon was blotted out. I was totally unprepared for the next minutes. Nothing I had read in literature had prepared me. Neither tales from the Arctic, the Cairngorms nor Dartmoor itself. In an instant we were enveloped in a freezing white hell and a screaming wind that tore at us in all directions. We were literally stopped in our tracks, unable to think or walk or speak or see. In an instant we were blinded, struck dumb by elemental forces beyond imagination. The real terror was the inability to breathe. I was sucking snow into my lungs and my nostrils were furred up with encrusting snow . . . But the Gods were with us. A lull came in the wind, although the atmosphere was still thick with choking powdery snow, and shapes could be discerned, and within a stone's throw was the blessed lump that could only be the car . . . then came the awful realisation that the lock was hopelessly frozen. It was Derek's turn to try . . . He silently took the key from me and put it in his mouth over the door lock with the key jutting out from the side of his mouth . . . Even inside the car the wind had forced snow through.'

-23.9C Aviemore 1895

18.3C Kings Langley 1945

It turns out to be one of the greatest blizzards of the twentieth century.

THE WHITE KING 'The weather in Moscow is permanently the 19th February,' says a character in Peter Moffat's television drama *Cambridge Spies*. This is not strictly accurate: Moscow enjoys a continental climate of cold, dry snowy winters and hot summers with showers – so summer isn't remotely like winter. Winter days, however, can be similar – and today in Britain has certainly, on occasions, been Moscow-like.

'At such a time as the King's body was brought out of St George's Hall, the sky was serene and clear, but presently it began to snow,' writes Thomas Herbert, the King's attendant, of Charles I's funeral today in 1649. The black hearse is covered in snow. It's been a bitterly cold month: bitter when he is beheaded (*9 February*); bitter today. When the use of the *Book of Common Prayer* is refused in St George's Chapel, Windsor (❓), Bishop William Juxon, remembering the King's strict adherence to the prayer book, says nothing throughout the ceremony. Charles is known as 'the white king' – at his coronation he chooses white robes, 'the colour of innocence', over the traditional purple. 'So went the white king to his grave,' finishes Herbert. 'In the 48th year of his age and the 22nd and 10th month of his reign.'

'The river Thames overflowed its usual bounds, and entered the grand palace of Westminster, where it spread and covered the whole area, so that boats could float there, and people went to their apartments on horseback.'
Matthew Paris, historian, 1236

'Your letter sounds spring like and here with a thrush singing in the garden the sun shining and the rooks beginning to discuss among themselves the prospects of the coming nesting season I feel as though spring were getting near. And although I see that Roosevelt is cutting short his holidays on account of disturbing rumours about the intentions of the autocracies, and though I am rather glad he is doing so I myself am going about with a lighter heart than I have had for many a long day. All the information I get seems to point in the direction of peace and I repeat once more that I believe we have at last got on top of the dictators.' *Neville Chamberlain, politician, 1939*

'The wind ... drove in long, icy gusts over the white empty miles of moor. It was what the Yorkshiremen called a "thin wind" or sometimes a "lazy wind" – the kind that couldn't be bothered to blow round you but went straight through instead.' *James Herriot, vet in the Yorkshire Dales, 1940*

-22.2C Braemar 1895 ❄

16.7C Rugby 1945 ☀

20 FEBRUARY

1.1C min / 6.7C max

WEATHER AND CATHEDRALS

A storm today in 1861 beats first on the north-east side of Chichester (*♀*) Cathedral's Gothic spire and then comes with unabated force from the south-west. Sir Christopher Wren's spire, complete with an interior swinging balance to counterpoise the winds, is creaking. The following morning, 'the spire was seen to incline slightly to the south-west, and then to descend perpendicularly into the church, as one telescope tube slides into another, the mass of the tower crumbling beneath it.'

THE MEDIEVAL WARM PERIOD
Climate historians have labelled the period from, roughly, AD 900 to 1300 the 'Medieval Warm Period' or 'Little Optimum'. Characterized by mild, wet winters, frost-free springs, warm summers and wet and windy autumns, these are some of the mildest centuries of the last eight thousand years. Harvests are regularly bountiful; vineyards flourish (*27 July*); the population rises sharply; knights go crusading; and – the greatest legacy of this climatic golden age – many cathedrals are built.

This contemporary account by engineering professor Robert Willis illustrates a theme that runs through the history of the English cathedrals: they have continually taken a pasting from the weather. In a gale in 1362, the Romanesque tower of Norwich Cathedral falls. In 1660, the spire on the central tower atop Ripon Cathedral collapses through the roof in a gale. And in 1714, the upper portion of St Asaph Cathedral in North Wales is whisked off. Most spectacularly of all, the great leaded spire of Lincoln Cathedral – at the time, the tallest building in the world – is nobbled by a storm in 1548. These cathedrals ('Luxury liners laden with souls,' as W. H. Auden put it) are, until the late Victorian age, Britain's largest buildings by far, and accordingly, the most exposed to the weather.

But the weather creates as well as destroys. Lincoln (called by John Ruskin 'the most precious piece of architecture in the British Isles') represents the zenith of our greatest era of cathedral-building: work on the triumphal Gothic masterpiece begins in 1192. The choir of Canterbury Cathedral is built in the 1170s. Salisbury Cathedral is rebuilt between 1220 and 1258. The thirteenth century also sees work on Rochester, Wells, Carlisle, Durham and Hereford. The central tower and choir of St Paul's are completed in 1240. This unprecedented burst of building comes towards the end of a climatic period known as the 'Medieval Warm Period' – four centuries of benign weather for which people gave thanks to God by building wondrous monuments.

-22C Keith 1978 ❄

17.1C Prestatyn 1998 ☼

♀

'A MACHINE CALLED A SNOW PLOUGH'

Heavy snowfalls in 1814 prompt the Secretary of the Post Office, today, to demand that regional postmasters find better ways for keeping the mail coaches running, recommending 'a machine called a snow plough for clearing the roads. This machine, which is made by a few boards in the form of a wedge, was first used some years ago at Wimpole, by Mr Wm. Oswald, then bailiff of the Earl of Hardwicke'. Although snow-ploughs are referred to in America as early as 1794, this is their earliest mention here.

The first 'snow plough' comes at a turning point for British roads. After the Romans depart around 450, their superbly designed roads steadily decline. By the Middle Ages, most roads are impassable in bad weather. Vehicular transport develops slowly as, until crude suspension appears in the seventeenth century in the form of leather straps or springs, horses are more comfortable than wheels. Indeed, until the turnpike (toll) system provides for maintenance, prejudice mounts against carts because of the damage they cause – as with juggernauts today. Before enclosure, farmers, too, resent travellers of all kinds, who tend, when weather makes the roads impassable, to cut across the open fields.

So until stagecoaches, around 1640, 'whereupon one may be transported to any place sheltered from foul weather and foul ways' (Edward Chamberlayne, 1669), roads just get worse and worse. In the wet, they are either flooded or a sea of mud. In frost, the iron ground is treacherously hard to wheel and hoof alike, and dangerously slippery. In snow, they are blocked. In dry weather, so much dust is kicked up by coaches that it's impossible to see. The mail service, inaugurated in 1784 (*2 August*), provides a much-needed impetus for improvement. By 1810, some 33,000 miles of tracks have become turnpikes – unpopular, at least to start with, but maintained.

-18.3C Braemar 1955
17.5C Banchory 1974

In the first decade of the nineteenth century Thomas Telford reinvents camber for drainage. In December 1815, John Loudun Macadam outlines his theory of road construction: essentially small, angular gravel, of a size 'small enough to fit in the mouth', compacted on to well-drained soil, thus preventing deep wheel ruts. Asphalt arrives after 1824, and thereafter all the technical improvements we know today are mere details.

22 FEBRUARY

1.2C min / 7.2C max

**BLOOD AND
SNOW IN
GLENCOE**

In a whiteout tonight in 1692 two companies of government soldiers led by Captain Robert Campbell receive a final turn of generosity from the MacDonalds of Glencoe (♥), before bedding down. The 120 or so soldiers (of whom a dozen are Campbells, the MacDonalds' sworn hereditary enemies) have been billeted in hamlets throughout the glen for twelve days, but it's tonight, as a storm blasts into the mountains, that they are most grateful for the Highland hospitality.

Then, at 5 a.m., a great bonfire is lit and the soldiers rise to carry out their order: 'Fall upon the Rebells . . . and putt all to the sword under seventy.' The Glencoe Massacre of the MacDonalds begins. Men, women and young boys are shot and stabbed in the dark in their own homes. Alexander MacDonald, clan chief, is shot twice in the back as he gets out of bed. Whether his wife is thrown naked into the snow and left to die, her rings bitten off her fingers by a redcoat soldier, may or may not be true – distinguishing history from myth is difficult with Glencoe. What is certain is that nearly forty MacDonalds are slaughtered. The storm prevents government reinforcements from blocking routes out of the glen and other MacDonalds escape into the mountains, their tracks quickly covered in the blizzard, where, it's thought, many more freeze to death.

A government enquiry concludes that the Secretary for Scotland ordered the massacre, under the pretext that MacDonald failed to swear the required oath of allegiance to the King. William III (*15 November*) fails to punish those held responsible by the inquiry – inaction later exploited by Jacobite propagandists. But it's the infamous betrayal of trust, the heinous breach of Highland hospitality – shooting in the back the man who offers you his own home in a storm – that fuels a tribal ill-will which still haunts the glen. To this day the Clachaig Inn at Glencoe carries the sign: 'No hawkers, no Campbells.'

17.3C Aberdeen 1897
-22.2C Braemar 1955

'A gret frost lastynge more than xi weeks' ends today in 1434, according to the *Chronicle of London*. One of a series of merciless winters in the 1430s, it fuels clan fighting and cattle raiding in Scotland. In England, the hardship merely adds to the rising discord – feuding nobles, disrespect for government – in the early reign of the child king Henry VI, eventually culminating in the Wars of the Roses (*11 February, 7 April, 23 April*).

MONET,
IMPRESSIONISM
AND THE CITY
OF FOG

The London fog this morning in 1901 delights the sixty-one-year-old French Impressionist painter Claude Monet. From his sixth-floor window in the Savoy Hotel he has a panorama of the Thames views he's painting: Waterloo Bridge, Charing Cross railway bridge, the Houses of Parliament. Entranced by the complex, ever-changing scene as the sun battles with the murk, by 9 a.m. he's already worked on five canvases, switching, in frenzied bursts, from one to another as the light changes.

It's Monet's third extended visit in three years. For him, as for writers before him (*5 October*), it's fog (*6 December, 16 December*) that makes London wonderful. 'The beautiful effects . . . are scarcely to be believed,' he writes. 'I can't begin to describe a day as wonderful as this. One marvel after another, each lasting less than five minutes, it was enough to drive one mad.' Without its fog, London is 'insufficiently London-like'. Without fog, 'London would not be beautiful.'

Monet, of course, is the artist who, three decades earlier, paints an orange sun rising – through fog, of course – over the cranes of Le Havre. He titles it *Impression, Sunrise* – and 'Impressionism' is born. It's the most dramatic leap in art history since the discovery of perspective: a new way of seeing 'from nature', spontaneously recording what he sees rather than what convention tells him he sees – the transient effects of light and colour. And that's the subtext of why he's here. For though Monet's the acknowledged master of atmospheric effects, there's a past master, dead fifty years, who just might be bigger: Turner (*13 November*). By choosing many of the same subjects, from the same views, Monet's deliberately inviting comparison, trying to show his way is better – trying to reclaim for France the supremacy in landscape that Turner took from Claude Lorrain.

-25C Braemar 1955

At Monet's exhibition in 1904, a French critic calls his London paintings 'marvellous . . . one of the most beautiful demonstrations of pure art'. Another agrees: 'Monet seems to have attained the extreme limits of art . . . on the fogs of the Thames!' Finally one writes the magic words: 'If it is true that Turner liked to compare certain of his works to . . . Claude Lorrain's, then one might place certain Monets beside certain Turners.' And with that, Monet goes home to France to paint water lilies.

19.2C March 1990

24 FEBRUARY

BROLLYWOOD 'Welcome to Brollywood,' says some wag as stars arriving at today's 2002 BAFTA awards in London get a dousing. The rain is so heavy that the red carpet starts foaming, turning 'movie idols into soap stars'. Versace dresses and suede Manolo Blahniks are ruined as Nicole Kidman, Kate Winslet, Russell Crowe et al. tiptoe through the chemical froth emitted by the soaking carpet. Ironically, it's the first time the BAFTAs are broadcast on American TV – confirming every American's suspicions about British weather.

'The fog so thick in London, that the illuminations for the Queen's birthday were not visible.' *Thomas Raikes, diarist and member of the beau monde, 1832*

'The coldest day of all; frost feathers, flowers, ferns all over the windows, giving a dim clouded light inside; my pen frozen ... Then broadest sunlight pouring on to me in bed, warm, melting the frost flowers so that a steam goes up, waving, wreathing its shadows across this page as I write.' *Denton Welch, novelist, 1947*

THE 1608 THAMES FROST FAIR MELTS

'Where, for three months and more, there had been solid ice of such thickness that it seemed permanent as stone, and a whole gay city had been stood on its pavement, was now a race of turbulent yellow waters. The river had gained its freedom in the night ... All was riot and confusion. The river was strewn with icebergs. Some of these were as broad as a bowling green and as high as a house; others no bigger than a man's hat, but most fantastically twisted ... But what was the most awful and inspiring of terror was the sight of the human creatures who had been trapped in the night and now paced their twisting and precarious islands in the utmost agony of spirit. Whether they jumped into the flood or stayed on the ice their doom was certain ... As they swept out to sea, some could be heard crying vainly for help, making wild promises to amend their ways ... For furniture, valuables, possessions of all sorts were carried away on the icebergs. Among other strange sights was to be seen a cat suckling its young; a table laid sumptuously for a supper of twenty; a couple in bed; together with an extraordinary number of cooking utensils.' *Virginia Woolf*, Orlando, *1928*

17.8C Nairn 1891
-20.6C Dalwhinnie 1955

25 FEBRUARY

1.1C min / 7.2C max

SNOW IN THE PENNINES

'Seven o'clock in the morning with the wintry dawn only just beginning to lighten the eastern rim of the moor was not time to be digging my car out of the snow.

'This narrow, unfenced road skirted a high tableland and gave on to a few lonely farms at the end of even narrower tracks. It hadn't actually been snowing on my way out to this early call . . . but the wind had been rising steadily and whipping the top surface from the white blanket which had covered the fell-tops for weeks. My headlights had picked out the creeping drifts; pretty, pointed fingers feeling their way inch by inch across the strip of tarmac . . .'

'The snow falls, and the sheep and lambs are disconsolate, the sea disappears. Then all is white, and the sea leaden and horrible. Then, in an hour, the snow is gone again, the earth is so warm.'
D. H. Lawrence, writer, Cornwall, 1916

'A week of intense cold. I have two large oil radiators in the library but by the time I have answered my morning letters I am too frozen to write any more and slink back to the drawing-room.'
Evelyn Waugh, author, Combe Florey, Somerset, 1956

In 1933 today, Harrogate lies beneath 2½ feet (75 cm) of snow. For three days, from 23 to 26 February, all roads across the Pennines are blocked. The *Royal Scot* train arrives at Euston over twelve hours late. No one evokes winter in the Dales better than the country vet turned bestselling author James Herriot:

'This was how all blocked roads began, and at the farm . . . I could hear the wind buffeting the byre door and wondered if I would win the race home.

On the way back the drifts had stopped being pretty and lay across the road like white bolsters; but my little car had managed to cleave through them, veering crazily at times, wheels spinning, and now I could see the main road a few hundred yards ahead, reassuringly black in the pale light.'

Meanwhile, there's fog in London tonight in 1832. 'The Duke of Devonshire's ball was held in the clouds,' reports Dorothea, Princess Lieven, wife of the Russian ambassador to London, evidently not yet used to London's dominant meteorological effect (*5 October, 6 December, 16 December*). 'So thick was the fog in the drawing-room that you could not recognize people at the other end of the room.'

16.6C Dyce 1976

-20.6C Woburn 1947

26 FEBRUARY

FISH IN THE STREETS Strong wind and flurries of snow today in 1461 make the Earl of Warwick's bombardment of the Lancastrian army completely ineffective at the second Battle of St Albans (⚔) during the Wars of the Roses (*11 February, 7 April, 23 April*). Warwick has a strong fortified position and plenty of artillery, but it's almost impossible to light matches and fire the handguns. Not only do arrows fall short, but several primitive cannons explode when, after they are thought to have fired, they are reloaded. The Queen, Margaret of Anjou (*9 April*), is victorious. Warwick's previously unblemished military reputation is damaged.

> 'A cold rainy day, a high wind and rain at night ... A great scarcity of coals; 8d a peck at the hucksters, and 14s a barrel at the quays.' *John Fitzgerald, schoolmaster, Cork, 1793*

In the twenty-first century, today is the average earliest flowering date for daffodils, thirteen days earlier than twenty years ago. Always a sign of spring, daffs can appear as early as the first week of January nowadays. This isn't due only to global warming. Extensive cross-breeding has produced ever earlier-flowering bulbs. In Sale (📍), near Manchester, retired schoolmaster Hugh Beggs has carefully plotted the changing fortunes of daffodils in his garden for more than two decades, from the once traditional flowering date of 11 March (*21 March, 10 April*).

In 1990, a pensioner in Towyn, north-west Wales (📍), wonders if a main has burst when, on a bright and breezy day, water starts pouring through the streets. She's then surprised to see that there are fish in it. Yes, it's sea water. Gale force winds and a very high tide combine to create mountainous seas which batter most of the west Welsh coast with 30-foot (9-metre) waves. Towyn is worst hit. Some 4 square miles (10 square km) are flooded and 2400 homes are damaged in Wales's worst-ever coastal flood. The town has to be evacuated: the emergency services and the army ferry people and pets in rowing boats to higher ground so that the mop-up can begin. Though numerous vegetable gardens are ruined, roses seem to thrive.

-18.3C Kielder 1963 ❄
16.7C Leeming 1953 ☼

27 FEBRUARY

1.4C min / 7.7C max

WINTER INSPIRES ELIZABETH DAVID

Marooned today in 1947 by ice storms and blizzards in a hotel at Ross-on-Wye in Herefordshire (♥), a young Elizabeth David begins to feel an 'agonised homesickness for the sun and southern food' she has recently left behind. Throughout the month – the coldest February of the twentieth century – David has been holed up in the hotel with her lover, George Lassalle, trying to escape the dreariness of post-war London. After six years living in Cairo, India and various parts of the Mediterranean, David is ill-equipped for the brutal winter of 1947 (*4 March*) – although roaring fires, hot water bottles and breakfast in bed every morning do help.

Nothing, however, can abate her horror at the hotel food. Used to the scents and flavours of simple, fresh food with 'life, colour, guts, stimulus', she finds the cooking in the hotel as unspeakable as the weather outside. Meals are, she later writes, 'produced with a kind of bleak triumph, which amounted almost to a hatred of humanity and humanity's needs'.

> 'Rose very early; had a most severe cold morning, and found the roads now very bad in some places, the ice being broke by the coaches that it bore not, and rougher than a ploughed field in others, yet hard as iron, that it battered the horses' feet.'
> *Ralph Thoresby, traveller and antiquary, 1709*

> 'How misty is England! I have spent four years in a gray gloom.'
> *Nathaniel Hawthorne, novelist and American Consul in Liverpool, 1857*

To ease her 'embattled rage', David begins to write descriptions of Mediterranean and Middle Eastern cooking. 'Even to write words like apricot, olives and butter, rice and lemons, oil and almonds, produced assuagement. I came to realize that in the England of 1947, those were dirty words that I was putting down.' David has been collecting and refining recipes for almost a decade. Now, in a creative rage, and as the ice and snow outside turn to heavy rain and floods, she begins to write a series of aromatic memories. The result is *A Book of Mediterranean Food*, published in 1950 and the most influential book in the British kitchen since *Mrs Beeton's Book of Household Management* (1861). Unlike anything the British have read before, it brings Mediterranean sunshine to reawaken our taste buds.

-21.2C Grantown-on-Spey 1986 ❄

19.4C Cambridge 1891 ☀

28 FEBRUARY

1.4C min / 7.6C max

'Everywhere full of brick battes and tyeles flung down by the extraordinary Winde the last night (such as hath not been in memory before) that it was dangerous to go out of doors,' writes Samuel Pepys of today's gale in 1662. 'Several persons have been killed today by the fall of things in the streets.' 'Windy Tuesday' is the best-documented and, in southern England, worst storm in 350 years – certainly between St Maury's Wind in 1362 (*21 January*) and the Great Storm of 1703 (*7 December*). Discussing the effects of 'this late, great wind' a week later, Pepys writes: 'We have letters from the Forrest of Deane, that above 1000 oakes and as many beeches are blown down in one walke there.' The final total lost in the Forest of Dean (🌳) is more than three thousand trees – a serious loss, as this is the largest nursery for naval timber in England.

The gale provides a crucial spur to the diarist John Evelyn, who's recently been asked by the Commissioners of the Navy to assess the state of British forests. The importance of timber in the seventeenth century is hard to over-emphasize: it's the most important natural resource after food – as fuel, as a building material and, most crucially, for defence. Britain's position as a trading power rests entirely on her navy. A third-rate, 74-gun ship of the line consumes up to 3800 trees – about 75 acres (30 hectres) of woodland. Dockyards like Chatham and Deptford are vast ship-building factories, voracious for wood. The timber, mostly oak, has to be seasoned and stored for years before it can be used. As the necessary trees take a century to replace, well-planned forestry management is vital; felling more than are being grown is suicide. During the English Civil War the destruction of landed estates, royal forests and other woodland in search of quick profits, combined with disasters like today's gale, creates a potential crisis for the restored monarch, Charles II. With intense commercial rivalry on the high seas, not to mention wars, it's essential to rebuild timber stocks – which is what prompts Evelyn's presentation to the Royal Society this September, and the publication early in 1663 of *Silva, or a Discourse of Forest Trees and the Propagation of Timber*, the first rigorous treatise on forest management.

-19.3C Keith 1986 ❄
146.1mm Loch Ailort 1955

19.1C Harestock 1891 ☼

MARCH

1 MARCH

1.4C min / 7.6C max

FIRST OFFICIAL
DAY OF SPRING

'Spring, I enjoyed that,' says Michael Flanders. 'Missed it last year. I was in the bathroom.' Today is the official start of spring, according to meteorologists, who tidy the year into four neat, equal-sized compartments, though most of us mark the moment from the spring equinox (*21 March*). Recently, however, the start of spring has been creeping forward, according to which indicators you choose to use (*9 March, 10 April, 13 April, 21 April*). Is spring getting shorter, as Flanders suggests? Gardeners certainly complain that it is, with winter tipping suddenly into summer so fast that the moments in between hardly count as a full season. March tends to be showery, often with large temperature variations. Snow falls in March as often as in February and mariners fervently believe in gales – the so-called 'equinoctial gales' – around the vernal equinox, though there is limited hard evidence for this. When the wind does blow, it is most often from the north-west.

'Still very cold. Snow last night. But there is no mistaking the growing power of a March sun. Clumps of yellow crocus are out, and the white-mauve ones beginning; green buds are appearing.'
J. R. R. Tolkien, academic and writer, Oxford, 1944

'The severest Frost that ever I had been sensible of.'
Thomas Hearne, Assistant Keeper, Bodleian Library, Oxford, 1716

'Embarkation at Bristol [for Ireland]. We set sail and drove between Cornwall and Wales ... At night, there was a storm and a lot of wind ... The women and children cried, there were some who prayed, others yelled, nearly all were so ill that some of them were vomiting and suffered from diarrhoea. I was disgusted by the stench and the howling and feared the roughness of the sea and its danger.' *Ludolf von Münchhausen, German traveller, 1590*

-20.0C Lagganlia 1986 ❄

'In the first week of March, alder catkins are ... scattering their pollen on the east wind ... Reed heads of last season still survive and bend before the wind ... The wind blows hard, swaying the reeds and bushes and bending the big trees ... to its cold breath.' *E. L. Grant-Watson*, The Ladybird Guide to What to Look for in Spring, *1965*

18.9C Greenwich 1959 ☼

'In alle the hevenes there was no clowde sayne;
 From other daies that day was so devyded,
 And fraunchisyd from mystys and from rayn;
 The erthe attempred, the wyndes smothe and playne.'
John Lydgate, monk and poet, describes Henry VI's entry into London, 1431

2 MARCH

1.3C min / 7.7C max

FIRST AND WORST HEATHROW AIR CRASH

In thick fog today in 1948, with visibility reduced to 65 feet (20 metres), a Sabena Airlines Dakota flying from Brussels to London misses Heathrow's concrete runway and nose-dives on to the grass. The DC3 ('the plane that changed the world') explodes in a ball of fire. All three members of the Belgian crew and sixteen passengers die. Airport crash tenders and the fire service manage to pull three passengers, including a former MP, Brigadier Nicholson, from the wreckage alive. London Heathrow airport – an army surplus tent with a bar and a post office – has only been open twenty-one months. Rescuers combing the wreckage report seeing a gentleman in a bowler hat and twill trousers asking about his briefcase, giving rise to the myth of 'the ghost of Runway One'. Sightings of the executive phantom, as well as radar detections of a figure on the runway, have been reported intermittently since. It's Heathrow's first-ever, and worst-ever, crash.

During the great winter of 1963, there are finally signs of a thaw. A family on Dartmoor, where snow has been piled up in 26-foot (8-metre) drifts, is rescued after sixty-five days marooned in a remote farmhouse. Up north, however, the freeze continues: Halifax Town FC finally give up all thoughts of football on their pitch and open it as an ice-rink.

FOG AND FLYING: BITTEREST ENEMIES

The first fatal accident to a scheduled British commercial flight is due to fog, when a Handley Page taking off from Cricklewood crashes in December 1920. When Croydon becomes London's official airport the same year, fog is the commonest reason for cancelled flights. Inbound flights are guided by 'smoke bombs and star shells' (*The Times*, 1929).

The first significant fog-flying technology arrives in 1931, in the form of a white line chalked on to grass to help pilots orientate themselves when taking off – it's regarded as a tremendous innovation. Landing is the real problem, however (*12 October*). Clearing fog by burning oil is tried in the Second World War (*20 November*), but is too expensive for commercial aviation. The first automatic blind landing is performed by the Hawker Siddeley Trident in 1966, and it's 1989 before British Airways are advertising: 'We can take off when other airlines haven't the foggiest,' due to new technology at Heathrow. Even today, fog remains the chief enemy to airline schedules – witness Christmas 2006, when more than 900 cancellations and delays over four days affect nearly 100,000 travellers. Gatwick airport, notoriously fog-prone, is another story (*17 February*).

-20.0C Braemar 1965

20.2C Exeter Airport 1977

3 MARCH

1.2C min / 7.8C max

COD / THE 500-YEAR COLD SNAP

As the Thames thaws today in 1684 (*19 January*), ships can reach the Port of London for the first time in weeks. And as ice along the Channel coast begins to break up, the cod fishing fleet, which usually leaves port in early February, can now finally prepare.

Relief, however, is tempered with fear. Cod are highly sensitive to temperature, and recently stocks have become scarcer. The fish has been a staple since Roman times. But throughout the seventeenth century sea temperatures around the British Isles and Norway have been falling – part of the 'Little Ice Age' (see below). Critically, temperatures have fallen below 2C for periods of twenty years. In the Faeroe and Shetland Islands, cod fishing has, in some years since 1675, failed completely. The English cod fleets have also traditionally fished along the Norwegian coast and off Iceland. Now there are no cod there either and British fishermen will have to make the perilous journey across the Atlantic, to fish off Greenland and the Grand Banks of Newfoundland.

THE 'LITTLE ICE AGE'

Know how winters from the past seem so much colder, with their frost fairs, famines and Dickensian winters? Well, that's because for roughly half a millennium, they are. Between the end of the 'Medieval Warm Period' (*20 February*) and the present sustained, incremental global warming is a cooler period between, very roughly, 1350 and 1850. This has been called the 'Little Ice Age'. And within this period there are three distinct minima, beginning around 1650, 1770 and 1850. The Little Ice Age is not one long freeze, with humans ranging across treeless tundra hunting mammoths, as the term might suggest. Rather it is a generally modest cooling, characterized by cycles of severe winters with easterly winds (one of the harshest being the 1690s – *7 November*), heavy spring and autumn rains, and occasional, intense summer heatwaves. Big storms also increase, in both intensity and frequency (*17 August, 19 October, 7 December*). Overall (especially between 1550 and 1700) it is cooler and damper than today. What lies behind climate change on this scale? Worldwide volcanic activity? Sunspots (dark, cool patches that appear in cycles on the sun's surface)? No one is sure. What's certain is that this zigzag of climatic movements has profound effects on society.

-21.7C Kinbrace 2001

17.2C Hunstanton 1939

SHIVER WITH SHINWELL

The blizzard which sweeps England and Wales today in 1947 buries a country already paralysed. For six weeks, snow has fallen somewhere every day. It's the snowiest winter of the twentieth century; the snowiest since records began. At Kew no sun is recorded for 22 out of 28 days in February.

'Economize in all fuels – even to the point of inconvenience,' implores the government. Post-war Britain is already in near-crisis, with chronic food shortages and unemployment approaching 2 million. Now, as coal barges freeze in the Thames, power stations have to cut back or close. Factories follow. As the freeze grips tighter, there are power cuts and reduced gas supplies. The little additional miseries pile on top: BBC programmes are cancelled, magazines cease publication and two hundred Football League matches are postponed. Water pipes burst by the hundred thousand. And still it freezes. And still it snows. With roads blocked for weeks, whole towns run out of food and lifeboats are used to deliver bread. Some 500,000 acres (200,000 hectares) of wheat are lost and 2 million sheep die, while farmers use pneumatic drills to dig turnips in Norfolk.

As the last glimmer of post-war optimism is snuffed out, Prime Minister Clement Attlee knows that his honeymoon is over. He can't be blamed for the weather, but for failing to anticipate the coal crisis he can – and is. Attlee institutes 'austerity' measures. Food rationing, so recently lifted, is reintroduced. Social services are cut. Later, wages are frozen. The most radical government of the twentieth century, the one that introduces nationalization of key industries and the National Health Service, is forced into economic orthodoxy.

Emanuel Shinwell, Minister of Fuel and Power, and John Strachey, Minister of Food, bear the brunt of the Tory jeers: 'Shiver with Shinwell and starve with Strachey.'

'There were two things I liked well in the Irish houses in the country: a pretty maid and normally, a pretty wind, sometimes also a pretty horse.'
Ludolf von Münchhausen, German traveller, 1590

'A beautiful day of early spring, quite perfect in light, colour, the shadow and tone of stone and lawn and blossom. We have started to mow with the sit-down mowers that roll and stripe. No more rotaries until September when the plantains grow.'
Alan Clark, politician, Saltwood Castle, Kent, 1990

-21.1C Houghall 1947

19.4C St. James's Park 1928

FILMING *RYAN'S DAUGHTER*

Under grey skies, in drizzle, John Mills shoots his first scene in David Lean's *Ryan's Daughter* at the end of the Dingle Peninsula, County Kerry (♀), today in 1969. Lean has insisted on building the fictional village of Kirrary atop the wild seascape near Dunquin, westernmost point of mainland Europe. Unfortunately, with the wild setting comes wild Irish weather. Within days things go awry and John Mills nearly drowns rowing through Atlantic rollers in strong winds, between the mainland and the Blasket Islands.

But Lean's problems have hardly begun. The utter unpredictability of the weather begins to haunt the crew. 'Time and time again we'd start shooting sunshine scenes and by midday it would cloud over,' producer Anthony Havelock-Allan later notes. 'Or conversely, we'd go out to shoot rain scenes . . . and the skies would be blue.' Sarah Miles, playing Rosy Ryan, recalls: 'Once we did half a scene and I remember waiting for three solid weeks before there was enough sun to finish the other half.'

It's all a cinematographer's nightmare. The gaps between the good weather grow so long that the seasons change. Slate skies interrupt spring sunshine in the love-making scene between Rosy Ryan and Major Doryan in a bluebell wood. By the time the sun comes out, it's autumn. The options are rewrite the script, complete the scene in New Zealand, or create a bluebell wood indoors. In a village hall, under lights, the crew do just that, with soil, trees, moss, birds and butterflies all brought in.

For the great storm scene, Lean wants meteorological menace on a biblical scale. Freddie Young, the cinematographer, even invents a spinning windscreen wiper mechanism to keep the camera lens clear when shooting into rain. But, with cast and crew on standby for a year (the schedule allows for six months), the sky resolutely refuses to darken – and the production goes £4 million over budget. After months of waiting, punctuated with fruitless storm-chasing up and down the coast, the scene has to be pieced together from five separate storms. The film is released in 1971 to savage criticism. Several films, pundits say, could have been made for the budget. David Lean goes into self-imposed exile and doesn't make another film for fourteen years. *Ryan's Daughter* wins two Oscars.

155.8mm Gobernuisgach Lodge 1983

-17.8C Braemar 1947

19.4C Leeming 1950

6 MARCH

2C min / 8.5C max

THE HIGHLANDS v ALPS COMPETITION

'It was a curious experience, this blizzard coming from below. Before we had been exposed to it for two minutes, we resembled two polar bears ... Our clothes were soon converted into coats of mail. Our eyes and noses were filled with drifted snow. Icicles depended from our hair, and even our cheeks were encrusted with ice. The fine dust penetrated every seam and opening in our attire. It drifted up our sleeves, and under coats and jerseys, and even formed snowballs in the sanctum of the watch pocket.'

The relish with which William Naismith recounts his ascent of Ben Lui (☛), today in 1897, perfectly reflects the new attitude to wild landscapes – and wild weather – that emerges around this time. In the wake of Edward Whymper's ascent of the Matterhorn in 1865, and the subsequent English 'discovery' of the Alps, a similar but less well-known discovery is occurring of our own wild spaces back home. The Scottish Mountaineering Club, founded in 1889 by Naismith, is the first club of its kind. Its members are an assortment of the new breed of hearty Victorian outdoorsmen – academics, vicars, country gents – who like nothing better than to pull on stout tweeds and nailed boots, and head into the hills in the teeth of a gale or blizzard, with the novel, eccentric aim of ascending mountains by awkward routes for no purpose other than pleasure. It's the beginning of a larger appreciation of wild spaces that eventually sees the creation of national parks and new 'rights to roam' (*14 March*).

The SMC's singularly Scottish approach to discomfort sets a tone that permeates British mountaineering for a century (still discernible, in some climbing books, in crag route designations like 'an ideal wet weather climb'). The enthusiasm for our home-grown winter landscapes soon spills over, in the pages of the *Scottish Mountaineering Journal*, into competitive swipes at the Alps: 'when rocks are coated with ice, when slopes of hard frozen snow often compel the use of the axe, when every *arête* has an overhanging cornice of snow, when, in short, our Scottish hills present most of the characteristics and many of the difficulties of mountains in the Alps' ... then who needs the Alps? But the SMC's delight in being on the hill in hard weather also culminates in the British invention of a new sport – ice-climbing (*22 April*).

-20.0C Braemar 1947

19.4C Redcar 1950

7 MARCH

THE NUMB MISERY OF SUPPORTING ARSENAL

'And the game went on, and the sky darkened, and Arsenal got worse, eventually conceding a goal, which in their hangover listlessness was one goal too many. And you stand there on the huge crumbling terrace, your feet stiffening and then actually burning in the cold, with the Chelsea fans jeering and gesturing at you, and you wonder why you bothered, when you knew, not only in your heart of hearts but with your head as well, that the game would be dull, and the players would be inept, that the feelings engendered on the Wednesday [when Arsenal beat Tottenham, at White Hart Lane, in a Littlewood's Cup semi-final replay] would have dribbled away to a flat nothingness before twenty minutes of the Saturday game had passed when, if you had stayed at home or gone record shopping, you could have kept the embers glowing for another week longer. But then, these are the games, the 1–0 defeats at Chelsea on a miserable March afternoon, that give meaning to the rest, and it is precisely because you have seen so many of them that there is real joy to be had from those others that come once every six, seven, ten years ... And yet as we were waiting to be let out ... the sheer awfulness of it all deepened and thus the experience was lent a perverse kind of glory, so that those of us there became entitled to award themselves a campaign medal.

Two things happened. First, it began to snow and the discomfort was such that you wanted to laugh at yourself for tolerating this fan's life any longer; and secondly, a man came out with a rolling machine and proceeded to drive up and down the pitch with it. He was not the irascible old git of football club legend, but an enormous young man with a monstrous skinhead haircut, and he obviously hated Arsenal with all the passion of his employers' followers. As he drove towards us on his machine, he gave us the finger, a delighted and maniacal smile on his face; and on his return visit he gave us the finger again, and so it went on – up, back, and the finger. Up, back and the finger.

And we had to stand and watch him do it, over and over again, in the dark and the freezing cold, while the snow fell on us in our concrete compound. It was a proper, thorough restoration of normal service.'
Nick Hornby, Fever Pitch, *1992, describing Chelsea v Arsenal, Stamford Bridge, 1987*

-18.9C Alston 1896 ❄

19.4C Norwich 1906 ☼

8 MARCH

1.9C min / 8.4C max

COLOURED RAIN / 'THIRTEEN DRIFTY DAYS'

A thunderstorm rains blue-black ink on the Shetland Isles (♀) today in 1935. Living as they do in a place where, the saying goes, 'it rains between showers', the Shetlanders know a bit about rain. But what's going on? In fact, coloured rain is not such an unusual phenomenon. Strong winds and convective currents (see *Definitions*) over the Sahara Desert lift vast quantities of sand into the atmosphere, which prevailing winds then carry across the Bay of Biscay to the British Isles, where it's deposited with rain, and occasionally fog. Usually it's red, occasionally yellow or inky blue. In November 1979, it's estimated that 60,000 tons of dust lands on County Cork. In 1902, pink snow falls across Cornwall and South Wales, coating washing hung out on lines with 'iron rust'. In July 1968, much of England and Wales gets a red dusting. Of course, understanding this phenomenon makes it seem a lot less interesting. Imagine how villagers in Somerset felt, in 1324, when for six hours it rains 'blood'.

In a thirteen-day blizzard – 'The thirteen drifty days' – most of the livestock in the Scottish borders perish in 1674. And in 1922 it's a case of 'Wind in the Channel – Continent cut off' when a gale on the south coast tears down telegraph and telephone wires to France. Instruments at many Met Office observatories are so overloaded they are destroyed or unable to report data, though at 4 a.m. a gust of 108 mph (170 kph) is recorded on the Scilly Isles – 'a speed which, since reliable observations began, has only once been exceeded,' *The Times* reports.

'I realized, quite suddenly, that spring had come ... in the lee of a small pine wood I leaned my back against a tree and was aware, all at once, of the sunshine, warm on my closed eyelids, the clamour of the larks, the muted sea-sound of the wind in the high branches. And though the snow still lay in long runnels behind the walls and the grass was lifeless and winter-yellowed, there was a feeling of change; almost of liberation.'
James Herriot in 1940, If Only They Could Talk

'It was the same jolly, clear spring weather, and I simply could not contrive to feel careworn. Indeed I was in better spirits than I had been for months ... Nesting curlews and plovers were crying everywhere, and the links of green pasture by the streams were dotted with young lambs. All the slackness of the past months was slipping from my bones, and I stepped out like a four-year-old.'
John Buchan, The Thirty-Nine Steps, *1915*

-21.1C Braemar 1947 ❄

149.2mm Honister Pass 1979

21.1C Colwyn Bay 1929 ☼

9 MARCH

1.7C min / 8.5C max

FIRST FROGSPAWN / SUN, SUN, SUN, HERE IT COMES

Today, on average (from the last five years), is the first day for the characteristic, giant tapioca of frogspawn to appear in ponds, ditches and slow-moving streams. Of the numerous 'indications of spring' (*10 March*), from blackbirds singing to swallows arriving (*13 April*) to grass growth kicking in (*21 March*, *18 April*), this is one of the most recognizable ones. Frogs breed only once a year, and will frequently do so early in mild weather (frogspawn has been reported as early as the first week of December). Although a hard frost will kill spawn, once tadpoles have hatched they can swim to the bottom of the pond where they are better protected.

'Out in a frightful snowstorm – the wind so bitter I had to take my hat off to guard my face, leaving my head to take care of itself.'
John Ruskin, critic, Brantwood, Coniston, 1876

'The spring is coming. On Sunday, the gorse, all happy with flower, smelled hot and sweet in the sun, and the lambs were leaping into the air, kicking their hind legs with a wild little flourish. But this morning all was white with snow. This evening, however, in the deep yellow sunset, the birds are whistling, the snow is nearly gone. Soon the very spring will conquer.'
D. H. Lawrence, writer, Cornwall, 1916

'[Brighton is] like a foreign town: the first springy day. Women sitting on seats. A pretty hat in a teashop – how fashion revives the eye! And the shell encrusted old women, rouged, decked, cadaverous at the teashop.'
Virginia Woolf, writer, 1941

'It seems as if winter in England goes on forever; by the time spring comes, you really deserve it,' writes George Harrison in 1969. 'So one day I decided I was going to "sag" off. I went over to Eric [Clapton]'s house.' There, seeking solace from the endless meetings at Apple, the Beatles' fledgling company, he wanders in the garden, the sun warming his face. It inspires one of the most unforgettable sounds of the *Abbey Road* album: 'Here comes the sun'. Elsewhere, he describes the effect of the afternoon: 'I found some sort of release and the song just came.' He completes the song on holiday in June, and it's recorded at Abbey Road studios in July.

-19.4C Braemar 1917

23.9C Wealdstone 1948

10 MARCH

1.9C min / 8.9C max

A NEW BRITISH WORD: 'BLIZZARD'

The snowstorm that rages in the West Country today, in 1891, is so dramatic that a new word has to be appropriated into the English language to describe it: 'blizzard'. Thought to derive from a German expression, '*Der Sturm kommt blitzartig*', which translates as 'The storm comes lightning-like', a blizzard is defined by the Met Office as 'a violent snow-storm with winds blowing at a minimum speed of 35 mph [56 kph] and visibility of less than a quarter of a mile [0.4 km] for three hours.'

This storm, which sets a benchmark for fifty years, ranks alongside the greats – 1881 (*18 January*), December 1927, 1963 (*6 February*), 1947 (*4 March*) and February 1978. Heavy, fine, powdery snow and strong easterlies tear across south-west England and Wales for five days. The average depth of snow is 2 feet [0.6 metres], with drifts of up to 30 feet (9 metres). Apart from innumerable incidents and private tragedies, over half a million trees fall, as well as many of the new 'telegraph poles': 220 die, many on 65 ships which founder in the English Channel. At least 6000 sheep are killed and 14 trains are stranded in Devon alone. The Great Western's *Zulu Express* leaves Paddington at 3 p.m. today, a Monday, runs into heavy drifts on the southern side of Dartmoor and reaches Plymouth four days late at 8.30 p.m. on Friday. There's even snow in the Scillies, the mildest part of the UK.

'I have been feeling cheerful today on account of the weather. Really it has been a delightful early spring day. The sun has been shining in the cold sky since dawn ... You can remember, I expect, days like this when every thing far or near at hand seems specially graced by the light. Sheep, railway engines, yards, lanes, distant hills, iron gates, drinking pumps.'
Philip Larkin, poet, Wellington, Shropshire, 1946

'Ted [Hughes] and I went for a walk on Primrose Hill this noon after lunch, and the air was mild and damply spring-like; the sun almost warm, and green buds out on all the lilac bushes.' *Sylvia Plath, poet, 1960*

'Slowly her wings curved back. She slipped smoothly through the wind, as though she were moving forward on a wire. This mastery of the roaring wind, this majesty and noble power of flight, made me shout aloud and dance up and down with excitement. Now, I thought, I have seen the best of the peregrine.'
J. A. Baker, The Peregrine, 1967

21.1C Roade 1929
-15.0C Rickmansworth 1931

11 MARCH

DROUGHT AND
THE WRONG
KIND OF RAIN

A meagre 0.3mm of rain falls in Canterbury (♀) today in 2006, and that's it for the next eleven days. By the end of the month, Kent has received less than its expected rainfall for the seventeenth month of nineteen. Six water companies will shortly impose hosepipe bans affecting 13 million, and the Environment Agency will claim that south-east England has less water per head than parts of Sudan and Syria. Remarkably, it's true.

Droughts are a recurrent feature of the British climate. They vary greatly in type, regional extent and intensity, and can be defined in several ways. The 2006 water shortage is a *hydrological drought*: two dry winters have failed to replenish the underground aquifers from which south-east England's water is drawn. Such a drought is remedied by steady rain, which seeps through soil into rock. Drastic downpours are no help, as water runs off quickly. Bad *hydrological droughts* in the south-east include those of 1890, 1934, 1964 and 1972.

An *agricultural drought* involves low rainfall during the growing season – spring and summer. During the spring of 1893, some parts of the south-east see no rain for 61 consecutive days, the dearth continuing into 1894, exacerbating the agricultural depression. In 1938, south-west Scotland has no rain for 38 days. The years 1934 (*11 July*), 1975 and 1997 are also bad for farmers.

The third kind of drought is a *meteorological drought* – straightforward rainfall deficiency. Rainless calendar months, or thirty-day periods, are not rare. Most occur in February and they can even affect Britain's wettest place – Seathwaite in Cumbria (*8 May*) is rainless in February 1932. And in February 1947 there are peat fires in western Scotland because it's so dry (while southern Britain freezes in the worst winter for half a century).

-19.4C Logie Coldstone 1958 ❄

22.8C Aber 1957 ☀

The 1976 drought (*3 July*) remains the benchmark, though the increased incidence of droughts in the last thirty years may have more to do with human wastage than with climate change. In 681, St Wilfrid arrives in the famine-hit south-east where there's been no rain for three years. On the day he converts the Saxons to Christianity, 'a soft but plentiful rain' falls.

12 MARCH

1.7C min / 8.7C max

FRENCH INVASION FOILED 'Decidedly the wind is not Jacobite!' Marshal de Saxe exclaims when the much-cherished scheme of the cabinet at Versailles – to invade England – is foiled, today in 1744, by gales in the Channel. The French have been planning an invasion along the east coast since the death of George I in 1727. The old enmity between the nations, each with developing empires overseas, is at a head.

Spurred also by James III, who has rekindled his dreams of reclaiming the thrones of England and Scotland, fifteen thousand French troops gather at Dunkirk in late February. Many have embarked by the time the gale intensifies today. But de Saxe, the illegitimate son of a Saxon noble who's risen to the top of the French army, is not easily deterred. He boards his flagship, but the swell is too great to put to sea and the fleet is badly battered.

> 'The sun shone while it rained, and the stones of the walls and the pebbles on the road glittered like silver.'
> *Dorothy Wordsworth, Lake District, 1802*

> 'Quite one of the horriblest black total eclipses we have had this year, with wild drifting snow.'
> *John Ruskin, critic, Oxford, 1876*

De Saxe is undaunted. Three days later, when the wind drops, he embarks his troops again. In an action replay, another severe gale hits the fleet. Virtually all the ships are driven ashore and many soldiers drown. The invasion is abandoned, and the Auld Alliance, between France and Scotland, is torn up. For Charles Stuart, James's son (*27 April, 8 July*), who's at Dunkirk and hoping to sail up the Thames as his father's supporters rise in Scotland and invade from the north, it's a crushing blow.

In 1957, the temperature reaches 23.3C at Haydon Bridge, Northumberland and 23C at Cape Wrath (♥) – record highs for the time of year. The reason is a *föhn*, a mountain wind that makes lowlands in the lee of hills unseasonably warm. Moist southerly airstreams rise over the windward slopes of high ground – the Pennines and Western Highlands in this instance – and descend the lee slopes, arriving back at low level, downwind, 3C or more warmer than their original state. The phenomenon is first studied in the Alps – the term is German – but *föhn* winds occur in many parts of the world. In Britain, the *föhn* effect is most noticeable in winter, on the North Wales coast (*27 January*), in Aberdeen and on the Moray coast.

-22.2C Grantown-oh-Spey 1958

23.3C Haydon Bridge 1957

13 MARCH

1.7C min / 8.7C max

BRITAIN'S BLEAKEST MAIN ROAD, THE A6

The snowstorm that begins today in 1947 causes 'the biggest road transport hold-up ever known', on Shap Fell in Westmoreland (now Cumbria). Until 1970, and the completion of the M6 between Kendal and Penrith, the A6, snaking up the flanks of the Lakeland fells to Shap summit (𝔓), is the main highway to Scotland – and, at almost 1400 feet (426 metres), is the highest main road in Britain. A plaque on the summit calls it 'this old and difficult route'. From the early eighteenth century the road, almost always busy, turns the village of Shap into a ribbon of prosperous shops, hotels and other businesses provisioning travellers. But in all its long history of bad weather and bad winters this is the bleakest moment.

'The worst section of the England to Scotland main road has a seven mile succession of the biggest drifts of the century and is likely to remain blocked for several days,' reports one newspaper. At each end of this stretch, hundreds of heavy lorries block the way as the police stop all vehicles joining the road. Drivers of stranded vehicles are advised to take the train home. The situation is worst for upland farmers, unable to get hay for their animals off the convoys of lorries sitting uselessly along the blocked road. While hundreds of vehicles are abandoned, many lorry drivers are instructed by their employers to stay with their vehicles and guard their loads. As they shiver through a long, dark night, they can enjoy the irony that British Summer Time begins. It's winter's last gasp, however. Following one more heavy snowfall tomorrow, the thaw creeps north from the south-west, bringing thick fog, heavy rain, a south-westerly wind, widespread flooding – and a clear road to Scotland.

'THIS OLD AND DIFFICULT ROUTE OVER SOUTH SHAP FELL'

'The tough, long-distance lorry men who traverse this road several times a week are not particularly worried about these conditions [snow and ice] but when they are met on the summit by a blinding blizzard and cannot distinguish the road ahead from the terrible drop over the edge, they have to stop and ride out the storm somehow. Sometimes the drivers have sat huddled in their draughty cabs waiting for the dawn; other men have left their lorries half buried ... and staggered knee- or waist-deep in the snow down the steep, twisting miles to the bright lights of the Jungle Café, which every driver knows.'
A. Harry Griffin, February 1952

-13.9C Logie Coldstone 1958

21.1C Peebles 1957

14 MARCH

1.9C min / 8.9C max

BIRTH OF
MODERN
MOUNTAIN
RESCUE

'Showers and sunny intervals' is the forecast for the fifty-first annual Derby Scouts 'Four Inns Walk' this Saturday in 1964, over the high moorland of the Peak District (📍). From the start, however, the rain is steady and increasingly heavy, with a severe, cold wind. In the rapidly deteriorating weather, a lightly clad nineteen-year-old collapses and dies. Two others, aged twenty-one and twenty-four, go missing. As rescue parties set out, mostly consisting of untrained volunteers, the weather turns into a blizzard. A second body is recovered on Monday, and a third – by this time 370 searchers are involved, desperately combing the snow-covered moorland – the next day. All three die from exhaustion and hypothermia.

'Being agog to see some Devonshire, I would have taken a walk the first day, but the rain would not let me; and the second, but the rain would not let me; and the third; but the rain forbade it – Ditto 4 ditto 5 – So I made up my mind to stay in doors.'

John Keats, poet, 1818

The tragedy is a landmark. It's the worst incident involving walkers since the Glen Doll disaster five years previously (*1 January*), and comes at a key moment in the outdoor leisure movement. After the Second World War, legislation enacted in 1949 (following the high-profile, mass trespasses of nearby Kinder Scout in 1932) has popularized walking and access to wild spaces. The Peak District, Britain's first designated National Park (in 1951), surrounded by densely populated conurbations, is already the most-visited wild area in Europe.

It becomes clear that the searches could have been far better organized. The disparate, well-meaning mountain rescue organizations that have sprung up after the Second World War (Keswick is first, in 1947) need to be far better equipped and coordinated. It also becomes clear how little is known, medically, about 'exposure' (as hypothermia is known) and 'wind chill'. (The term is coined by an Antarctic explorer, Paul Siple, in 1939, highlighting the fact that almost the only available information for those undertaking severe exercise in adverse conditions is for cold but dry regions such as the Himalayas or Antarctic, rather than Britain's predominantly wet, windy conditions.) By 2007, there are seventy-five well-equipped mountain rescue organizations serving the wild upland areas of Britain – and 75 per cent of mountain rescue incidents are still due to the weather.

-22.8C Logie Coldstone 1958 ❄

20.6C Gatwick 1961 ☀

15 MARCH

2.2C min / 9C max

FIRST LIFEBOAT / CUP FINAL IN THE MUD

When a violent storm sinks a Newcastle sailing ship, the *Adventure*, just offshore at South Shields (⚓) today in 1789, hundreds watch helplessly from the cliffs. Everyone aboard is drowned. The local community is so appalled by their inability to help that the members of a local social club institute a competition to design a purpose-built rescue vessel. A prize of two guineas is offered. Willie Wouldhave, the parish clerk, comes up with the design (below), which is constructed by local boatbuilder Henry Greathead. The 30-foot-long (10 metres) boat, with twelve oars and a curved keel, carries 784 lb (356 kg) of cork for buoyancy. It's called the *Original* as it's the first lifeboat built for the job, rather than a conversion, and launches ten months later – the most important ever development in the history of sea rescue.

'I began my tour through England and Scotland; the lovely weather continuing, such as the oldest man alive has not seen before, for January, February, and half of March . . . I preached in Mr Stephen's orchard to far more than his church would have contained. And it was no inconvenience either to me or them, as it was a mild, still evening.'
John Wesley, founder of Methodism, Tewkesbury, 1779

'The air is so mild, I have been sketching out of doors all morning, and felt almost too hot in the sun in spite of the snow on the hills. I have been drawing the stump of a hollow tree for another hedgehog drawing; there is not much sign of spring yet but the moss is very pretty in the woods.'
Beatrix Potter, author and artist, Denbigh, 1904

After days of rain, the pitch is 'like a cabbage patch' at Wembley today in 1969, when Arsenal lose 3–1 to Third Division Swindon Town in the League Cup Final. Mud, like laughter, is a great leveller: there are plenty of both as Arsenal, fielding household names like Frank McLintock and George Graham, lose their way in the mire. In extra time, twenty-three-year old Don Rogers flicks the ball round Bob Wilson to seal the Gunners' fate. The image of the Arsenal goalkeeper, left flailing in the mud, says it all.

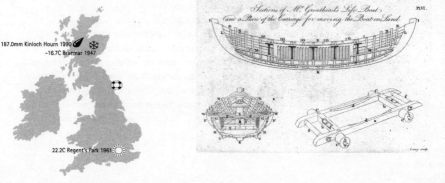

187.0mm Kinloch Hourn 1990
-16.7C Braemar 1947

22.2C Regent's Park 1961

16 MARCH

2.3C min / 9.4C max

THE GREAT SUCK

The gale begins as gangs of men are out on top of the banks of the Great Ouse (♥), in failing light, desperately trying to repair breaches in the dykes. It's a Sunday, in 1947, in the Fens. The year has already brought the severest winter, with some of the heaviest snow, of the twentieth century (*4 March*). (And this is in a country still picking up the pieces after the Second World War.) In the second week of March the thaw kicks in. Several feet of snow melt and the ground, still frozen brick-hard, is unable to absorb the water. As thawing snow releases water both faster, and in far greater quantity, than even the heaviest rainstorm, tracts of low-lying Britain are poised for disaster.

'I am now ... IN but not "settled" in. The weather (which seems a slice of our normal "wedding-day weather" come too early) contributes to my comfort. The great bank in the Fellows' Garden looks like the foreground of a pre-Raphaelite picture: blazing green starred like the Milky Way with blue anemones, purple/white/yellow crocuses, and final surprise, clouded-yellow, peacock, and tortoiseshell butterflies flitting about.'
J. R. R. Tolkien, academic and writer, Oxford, 1972

'Most bitter cold, the King saying today that it was the coldest day he ever knew in England.'
Samuel Pepys, diarist, 1667

In the Fens, following centuries of drainage and shrinkage, much of this reclaimed marshland is below sea-level. As the warm air starts arriving, and the flooding begins, the worst possible thing happens. It starts raining. Nothing – not even the warm air and bright sunshine – melts snow like rain. Then, today, as the rivers begin bursting their banks, the south-westerly gale gets up. Rapidly reaching 65 mph (104 kph), gusting 100 mph (160 kph), it could not blow from a worse direction: directly along the flooded rivers, piling the water into waves. Local authorities, fire brigades, the army, even remaining prisoners of war, get to work with armadas of pumps. The Great Suck begins. Not all the land will re-emerge until June.

-16.1C Braemar 1947 ✳

The massive thaw gives the writer John Wyndham – author of *The Day of the Triffids* and *The Midwich Cuckoos* – the idea for a story about aliens melting ice sheets to cause a rise in sea levels. He calls it *The Kraken Wakes*. In 1801, gales wreck the 74-gun HMS *Invincible* off the Norfolk coast as she sails to attack the Danish fleet, drowning 400. In 1978 the Cheltenham Gold Cup is cancelled because of 6 inches (15 cm) of snow; although a rapid thaw follows, it's too late to reinstate the race. The Gold Cup is snowed off in 1931 and 1937 too.

22.8C Goldington 1961 ☼

THE 'FATHER OF
METEOROLOGY'

'Hail with snow and light wind four or five times in the day,' the Reverend William Merle records in his diary today in 1340. Merle, rector of Driby in Lincolnshire (🌑) and a fellow of Merton College, Oxford, is a pioneer: he is the first person to keep a daily weather record. He does so scrupulously for sixty-seven years, from January 1277 to January 1344 – in Latin. Merle, accordingly, is the father of meteorology.

'The severest winter that any man alive had known in England. The crowes feete were frozen to their prey. Islands of ice inclos'd both fish and fowl frozen, and some persons in their boates.'
John Evelyn, diarist, 1658

Of course, monks, chroniclers and other assorted scribes note weather events long before the fourteenth century. But their recordings are fragmentary, and often anecdotal. Merle's journal is systematic (hinting perhaps at a national trait that will emerge six centuries later on railway platforms). His notes are still of value, in the twenty-first century, to climatologists trying to comprehend what's happening at the beginning of the 'Little Ice Age' (*3 March*). The literary value to be wrung from his slim, vellum volume is more questionable. It's been described as 'the dullest narrative ever containing an earthquake': 'June 1340 – great heat and light rain on the 2nd, and on the 11th there was rain, but less than on the 2nd. From the 13th onwards there was moderate wind occasionally throughout the remainder of the month, and it was very strong on the 14th, and stronger on the 23rd than on the 14th.' You get the gist.

Sadly, we know little about Merle and we can only guess at what inspires him. This is the end of a long period of relatively benign weather called the 'Medieval Warm Period' (*20 February*). After decades of abundant harvests, suddenly famines start to become commonplace (*12 May*). The climate is changing. Is he inspired to start detailing the weather because he's worried?

-16.4C Braemar 1888 ❄

Jacobite sympathisers are quick to interpret a mysterious flaming illumination in the sky tonight in 1716 as 'an omen of God's displeasure for beheading the rebel lords' – the Earl of Derwentwater and others have just been executed for rebellion. In fact the auroral display, witnessed over much of the country, is the Northern Lights (*23 March*).

22.2C Enfield 1990 ☀

DULLEST WEATHER DAY / RAINDROP WAGER

Weatherwise, today is the dullest day of the year. By which we mean that nothing – literally nothing – that we can find happens due to the weather. Even the disasters which *ought* to be due to the weather – such as the *Torrey Canyon* – are not. The *Torrey Canyon* is the world's first supertanker, and in 1967 she ploughs into Pollard's Rock in a reef between Land's End and the Scilly Isles (✪), en route to Milford Haven. As she breaks up she spills 120,000 tons of crude oil, laying waste to tens of thousands of seabirds and miles and miles of the marine ecology along the coasts of Cornwall and Normandy. Along with Windscale (*11 October*), it shares the honour of taking man-made environmental damage into a new league in the space of an afternoon. The shorelines are still black today. Appalling weather might just be an excuse. But it's a perfect day: flat calm, windless, clear – the crash is a straightforward navigational cock-up by a tired captain running late.

With so little to report, we'll tell you about the wet day – it *might* have been today – when Lord Alvanley (*2 August*) makes his famous raindrop wager in White's Club in 1817. White's is the oldest and grandest of the St James's gentlemen's clubs, the social hub of Regency London. Within White's, the table directly in front of the bow window on the ground floor is the throne of the most socially influential members. The *arbiter elegantiarum* of this table is Beau Brummell, until he moves to the Continent in 1816, when his place is taken by the aforementioned Lord Alvanley. White's at this time is a byword for high stakes gambling and here, this rainy day, Alvanley bets another club member £3000 – a titanic sum at the time – that one of two raindrops will reach the bottom of the window pane first.

-16.7C Grantown-on-Spey 1985 ❄

Some accounts say that the two drops merge, nullifying the bet and enraging Alvanley; others that, in a few seconds, a fellow member is completely ruined. Sadly there is no record in the White's Wager Book. This doesn't mean the bet didn't happen – rather that, because the result was apparent almost immediately (rather than being contingent on some future development), there was no point writing it down. Anyway, the idea has been borrowed regularly ever since, by children's authors (A. A. Milne's *Now We Are Six*), film-makers, novelists and even car commercials.

22.3C Cambridge 1990 ☀

2.6C min / 9.5C max

EMLEY MOOR MAST / WUTHERING HEIGHTS

So much ice accumulates on the Emley Moor television mast in Yorkshire (☂), today in 1969, that the weight becomes too heavy for the 1260-foot (385-metre) steel tower to bear, and it collapses. No one is hurt, but several million have to live without *Coronation Street* for weeks. A report on the incident notes it is the 70 tons of icicles on the stay cables that topple the mast, leading to the first serious research into icing effects on cables, and improved designs for power masts.

Some 20 miles (32 km) north-west of Emley Moor also today, in 1784, a classic scene from *Wuthering Heights* takes place. It's a spring day on the moors, 'so warm and pleasant' that the doors of Thrushcross Grange are left wide open, 'too tempting for Heathcliff to resist walking in' to see Catherine who is sick. After a violent exchange, they embrace passionately for the last time as 'the shine of the westering sun up the valley' fades. That night, Catherine gives birth; two hours later, she is dead. Next morning, the soft glow of sunshine strikes the house; Catherine has found peace.

'Such a Tempest of raine and Thunder that all in the Church were readie to fall to the ground, and such flashes of lightning entered the Church that each man thought it had beene set on fire; and such a filthie stench arose withal, that manie of the companie fell sick thereof, and hardly escaped death.'
Annals of Cambridge,
Barnwell, 1223

Weather permeates every page of Emily Brontë's dark novel – as metaphor, and simply as weather, wild and wet. There's even weather in the title: '"Wuthering" being a significant provincial adjective,' Mr Lockwood says, 'descriptive of the atmospheric tumult to which its station is exposed in stormy weather . . . one may guess the power of the north wind, blowing over the edge, by the excessive slant of a few stunted firs at the end of the house; and by a range of gaunt thorns all stretching their limbs one way, as if craving alms to the sun.' Early on, Brontë emphasizes that this storm-tossed house ('Out on the wiley, windy moors,' as Kate Bush sings), is appropriate for Heathcliff; when Lockwood first visits the 'bleak hill-top . . . hard with a black frost', the 'first feathery flakes of a snow shower' begin to fall at the gate of the house. The place is as cold and desolate as the protagonist's heart.

-16.5C Braemar 1979 ❄

19.4C St. James's Park 1972 ☀

20 MARCH

2.8C min / 9.8C max

EQUINOX, TRUE START OF SPRING / SPRING CLEANING

What day does spring really start? Meteorologists may date the season from the beginning of the month (*1 March*), but most people tradition-ally agree that the real time is the Vernal or Spring Equinox, around now. While the equinox usually falls between 21 and 23 March, day and night length *seem* to become equal a day early – today – because the sun's rays 'refract', or bend, as they come through the atmosphere. After the equinox, temperatures usually rise, as the input of the sun's heat becomes greater than the loss by cooling at night.

'Long walk through Regent's Park. Sunshine and the first ghastliness of Spring ... Sat in the blazing light and noticed how hideous the bright sunshine made everyone (including myself) appear.' *Joe Orton, dramatist, London, 1967*

The tradition of 'spring cleaning', which follows from these astronomical developments, is firmly rooted in weather. Today is, on average, the day that tempera-tures first rise above 15.5C (60F). The arrival of warmer temperatures signals that the home no longer requires heating, and as traditional open fires or wood- or oil-burning stoves and ranges deposit dust, ash and soot over everything, now's the time, with the heating off until the autumn, for curtains, carpets and other domestic surfaces to get their thorough annual 'spring clean'.

'A lovely day here, starting with a thick white fog and coming out into a cloudless blue sky and hot spring air. Steady, perfect electioneering weather. It really is getting uncannily like the autumn of 1959 when Gaitskell was fighting his valiant, hopeless campaign against Macmillan and the country had never had it so good and would have nothing said against him. All this week we have been fighting 1959 in reverse. Now it is we who are on top of the world, we who are the Government being given credit for the weather.'
Richard Crossman, Labour Leader of the Commons, at Prescote, his country house near Oxford, 1966 – and he's wrong, of course; Labour lose the election

172mph Cairngorm summit 1986 (record high-level gust)

-16.1C Newport 1930

21.4C Rickmansworth 1938

20 MARCH

THE SPARKLE, THE RIPPLE, THE SCENTS AND THE SUNLIGHT

'The Mole had been working very hard all the morning, spring-cleaning his little home. First with brooms, then with dusters; then on ladders and steps and chairs, with a brush and a pail of whitewash; till he had dust in his throat and eyes, and splashes of whitewash all over his black fur, and an aching back and weary arms. Spring was moving in the air above and in the earth below and around him, penetrating even his dark and lowly little house with its spirit of divine discontent and longing. It was small wonder, then, that he suddenly flung down his brush on the floor, said "Bother!" and "O blow!" and also "Hang spring-cleaning!" and bolted out of the house without even waiting to put on his coat. Something up above was calling him imperiously, and he made for the steep little tunnel which answered in his case to the gravelled carriage-drive owned by animals whose residences are nearer to the sun and air. So he scraped and scratched and scrabbled and scrooged and then he scrooged again and scrabbled and scratched and scraped, working busily with his little paws and muttering to himself, "Up we go! Up we go!" till at last, pop! His snout came out into the sunlight, and he found himself rolling in the warm grass of a great meadow.

"This is fine!" he said to himself. "This is better than whitewashing!" The sunshine struck hot on his fur, soft breezes caressed his heated brow, and after the seclusion of the cellarage he had lived in so long the carol of happy birds fell on his dulled hearing almost like a shout. Jumping off all his four legs at once, in the joy of living and the delight of spring without its cleaning, he pursued his way across the meadow ... intoxicated with the sparkle, the ripple, the scents and the sounds and the sunlight.'

Kenneth Grahame, The Wind in the Willows, *1908*

21 MARCH

2.7C min / 9.6C max

MR GRISENTH-WAITE'S LAWN / 'AMAZING GRACE'

Today (on average, over the last twenty years) is when you are most likely to mow your lawn – if you have one – for the first time in the year. This says much about twenty-first-century weather, because it's thirteen days earlier than twenty years ago. In short, without global warming you wouldn't be risking your back on the pull-start for almost another fortnight.

We have this remarkable information thanks to Mr Grisenthwaite, a seventy-seven-year-old pensioner from Kirkcaldy, Fife (🎵). Mr Grisenthwaite starts a lawn-mowing diary (and why not?) in 1984. He notes the date of his first cut and last cut of the year, and all the cuts in between. Then he continues to record these facts for the succeeding twenty years. As the dates of first and last lawn cut are excellent indicators of the start and end of the growing season, the result is a uniquely precious record of climate change – so important, in fact, that in 2006 academics borrow Mr Grisenthwaite's mowing diary to write a paper for the journal of the Royal Meteorological Society. As a result, we have hard evidence that, with the average last cut seventeen days later, the growing season has grown by a month.

Obviously, every year's weather is different, so the dates vary considerably. Grass only grows when the ambient temperature is 5C (40F) or higher. A cold (or dry) February or March delays things; a mild, wet February and March brings them forward. For farmers, milk yields in April depend largely on the 'spring flush' of growth now (*18 April*) – and the crucial moment when grass grows faster than stock can eat it. This tends to be around 20 April, but can, if spring is cold and late (as in 2006), be up to a month later.

A STORM INSPIRES BRITAIN'S FAVOURITE HYMN

A furious Atlantic storm today in 1748 so frightens John Newton, aboard a slave-trader returning to Ireland, that he prays for divine mercy. Referring to his 'great deliverance', twenty years later, Newton – now a minister and ardent abolitionist – writes of his salvation in a hymn (co-written with the poet William Cowper). 'Amazing Grace' becomes the most popular hymn in the English language:

'Amazing grace! (how sweet the sound) That sav'd a wretch like me! I once was lost, but now am found, Was blind, but now I see.'

-16.1C Braemar 1899 ❄

21.1C Cambridge 1938 ☀

SEX IN THE RAIN

A late afternoon walk through the woods, in drizzle, in 1920. Sounds sexy? Now read this: 'The drizzle of rain was like a veil over the world, mysterious, hushed, not cold. She got very warm as she hurried across the park. She had to open her light waterproof. The wood was silent, still and secret in the evening drizzle of rain, full of the mystery of eggs and half-open buds, half-unsheathed flowers. In the dimness of it all trees glistened naked and dark as if they had unclothed themselves, and the green things on earth seemed to hum with greenness.'

From *Lady Chatterley's Lover* (1960), of course. Like so many British novelists, D. H. Lawrence uses weather incessantly to evoke and emphasize moods and feelings, the so-called 'pathetic fallacy'. In *Chatterley*, today's weather is also integral to the plot. If it hadn't been raining, making the ground too soft for her husband's wheelchair, Lady Chatterley would not have gone down to the woods alone today. And if she hadn't done that, literature would have missed out on its most famous sex-in-the-rain scene:

'She slipped on her rubber shoes again and ran out with a wild little laugh, holding up her breasts to the heavy rain and spreading her arms, and running blurred in the rain with the eurythmic dance-movements she had learned so long ago in Dresden. It was a strange pallid figure lifting and falling, bending so the rain beat and glistened on the full haunches, swaying up again and coming belly-forward through the rain, then stooping again so that only the full loins and buttocks were offered in a kind of homage towards him, repeating a wild obeisance.

He laughed wryly, and threw off his clothes . . .

She was nearly at the wide riding when he came up and flung his naked arm round her soft, naked-wet middle. She gave a shriek and straightened herself, and the heap of soft, chilled female flesh became quickly warm as flame, in contact. The rain streamed on them till they smoked. He gathered her lovely, heavy posteriors, one in each hand and pressed them in towards him in a frenzy, quivering motionless in the rain. Then suddenly he tipped her up and fell with her on the path, in the roaring silence of the rain, and short and sharp, he took her, short and sharp and finished, like an animal.'

-12.8C Laing 1899

20.0C Lincoln 1918

23 MARCH

2.5C min / 10.3C max

THE NORTHERN LIGHTS 'The glow kindled rapidly into brilliance and, ascending slowly from the horizon, assumed the shape of a huge arc,' records one Cicely Botley in Abernethy, Perthshire (♥), this night in 1946. 'Suddenly the arc broke into feverish activity along its whole length, dividing here into bundles of short rays, and there into diffuse pulsating patches of light. Tinges of red and green sparkled to enhance the yellow-white ... The rays were leaping upwards, one bundle subsiding as an adjacent one leapt ahead, and eventually the whole northern half of the sky was filled with streamers.'

'Last night the moon through the mist on our meadow [*sic*] – and time had disappeared – and night birds whirring so that past and present and future were one. In the Vale of Avalon, the waters have receded, but in these Mendip hills nothing has changed ... The moment the wind switched to the west, the sweet warm air of the Gulf Stream came in. Today is gloriously sunny.'
John Steinbeck, writer, at Bruton, Somerset, where he rents a cottage in 1959 to explore Arthurian sites

Miss Botley is witnessing that most legendary crepuscular orgasm, the Northern Lights or aurora borealis (in the southern hemisphere, aurora Australis), which occur when solar particles collide with atmospheric gases in the earth's magnetic field, creating light particles in the sky. 'The coronal rays now merged to form what appeared to be a bluish-white vapour, sometimes like the smoke of the straw which is burned in the country, at other times remarkably like cirrus clouds. Before long, however, the "vapour" resolved itself into rays, patches of green light waxed and waned in the east, and a quite fantastic cloud of brilliant red appeared and remained fixed in the western sky for almost an hour – so unreal that one imagined it to be a reflection of a gigantic fire below the horizon. It was such an illusion which caused fire engines to race towards the horizon in many parts of Southeast Europe during the great aurora of January 1938.'

The Northern Lights occur mostly in March, April, September and October (around the equinoxes). In far north Scotland, they occur roughly once a month. Across the rest of the British Isles, especially in areas with light pollution, this spectacular celestial phenomenon is rare, though the *Anglo-Saxon Chronicle* records an instance in 1122: 'Many shipmen said that they saw in the north-east a great and broad fire near the earth, which at once waxed in length up to the sky: and the sky ... fought against it as if it would quench it; but the fire nevertheless waxed up to the heavens.'

-14.2C Braemar 1883

22.2C Geldeston 1918

THE END OF
THE AGE
OF SAIL /
AMOCO CADIZ

In deteriorating weather, 'moving fast under plain sail', HMS *Eurydice* passes Ventnor, Isle of Wight, today in 1878. The Royal Navy training frigate is bound for Portsmouth, after a swift passage across the Atlantic. Two miles (3 km) offshore in Sandown Bay (✛), she's hit by a squall accompanied by a blinding snowstorm. In eleven fathoms (20 metres) she heels, topples on to her starboard side and sinks. Within minutes the storm clears and the sun breaks through again. All but two of the 370 or so crew, trainee sailors and passengers are drowned. It's one of the worst peacetime disasters in British naval history. The inquiry finds the crew innocent – orders to take the sails in are given, but the snow is so thick it is impossible to see – and the weather guilty.

'Too proud, too proud, what a press she bore!
Royal, and her royals wore.
Sharp with her, shorten sail!
Too late; lost; gone with the gale'
Gerard Manley Hopkins, 'The Loss of the Eurydice'

'A grey swirl of snow with the squall
at the back of it,
Heeling her, reeling her, beating her down!
A gleam of her bends in the thick
of the wrack of it.
A flutter of white in the eddies of brown'
*Sir Arthur Conan Doyle, 'The Home-Coming of
the Eurydice'*

When the *Eurydice* is launched, in 1843, her shallow draught is an experimental design, but she sails well and, in 1877, the frigate is recommissioned for training young seamen. However, following her loss (and the disappearance of HMS *Atalanta*, her sister ship, two years later), the Royal Navy abandons sail-training. The front-line fleet is now built of iron: learning how to hand and reef sails is irrelevant. It's the final recognition that the man-o'-war days of sail are over (*6 September*). Not that the *Eurydice* is quickly forgotten. Arthur Conan Doyle, Gerard Manley Hopkins and a youthful Hilaire Belloc all compose poems about the disaster. The phantom of a three-masted ghost ship that vanishes when approached has been seen off the Isle of Wight intermittently ever since.

-13.3C Newton Rigg 1916 ❄

22.2C Hodsock Priory 1918 ☀

In raging seas the wreck of the supertanker *Amoco Cadiz* splits in two, today in 1978, after running aground off Cape Finisterre. Strong winds have crippled the salvage operation. In all, 223,000 tons of oil spill into the English Channel forming an 800 square-mile (1287 sq-km) 'chocolate mousse' slick. It's the world's worst oil spill (until the *Exxon Valdez* in 1989), and the damage to the Breton and British coastline is unprecedented.

25 MARCH

2.8C min / 10.3C max

When does summer begin? An odd question, perhaps, just five days after making the same enquiry about spring (*20 March*), but if you believe it's the moment when the evenings suddenly seem lighter and longer – in short, when the clocks change – then it's around today. The rule defining the start of Summer Time is that it's the last Sunday in March (and it ends on the day following the fourth Saturday in October). If you can't remember whether clocks go forward or back, the rule is: 'Spring Forward, Fall Back'.

British Summer Time (BST) is introduced in May 1916 as a wartime measure to allow exploitation of the maximum number of hours of daylight, giving darker mornings but lighter evenings – thereby saving hundreds of thousands of tons of coal and making summer days seem even longer. BST remains all the year round in the war years, from February 1940 until October 1945 (and again from February 1968 until 1971) – highly unpopular with early risers, such as farmers and builders, and in Scotland, where in the far north sunrise could be as late as 10 a.m. in midwinter.

HOW A STORM LEADS TO THE PENCIL

When a storm blows down dozens of trees on Seathwaite Fell (8 May) near Borrowdale (♥) in 1564, the roots of one tree, a mighty oak, reveal some unusual grey-black 'stones'. Shepherds in the area find that if these stones are rubbed against another surface, they leave a dark mark. They start to use the stones to mark their flocks. Because the material is very messy to handle, after a time they cut it into square pieces which they encase with wood. Bingo! The pencil!

The material, of course, is graphite, and the Borrowdale deposit is unusually pure. Initially called 'plumbago', or 'that which writes like lead', it in fact has nothing to do with lead; but the misnomer sticks, even after a Swedish chemist in 1779 renames it 'graphite' after the Greek word for writing. An industry springs up around Borrowdale's graphite which becomes fantastically valuable. It is mined only six weeks of the year and armed guards escort the wagons that carry it to London. The English Guild of Pencilmakers sells the 'writing sticks' in hand-carved wooden cases, enjoying a world monopoly until the Chinese discover graphite deposits in the 1800s.

-11.7C Balmoral 1919 ❄

21.7C London Road Station 1953 ☀

FRIDGE
MAGNETS
GALORE!

Thick fog shrouds the Scillies today in 1997 (and the airport's been closed for two days). Regardless of this, the container vessel MV *Cita* proceeds at a brisk 19 knots through the murk en route from Southampton to Belfast. On the bridge it's a peaceful scene. Lights wink from the automatic pilot, the engines murmur their rhythmic throb, and the watch officer snores peacefully. Until, with a loud crunch, at 3.30 a.m. she arrives on the rocks of Newfoundland Point, south of the Scillies (✪). In many ways the remarkable thing is that, in one of the world's busiest shipping lanes, she doesn't collide with something sooner.

'Snow fallen upon the leaves had in the night coined or morselled itself into pyramids like hail. Blade leaves of some bulbous plant, perhaps a small iris, were like delicate little saws, so hagged with frost.'
Gerard Manley Hopkins, poet, 1872

'A cold cloudy day with a piercing wind ... I've never known so much sunshine, so many clear blue skies with sailing white clouds, so many magnificent sunsets with the evening star coming out over the poplars. It's this winter weather which has kept me going during these ghastly eight weeks of failure, frustration and fiasco up in London.'
Richard Crossman, Labour Leader of the Commons, Prescote, Oxford, 1967

Once the immediate panic of an oil disaster passes – it's thirty years almost to the day since the *Torrey Canyon* goes down a few miles away (*18 March*) – fear switches to *Whisky Galore!*-style glee (*5 February*) as locals pounce on the first of two hundred 40-foot (12-metre) steel containers washing up along the rocky shore. Under present-day rules, salvors have to report 'salvaged' wreckage to an officially appointed Receiver of Wreck, but may retain salvage rights (in the form of compensation) if the receiver wishes to repossess their haul. The *Cita*'s cargo becomes, as with the *Napoli* in 2007 (*22 January*), a glorious free-for-all for 'wreckers'. It's like the good old days (*4 January*). There's an eclectic assortment of merchandise: computer mice, house doors, golf bags, granite headstones, fridge magnets, fork lift trucks, Ben Sherman shirts, bales of raw tobacco, carrier bags and trainers (left and right feet packed in separate containers). While newspaper headlines talk of 'Tide of Fortune Turns for Islanders Whose Boat Came In' and 'Scilly Isles Clean Up', locals are more sanguine: 'It's like unwanted Christmas presents from distant relatives,' complains one. Wreckers can't be choosers, of course.

164.3mm Glen Etive 1968

-11.1C Eskdalemuir 1919 ❄

22.8C Whitby 1907 ☀

HELICOPTER'S MOMENT / BEVERLEY NICHOLS' GARDEN

A North Sea storm today, in 1980, causes the worst-ever weather-related oil rig disaster – and the coming-of-age of the helicopter. A North Sea accommodation platform ('flotel'), the *Alexander Kielland* – 10,000 tons and the size of a football pitch – collapses in high winds, drowning 123. In appalling conditions, 180 miles (290 km) from the Scottish coast (✪), the international rescue effort involves 47 ships, two planes, three diving vessels and 23 helicopters. RAF crews in British-built Westland Sea Kings employ every avionics feature the helicopters possess, in particular the automatic hovering facility which allows them to move down to 40 feet (12 metres) above the life-rafts in 30-foot (9-metre) waves, in darkness and dense fog. 'They were the worst conditions you could get – bad visibility, cross winds, high seas and a very small target,' recalls Flight Sergeant John Moody. But 89 men are saved. 'Within the closely-knit circle of helicopter rescue crews the *Alexander Kielland* affair is regarded as . . . the ultimate test since this once-despised type of aircraft started saving lives in the Burmese jungles,' reports *The Times*. In the rivalry between RAF and Royal Navy rescue crews, the RAF regard the event as their answer to the Navy's 'finest hour' – the Fastnet race rescue (*13 August*).

A GREAT GARDEN WRITER'S FIRST GARDEN

'It was a cold evening in late March, and the shadows were falling. No garden can be expected to look its best in such circumstances. But this garden did not look like a garden at all . . . Even in the grimmest winter days a garden can give an appearance of discipline . . . no matter how wild the winds or dark the skies. But this garden was like a rubbish heap. Nothing. Earth. Sodden grass. Rank bushes. A wind that cut one to the marrow. I shivered.

'Then I pulled myself together. It was unreasonable, surely, to be shocked by this prospect. One could not expect summer glory in the middle of winter.

But . . . it was not the middle of winter. It was the beginning of spring . . .

I pulled the collar of my coat up and strode into the garden. Everywhere, there was the evidence of appalling neglect . . . Why – there were even old newspapers lying sodden in the orchard! . . .

Such was the garden, when I entered into my inheritance.'
Beverley Nichols, Down the Garden Path, *1932*

-11.7C Barr 1969 ❄
22.8C West Witton 1929 ☀

28 MARCH

2.5C min / 10.3C max

FOXHUNTING
MEMORIES

'On one of my expeditions, after a stormy night at the end of March, the hounds drew all day without finding a fox. This was my first experience of a "blank day". But I wasn't as much upset about it as I ought to have been, for the sun was shining and the primrose bunches were brightening in the woods. Not many people spoke to me, so I was able to enjoy hacking from one covert to another and acquiring an appetite for my tea at the "Blue Anchor". And after that it was pleasant to be riding home in the latening twilight; to hear the "chink-chink" of thrushes against the looming leafless woods and the afterglow of sunset; and to know that winter was at an end. Perhaps the old horse felt it, too, for he had settled into the rhythm of an easy striding walk instead of his customary joggle.

I can see the pair of us clearly enough; myself, with my brow-pinching bowler hat tilted on to the back of my head, staring, with the ignorant face of a callow young man, at the dusky landscape and its glimmering wet fields . . . I can hear the creak of the saddle and the clop and clink of hoof as we cross the bridge over the brook by Dundell Farm (♀); there is a light burning in the farmhouse window, and the evening star glitters above a broken drift of half-luminous cloud. "Only three miles more, old man", I say, slipping to the ground to walk alongside of him for a while.

It is with a sigh that I remember simple moments such as those, when I understood so little of the deepening sadness of life, and only the strangeness of the spring was knocking at my heart.'
Siegfried Sassoon, Memoirs of a Fox-Hunting Man, *1928*

'Wembley. Snowing. No heat. No-one else here. Why am I doing this?'
Brian Eno, curating 'Self Storage', his art installation, 1995

'And now I shall rise from the lovely fire . . . & go out into the grey day . . . I am going to walk, alone and stern, over the miles of grey hills . . . I shall call at a public house & drink beer with Welsh speaking labourers. Then I shall walk back over the hills again, alone & stern, covering up a devastating melancholy & a tugging, tugging weakness with a look of fierce & even Outpost-of-the-Empire determination & a seven-league stride.'
Dylan Thomas, poet, 1934

-10.0C Alwen 1969 ❄

23.3C Earls Colne 1965 ☀

29 MARCH

2.9C min / 10.4C max

GRAND
NATIONAL FOG
/ BOAT RACE
SNOW

Does thick fog hide skulduggery at today's Grand National in 1947? Emerging from the murk to win, way ahead of the field, is not the 8:1 favourite, ridden by Irishman J. J. MacDowell, but 100:1 outsider Caughoo, ridden by fellow Irishman Eddy Dempsey. Five years later, a sinister accusation surfaces. Does Dempsey wait out, unseen in the fog, at a remote part of the Aintree course (♥) during the first circuit, before rejoining, in the lead, when the field comes round for the second time? MacDowell confronts Dempsey – both now retired – with the accusation during a Dublin drinking spree. Before the local magistrate, following fisticuffs, the story leaks out. It makes headlines across the world. Meanwhile, in 1901, also today, the snow is so thick that the jockeys request the Grand National be postponed, a request rejected by the stewards. The enterprising trainer of the well-fancied Grudon promptly sends out for two pounds of butter. The butter is pressed into Grudon's hooves to prevent the snow clumping up – and Grudon duly wins.

SINKING BLUES

It's also snowing, with a sharp easterly churning the Thames, for today's 1952 Boat Race – regarded by many as the greatest ever. Rowing from behind all the way round the outside of Barnes bend – a feat never repeated – the Dark Blues win by 6 feet (1.8 metres). It's the closest result there's been, save for the 1877 dead heat.

A strong west-north-wester blowing over a slack tide creates preposterous rowing conditions for the Boat Race in 1912. Both crews sink.

In 1951, a squall sweeps up the river at the start and Oxford sink after only two minutes. In 1978, in horizontal, driving rain and rough water, Cambridge sink shortly after Barnes Bridge. These days, hi-tech boats fitted with self-baling pumps make sinking in rough weather far less likely.

-17.2C Braemar 1901

139mm Doune 1993

25.0C Santon Downham 1968

At Twickenham, also today in 1952, England and Ireland ignore the snow and the 'biting blast' (rather than postpone again) and play, confounding 'all those who believed that a presentable game of Rugby football was impossible in such conditions,' The Times reports. Spectators, 'many of them dressed for a visit to the front line in Korea', watch England win 3–0.

30 MARCH

2.8C min / 10.7C max

EASTER WEATHER / BUCHAN'S SPELLS

A blizzard arrives today in 1952, with gale-force easterlies. Villages in the south-east are cut off, with 8-foot (2.4-metre) drifts in the Chilterns (♀). Many are surprised to learn that, for all its springtime associations, notable snowstorms are more frequent in March than in any other month. Many are also surprised to learn that there's a considerably higher chance of a white Easter than a white Christmas. This is less surprising when you consider that Easter moves – by anything up to thirty-five days, from 22 March (earliest) to 25 April (latest). The fiendishly complicated arrangement for calculating the date of Easter is settled in the fourth century. It states that Easter Day is the Sunday following the first full moon after 20 March. Generally, the earlier Easter is, the greater the chance of snow; the later, the more chance of sunshine. (Not to say that several late April Easters haven't been very snowy, such as in 1983.) Later Easters also feel more spring-like because (at the latitude of London) there is two hours and twelve minutes more daylight than on the earliest date, and the clocks have changed.

Freezing or snowy early Easters invariably prompt cries to fix the date of Easter, as Harold Wilson's 1966–70 government fixes the date for the Whitsun Bank Holiday (to the last weekend in May). In 1928 a bill is presented to fix Easter to the middle weekend of April (the first Sunday after the second Saturday – got that?). However, chiefly for weather reasons (see below) it never happens.

ALEXANDER BUCHAN AND HIS WEATHER 'SPELLS'

In discussing the possibility of 'fixing' Easter, in 1927, Parliament takes into account the ideas of the President of the Scottish Meteorological Society, Alexander Buchan. Following careful study of long-term weather records, Buchan evolves a theory which holds that at certain, regular times of year the weather is consistently warmer, or consistently colder, than might be expected. He identifies six cold periods, and three warm ones. These tendencies – or 'singularities', as the Bavarian climatologist August Schmauss calls them in 1938 – continue to be argued about today. Can such a tempting notion, promising regular, reliable interludes, really be the case? Whether they exist or not, the date proposed for the fixing of Easter happens to correspond with a Buchan 'cold spell' – 11–14 April – and the prospect of cold Easters for ever is one reason why the legislation is never enacted.

-10C Balmoral 1941 ❄

23.9C Cullompton 1929 ☼

31 MARCH

3C min / 10.8C max

THE GALE AND THE LONELY BOMBER

The last two days of March brought equinoctial gales. The wind howled through the wood near our barracks (●), scattering small branches along the road. Every few hours the tannoy uttered warnings of storms and cancellation of flights. Only one 'plane operated. The circumstances were unusual and more than usually tragic. We had a new Squadron Leader named O'Donoghue, a permanent RAF man who had operated in India by daylight. His crew once remarked that he disliked night operations strongly. We ... were amazed when one night he asked permission to go out alone at dawn. When he made this request, operations had been cancelled owing to weather. Very early in the morning the sound of a single 'plane roused us. By the time we were up, O'Donoghue was back. The papers announced that night that a lone Lancaster had raided a town not far inside the German coast.

During the gale weather O'Donoghue was granted a second attempt. Again we heard him depart in the most melancholy hours of the morning, the roar of his engines torn by the wind. But by the time we were up on this occasion he had not returned ... At 9 o'clock the tannoy called all crews to the briefing room. A listening post in England had heard a Lancaster calling for help. It had been hit as it crossed the enemy coast, homeward bound. Group believed that it may have ditched in the North Sea and ordered us to search an area east of the Wash. Our chances of success were obviously negligible, as met had forecast waves of 30 to 40ft high. All the morning we scanned the changing mountains and valleys of the North Sea. Whitecaps rose and broke, lashed by squall after squall. Eventually we flew back low across the scudding windmills of Norfolk with little hope that anyone would have anything to report. When we reached Elsham it was to hear that the German radio had claimed a single heavy bomber shot down.
Don Charlwood, No Moon Tonight, *1956*

'Curiously silent after a week of wind and rain ... The sun comes out, the bees with it ... March is gone, mad-haring it.'
Derek Jarman, film-maker, Prospect Cottage, Dungeness, Kent, 1992

'There came a day of full sun ... young foliage peeped; water was bright as a blade. This was spring's ambassador. The day showed in all our faces. The men sat under a wall to breakfast and basked.'
Adrian Bell, Corduroy, *1930*

-9.2C Braemar 1897 ❄

22C Wryde 1907 ☼

APRIL

1 APRIL

3.3C min / 10.9C max

APRIL'S HERE /
SOIL WATER
CAPACITY /
SEVERN BORE

Showers, confusingly, are the one thing *not* to expect, now that April's here. Contrary to folklore, April is one of the drier months. Thunderstorms are rare, and with our surrounding seas at their coldest, and the land just beginning to warm up, near the coast breezes off the sea become common between now and June. Snow is not uncommon (one day a year on average in the south; three or four in the north, seven in Scotland).

Today is a critical day for both farmers and commercial fruit and vegetable growers, as the amount of water measured in the soil on 1 April is a guide to watering requirements for the year ahead. This figure, known as the 'field capacity' – that is, the amount of water that the soil naturally holds when all the pore spaces are full but it is not waterlogged – varies greatly from soil to soil (clay holds masses of water, sand none). From this figure, growers calculate how much water they need to provide for particular crops (some plants, usually those with large leaf areas, such as lettuces, cabbages, potatoes and rhubarb, need much more water than others) on that particular soil, and so know their irrigation requirements. Why today? Because snow has usually thawed, winter rain has usually drained, yet it is before the temperature has risen sufficiently for evaporation to begin.

A week of steady rain, culminating in a south-westerly gale blowing exactly as the spring tide reaches its height – these are the conditions hoped for by surfers on that curious tidal phenomenon, the Severn Bore. The Severn has a huge tidal range (the second-largest in the world, after the Bay of Fundy in Canada) and a narrowing bay, so that the leading edge of the incoming tide is funnelled, as the bay narrows, forming a surfable wave moving at up to 15 mph (24 mph). Why do surfers want such specific conditions? The rain is needed so that a full downriver freshwater flow hits the rising flood tide; and the wind helps drive the 'wave' upriver. Amazingly, these conditions *do* perfectly coincide today in 2006, allowing Steve King, a railway engineer from Gloucester, to set the world distance surfing record. It's accepted by the media, but later rejected by the *Guinness Book of Records* and the British Surfing Association because the exact distance (about 7 miles/11 km) is never verified.

-11.7C West linton 1917

22.6C Wryde 1907

2 APRIL

3.1C min / 10.9C max

PUNCH-UP IN
THE PUDDLES /
THE WELSH
FRET

'It was paradise in the puddles, frenzy in the frozen wind, and the mother of all games on Mothers' Day,' says former Scotland rugby international-turned-BBC-journalist John Beattie of Scotland's unexpected 19–13 victory over England in Edinburgh today in 2000. Despite pre-match forecasts of an overwhelming English victory, the elements step in to even things up. Curtains of sleet sweep across Murrayfield and, as the ball becomes increasingly slippery, the fast, open rugby England have played with such success all season falls apart. The win passes into Calcutta Cup lore, alongside that greatest of showdowns ten years earlier, also at Murrayfield, also in atrocious weather. That time, again Scotland are written off. But in a gale, England lose their composure and fail to take their penalties – and Scotland steal the Grand Slam from under the Auld Enemy's nose. So, rain, wind and mud – it means sunshine in Scotland.

'If you choose a misty morning to take a stroll across the bridge over the lake in St James's Park, you will see to the east one of the most magical views in London. For the Horse Guards, the War Office, and the black pyramidal roofs of Whitehall Court beyond form a composite group which mounts, with all the majesty of some legendary castle cupola above cupola, roof over roof, up from a frame of mist and plane trees to the palest of pale heavens. It has, on such a morning, all the mystery of some faerie citadel.'
Cecil Lewis, Sagittarius Rising, 1936

It's -15C this morning in 1917 at Newton Rigg in Cumbria – the coldest day of the coldest April on record. In 1995, during the first three days of April, South Wales and the south-west see less than an hour of sunshine, so buried are they beneath sea fog. Such sea mists or 'frets' are unusual on the west coast, where the sea surface temperatures are higher because of the Gulf Stream. The east coast is their usual home (*29 August*), with east Scotland particularly prone in the late afternoon when, like clockwork, the 'haar', as it's locally known, slides up valleys such as the Spey. Haars are most common between now and June, as the land begins to warm up. In this instance, in Wales, it's a westerly flow of moist air blowing in across the Atlantic from the tropics.

165.1mm Loch Carron 1933

-15.0C Newton Rigg 1917
23.9C Leeming 1946

3 APRIL

3C min / 10.8C max

BEACHY HEAD COLLAPSES

Twelve months of interminable rainfall – the heaviest since records begin in 1766 – takes its toll, today in 2001, on one of Britain's greatest natural landmarks. A 525-foot (160-metre) section of the white cliffs at Beachy Head in Sussex cracks and slips into the sea. The Devil's Chimney, a 230-foot (70-metre) high chalk outcrop jutting into the Channel where the South Downs meet the sea, near Eastbourne (🖉), is among the thousands of tons of chalk to collapse. Rain is the culprit one way or another: either it seeps through fissures in the chalk and freezes, expanding as it does so, to fracture the cliffs, or the immense weight of water absorbed by the porous chalk is simply too much for the cliffs to bear.

> 'The wind continues as foul as it can blow, but as I am now fixed on board, it is my intention not to move out of the ship, to which I begin to be reconciled.' *Horatio, Lord Nelson, Scilly Isles, 1798*

> 'Jeremy Geany, Dan. Shea, and M. Rourke were executed on a wooden gallows near Blackpool bridge for murder and robbery ... It was a charming evening, dry, and the sun shining, though there were several showers in the afternoon.' *John Fitzgerald, Cork school teacher, 1793*

The record rainfall causes numerous landslides across southern Britain during the winter and spring of 2001. At Charmouth in Dorset, 1.6 miles (2.5km) of cliffs collapse; 200,000 tons of rubble drop from the palisades at St Margaret's Bay, near Dover; and a huge landslip in Brighton stops just short of a supermarket. Despite the alarmist headlines, there's nothing unusual about this. The continuous cycle of the weather drenching and drying, freezing and thawing, sun-baking and storm-battering means the coast is always retreating in some places and accumulating in others. It's only because the space of a human life is so short that we view our coastline as immutable – with certain sections, like Beachy Head, sacrosanct.

On a damp, murky day in 1930, Vita Sackville-West drops in to view a property for sale near Sissinghurst in Kent (📍). 'Fell flat in love with it,' she writes in her diary. She returns the following day, and the day after that, but each time it's grey and gloomy and over-cast. Her family is very unimpressed. On her next visit, it's pouring with rain and the 7-acre plot is a muddy wilderness. Vita and her husband Harold buy it anyway, even though they still haven't glimpsed the views across the Kent Weald. It will become the most famous garden in the world.

-12.2C Newton Rigg 1917

25.6C Hunstanton 1946

CASTLETOWN HAIL AND 'THE VOICE OF THE LORD'

'Every stone of it was the size of four inches in circumference. A hen was killed in Ballykilmurry, and her two legs were broken by one piece of it ... Every stone went two inches in the earth, and what fell in the water went to the bottom like a natural stone.' Thus is described today's hailstorm at Castletown, in County Offaly () in 1635.

Many indicators suggest that hail and thunder are both more frequent and more extreme between 1631 and 1680 than in any previous or subsequent period. Certainly the percentage of space devoted to them in European chronologies is almost twice as great during this time – partly explained by the fact that many pamphleteers still see thunderstorms as the 'Voice of the Lord'. The thirty years from 1651 to 1680 especially, when Londoners are enjoying the great 'Frost Fairs' on the Thames (*4 February, 24 February*) stand out as one of the coldest periods of the Little Ice Age (*3 March*); the increase in hailstorms may well be caused by this climatic deterioration.

> 'Dry, windie, indifferently cleare in the nights very cold, and much given to frosts ... good to purge the aire, and to prevent infection, the small poxe is about in the country ... oh lord watch over mee and mine.'
> *Ralph Josselin, Puritan minister and diarist, Essex, 1649*

> 'The spring, which appeared so early, lingers on, and seems at a stand. An easterly wind, dry, acrid, and cold, suspends vegetation.'
> *Louis Simond, traveller,* An American in Regency England, *1811*

In 2000, two weeks into British Summer Time and just a month after meteorologists pronounce the winter the sunniest of the century (with a daily average in England and Wales of 2½ hours' sunshine per day), freezing weather brings transport chaos as roads, railways and runways become snow-bound. Luton airport closes for ten hours. A 'white-out' blocks the Snake Pass (⚡) on the A57, the famously scenic road (height 1679 feet/512 metres) through the Peak District (*14 March*). Thousands of newborn lambs freeze to death or starve as gales blow snow into drifts 5 feet (1.5 metres) deep, and 65 mph (104 kph) winds force motorists on Dartmoor to abandon their cars. Meanwhile, heavy rain floods Cambridgeshire and the Somerset Levels.

-11.2C Glenlivet 1961 ❄

26.5C Greenwich 1946 ☼

5 APRIL

WINDSURFER INVENTED / UK'S BIGGEST TREE FALLS

A light breeze blows over Hayling Island on the Hampshire coast (☀) this Saturday in 1958. Away from the sea front, up a quiet creek, twelve-year-old Peter Chilvers is down as usual for Easter. He is busily experimenting with his latest brainchild: a plywood board with improvised mast and sail. Fortunately the wind is light, and the water calm, as mastering his curious water-scooter is by no means straightforward. It requires balance, perseverance and coordination as he tries to remain upright, catch the wind and steer using the foot-operated tiller connected to the rudder on the back of the board.

'Children and grandchildren arrive. Cold wind, hot sun, blue sea. Read Ruskin, walk on the beach, play snakes and ladders. Feel well blessed.'
Hugh Casson, architect, artist and broadcaster, 1980

'The gale arrived in time for breakfast ... The boat thrummed and shivered on its moorings. Below, it felt as if one was squatting in the sound box of an out-of-tune harp ... It was all quite safe, but it meant a spell of enforced retirement ... So I gave myself up to a bungalonely life of pipe and slippers, pottering about the library, feeding the stove with charcoal, making a ritual to-do about having elevenses at eleven and lunch sharp at one. At intervals I went upstairs to watch the wind harrowing the water.'
Jonathan Raban, Coasting, 1987

On a later visit, watching the weathervane on a local church spire gives him a further idea: the arrow is on a pivot that rotates, so the vane always points into the eye of the wind. If his 'sail board' were the arrow, he realizes, with the keel as the pivot, then his board would react in the same way. Eureka! If he can only make his sail move, he needs no rudder. There are plenty of details that the twelve-year-old has not yet fully understood, to do with the dynamics of using the sail to counterbalance body weight. But these are no more than details. Although legal shenanigans muddy the water a decade later, when two Californians file patents and try to take the credit, the fact is that young Peter Chilvers has invented the windsurfer.

A snowstorm in 1911 blows down the largest tree in Britain: the Huntingdon wych elm at Magdalen College, Oxford. It's 142 feet (43 metres) high, with a trunk circumference of 27 feet (over 8 metres) – the length of a double-decker bus. The 'gloom of the wych-elm's hollow', as Oscar Wilde describes it in a poem, has shaded undergraduates for more than four centuries.

22.8C Cambridge Observatory 1892
-9.0C Grendon Underwood 1990

3.5C min / 11C max

CRUELLEST DAY
OF THE
CRUELLEST
MONTH

'Suicide reaches its maximum during the fine season,' writes Emile Durkheim, the father of sociology, in his landmark monograph, *Suicide* (1897). 'When nature is most smiling and the temperature mildest.' According to Durkheim, the rate peaks in April (and particularly today, according to an early twentieth-century study building on his work), though his explanation is social rather than meteorological: in winter, when communal life and activities are strongest, it is easier to share problems. Come April, you're on your own again.

Morselli, the other great early epidemiologist of suicide, highlights the biometeorological aspects of the April peak, namely the change in temperature. In the first thirty years of the twentieth century various weather conditions – falling barometric pressure, humidity, cloudiness, and rainfall – are all thought to influence suicide rates. A third explanation, suggested by some psychologists more recently, is that April quite simply fails to meet expectations: potential suicide cases hang on for the promise offered by the new beginnings of spring; then, around today, they feel let down when their life does not improve and reach for the razor blades. Is this why T. S. Eliot calls April 'the cruellest month' in the first line of *The Waste Land* (1922)?

For several decades, seasonality in suicides is viewed as fact. Recent research, though, into suicide rates in England and Wales between 1982 and 1996 has shown that the seasonal effect, still prevalent in the 1970s, is now greatly diminished. In fact, it's almost vanished. Speculative explanations for this include societal changes, and the gradual disappearance of our traditional seasons due to global warming.

The temperature hits 20C in the Solent today, in 2007, as a belt of high pressure settles over much of the UK, delivering clear skies and sunny days: a theme that persists for the rest of what turns out to be the warmest April since records began, 350 years ago.

-12.2C Wolfelce 1917
23.3C Southampton 1892

7 APRIL

3.4C min / 11.1C max

BLOOD AND SNOW IN THE WARS OF THE ROSES

In a blinding snowstorm, today in 1461, the bloodiest battle ever fought on British soil takes place at Towton, Yorkshire (⚔). Over fifty thousand men (including half of the peerage) take part in this decisive encounter between the House of York and the House of Lancaster. Around 11 a.m., as snow begins to fall, the Yorkist commander, Lord Fauconberg, takes advantage of the strong southerly wind: calling forward eight thousand longbowmen, he rains 120,000 arrows a minute on the frozen Lancastrians facing into the blizzard. When they return fire, unable to see a thing, they are unaware that their arrows fall short. Fauconberg's archers then move forward, refilling their quivers with the spent arrows.

> 'You never know what the weather will do in Yorkshire, particularly in the springtime. [Mary] was awakened in the night by the sound of rain beating with heavy drops against her window. It was pouring down in torrents and the wind was "wuthering" round the corners and in the chimneys of the huge old house. Mary sat up in bed and felt miserable and angry.'
> *Frances Hodgson Burnett, The Secret Garden, 1909*

In this plight, the Lancastrians have no choice but to attack, and so the murderous mêlée of hacking and hewing begins, raging through the night in atrocious weather. 'This battle was sore fought . . . and taking of prisoners was proclaimed a great offence,' a witness records. The front lines, it is said, have to stop fighting at times to clear bodies, simply in order to engage each other.

The battle turns into a rout by the Yorkists. Thousands die trying to cross the swollen, freezing rivers surrounding the dale. Piles of bodies form bridges over Cock Beck. Lancastrian soldiers are chopped down miles from the battlefield in every direction. With no prisoners taken, the slaughter is on an epic scale: 28,000 die – as many fleeing the field as fighting. The image of a vast, snowy upland strewn with bloody corpses comes to epitomize the bleakest horrors of this civil war. The strength of the great northern families is shattered and Edward IV is crowned king on returning to London.

22.0C Achnashellach 1974 ☼

-8.9C Balmoral 1935 ❄

The ground is so heavy with rain (or 'bottomless' as jockeys say) at Aintree today in 2001 that only four horses complete the Grand National. Many simply retire with exhaustion. For millions of once-a-year punters who don't understand the importance of 'going' (*22 May*), it's a huge anti-climax.

3.1C min / 11.2C max

LIGHTNING
STRIKES A
FOOTBALL
TEAM

Lightning strikes eleven soccer players simultaneously this afternoon in 1979 as they run for cover during a thunderstorm in Caerleon, Gwent (♀). How can a single bolt zap eleven simultaneously? By striking the waterlogged ground is the answer, creating an electrical gradient felt by everyone – including the spectators. Only one player is seriously hurt.

Such group strikes at outdoor sporting events are not uncommon: at the Army cup final, in April 1948, eight players, two spectators and the ref are felled; two later die. *The Golfers' Handbook* includes guidance for players' safety ('raising golf clubs or umbrellas above the head is dangerous' – see below). Between thirty and sixty people are struck each year in Britain, out of 300,000 or more annual ground strikes. On average, about three are killed – a sixfold decrease from the nineteen per year killed, on average, between 1852 and 1899. This is because far fewer people work outdoors now. A strike only lasts a fraction of a second, and surviving it depends on the route it takes past the vital organs (*20 June*). It can, for example, run down a wet coat, missing the body, and end up in the shoes.

HOW NOT TO DIE FROM A LIGHTNING STRIKE
- Avoid open spaces and beaches.
- Don't shelter under a tree, especially a lone tree, however tempting.
- If in water, get out.
- If caught in the open, keep away from metal objects (golf clubs, fishing rods, mobiles, iPods, bicycles, wire fences, even coins). Crouch as low as you can, feet together, preferably in a ditch or hollow.
- Get in a car. The superstructure acts like a giant Faraday cage, guiding the charge harmlessly round the outside of you.
  - If your or anyone else's hair stands on end, or objects around you begin to buzz, run – lightning may be about to strike.

-9.4C Caldecott 1968 ❄
24.4C Brixton 1894 ☀

9 APRIL

3.1C min / 11.3C max

'IT'S THE SUN WOT WON IT'

Does weather affect voting in General Elections? Everyone believes it does today, in 1992. The electorate has to choose between the incumbent Conservatives, under John Major, or the Labour opposition, under Neil Kinnock – equally unappealing choices for many. 'LABOUR'S TAX BOMBSHELL' bawl the Conservative posters, while Neil Kinnock rabble-rouses Sheffield with a triumphalist speech – 'We're aaallrright! We're aaallrright!' – now generally hailed as one of TV's all-time embarrassing moments. The *Sun*, the day before polling day, runs eight pages of copy beneath the headline: 'NIGHTMARE ON KINNOCK STREET'.

Polling day dawns, dry, sunny and warm, just about everywhere. The *Sun* runs the now immortal front page, showing Kinnock's head inside a light bulb, alongside the headline: 'IF KINNOCK WINS TODAY, WILL THE LAST PERSON IN BRITAIN PLEASE TURN OUT THE LIGHTS.' The turn-out is a spectacular 78 per cent, the highest since 1974. It breaks several records, and brings a fourth Conservative victory, with more votes cast than ever before. For the first time, one party receives more than 14 million votes. Next day, ever-bashful, the *Sun* bawls: 'IT'S THE SUN WOT WON IT'. It probably is – just not in the way the newspaper claims.

Weather, the evidence suggests, has affected several recent elections. In 2001, when it's cool, with a brisk wind and blustery showers over most of the UK, there's a turn-out of just 59 per cent – the lowest since 1914 (partly explained by the fact that most people regard a Labour victory as a foregone conclusion). And dry, sunny weather with temperatures several degrees above the average at the 1997 election sees a turn-out of 71.4 per cent. But if rain and cold reduce turn-out, and good weather increases it, can weather influence the *way* people vote? Here again, the evidence suggests that yes, fine weather on polling day makes people feel better, and so more likely to give the benefit of the doubt to the incumbent government. Richard Crossman, Labour Cabinet Minister (under Harold Wilson), Leader of the House of Commons and political diarist, writes in buoyant mood of 'perfect electioneering weather' when fighting the Heath government in the 1970 election (*14 June, 18 June*). The weather is indeed wonderful – which probably affects Britain's decision to return Edward Heath.

-8.3C Leadhills 1923 ❄

23.9C Southend-on-Sea 1969 ☼

10 APRIL

3.6C min / 11.5C max

ROBERT MARSHAM'S 'INDICATIONS OF SPRING'

This spring day, in 1736, the first swallow (*13 April*) arrives at the village of Stratton Strawless in Norfolk (♀). It's noteworthy only because of what the twenty-eight-year-old squire who watches its arrival, Robert Marsham, does next. It's something no one has ever thought to do before. *He writes the fact down.* Soon, he's noting down other seasonal occurrences, along with their dates: earliest singing of cuckoo and nightingale, dates different trees come into leaf, first flowering or blossoming dates, first time he hears frogs croaking and rooks nesting, first appearance of butterflies. In all, he records twenty-six of what he calls his 'Indications of Spring'.

Marsham has invented a new field of study, 'phenology': the study of the times of recurring natural phenomena. Although his initial purpose is merely to find the best time to cut timber, it becomes his life's work. He maintains his record, without interruption, for sixty-two years. More remarkably still, succeeding generations of the Marsham family continue it – until 1958. As variation in the dates of phenological events can be shown to correspond with warm and cold years, we know that phenology is related to temperature: as a rule of thumb, flowering and leafing occur six to eight days earlier for every degree C rise. Accordingly, with concerns about global warming, Marsham's record provides one of our most precious sources of evidence of its consequences. '150 to 200 species may be flowering on average 15 days earlier in Britain now than in the very recent past,' states a paper in *Science* in 2002, concluding that these earlier first flowering dates will have 'profound ecosystem and evolutionary consequences'.

Marsham, along with the Reverend Gilbert White (*27 January, 10 December*), with whom he corresponds, and White's brother-in-law, the Rutland squire Thomas Barker (*2 May*), are the first of a new breed of gentleman amateur naturalists that appeared in the eighteenth century. Their legacy lives on in the colossal popularity of the BBC series *Springwatch* and *Autumnwatch*.

-10.1 Grantown-on-Spey 1978 ❄

Nine weeks' rain falls in forty-eight hours in 1998 over central England. The rivers Avon, Leam, Cherwell, Ouse, Nene, Soar and Wreke all burst their banks in a once-in-a-150-year event. There is little furore: water just creeps up and up, leaving, when it retreats, a layer of soft mud – and an insurance bill of £700–800 million.

23.3C Cullompton 1909 ☀

11 APRIL

3.6C min / 11.7C max

BACON'S CHICKEN

On this cold, snowy day in London in 1626 the Elizabethan philosopher and scientist Sir Francis Bacon is travelling from Gray's Inn to Highgate with Sir John Wedderburn, the king's physician. At the foot of the hill, the ever-enquiring Bacon buys a chicken. He proceeds to kill it and stuff it full of snow with his bare hands. He is, he later writes from his deathbed, 'desirous to try an experiment or two, touching *the conservation and induration of BODIES*' – to see if snow can preserve flesh as well as salt does. Unfortunately, as a result of this experiment Bacon is chilled to the marrow and falls ill. For some reason he's taken to Arundel House, an empty summer mansion on Highgate Hill, where he's put in a damp bed. Within days, the great man is dead from bronchitis.

Or is he? Such are the particulars of Bacon's death that it's been suggested the whole thing is a hoax. None of his friends come to see him while he's ailing in Highgate. He has already published his *Experiments on the Preservation of Bodies by Cold*. And he dies on 'Resurrection Sunday' – nudge, nudge. Is a more plausible story that 'sick of the struggle against penury, ill fame and ingratitude', and uncertain how the new King, Charles I, and court might use him, Bacon takes the opportunity of an unseasonably cold day in April to 'do a Reggie Perrin'? As the nation mourns, does he slip off to Germany to live for another twenty-five years? Either way, half a century of English Renaissance culture ends following today's snow. As Ben Jonson notes after the event: 'Wits daily grow downward.'

'Today has turned suddenly oppressively warm, like a day in early summer, and through the hot dusk green trees loom hugely. Isolated rain-drops spot the pavements ... After work tonight I had a gin & lime & soda, my favourite drink at present. 1/8 [about 8p] a mouthful, though. Holy God.' *Philip Larkin, poet, Wellington, Shropshire, 1945*

'Today is clear and of course cold. Spring has come. Outside my window white lambs leap and jump. There's faint green in the trees ... I said to the cats the other night, "I'm tired of being an under-rated writer, I want to be an over-rated writer like everybody else." But also, does it matter? Just now I'm more interested in weather.' *Martha Gellhorn, war correspondent, Monmouthshire, 1982*

-13.3 Braemar 1917 ❄

26.0C Wryde 1894 ☀

3.8C min / 11.7C max

'COASTING'
WITH
JONATHAN
RABAN

'The weather and the tides ... kept me in a state of dazed preoccupation ... Living in a city, I'd hardly bothered to notice whether it was raining or shining. Weather was something that just was, and I couldn't have been less interested in its whys and wherefores.

Now I studied it as intently as any text that I'd pored over in the past. I watched the Atlantic lows winging their way in from south Greenland – unstable, whirling cones of disturbed air, filling, deepening, changing track, spawning more depressions in their wake. Spinning against the clock, they brought the powerful, salty southwesterly winds that whipped the sea up into untenantable hills of froth and spume, took slates off roofs and made the water in even sheltered harbours slop and gurgle round the quays, slamming boats into walls and tossing them frivolously about on their moorings.

I learned Buys Ballot's Law. Face the wind, and you'll find low pressure to your right and high pressure to your left (the reverse, of course, applies in the Falkland Islands). It was the high pressure to the left of England that I began to dream of wistfully as I'd once dreamed of unattainable girls. Please, God, give me a kindly ridge of it, just from, say, north of the Azores to somewhere a little west of Ireland. The northwest wind would be cold, but it would come from the shore, the sea would be flat and *Gosfield Maid* would whistle up-Channel to the Dover Straits with her sails wide out to starboard.

I watched the drum on the barograph in the saloon revolve at a tenth of an inch to an hour, its inked stylus leaving a thin blue line on the paper as the vacuum cylinder swelled and contracted with the changing atmospheric pressure. Very soon I found myself subscribing to a theory of natural magic. On the rare morning when the barograph needle had climbed overnight and was holding steady, so I discovered a buoyancy of spirit in myself, a sudden rush of cheerfulness and hope for the day. As the needle dipped, my mood darkened in sympathy, and I could feel myself sinking down the inky slope on the graph paper. Galebound at 995 millibars and falling, I sat up in the wheelhouse under a sky of heaped slag and asphalt, lost in the small print of *Reed's Nautical Almanack'.*
    Jonathan Raban, on his four-year perambulation in the ten-ton
        Gosfield Maid *around the coast of the British Isles in 1982,*
        Coasting, *1987*

-9.4C Alwen 1958
25.6C Cromer 1939

13 APRIL

3.8C min / 11.6C max

SWALLOWS ARRIVE

It may not make a summer, but the sight of a single swallow is probably everyone's favourite indication that the season's on its way – and according to the great eighteenth-century amateur naturalist, the Reverend Gilbert White (*10 April, 10 December, 27 January*): 'The house-swallow . . . appears in general on or about the thirteenth of *April*, as I have remarked from many years' observation.' There is a strange satisfaction in knowing that, despite inhabiting worlds different in almost every conceivable way, the arrival of the first swallow is a pleasure identical to us and to someone living 250 – and, presumably, 2500 – years ago.

> 'The eleventh day of sunshine . . . A lot of my time is being spent assisting indignant bumble-bees trapped behind window panes: their bigness, noisiness and silliness is somehow appealing.'
> Alec Guinness, actor, Hampshire, 1995

Obviously, there's a good chance we'll see swallows before today. 'When I was a boy I observed a swallow . . . on a sunny warm *Shrove Tuesday*; which day could not fall out later than the middle of *March*, and often happened early in February,' White goes on to say. In fact, according to figures from the Woodland Trust and BBC's *Springwatch*, nowadays our first sighting of a swallow (averaged from the last five years) is six days later than White's estimate, on 19 April. This is not because swallows' habits have changed, but because, while White is averaging only for his own parish of Selborne in Hampshire (✿), the recent figures are averaged from the whole country – and swallows arrive in the north later than in the south.

Changes in day-length, it's thought, monitored by the swallow's 'internal clock', are what trigger the phenomenon of migration. The birds are astonishingly savvy about picking clear skies and tail winds. British swallows arrive here having completed a 6000-mile (9600-km) journey from southern Africa, via Namibia, the Congo rainforest, the Sahara, Morocco, eastern Spain and western France. They fly in large flocks, low, during daylight (eating on the wing), covering about 200 miles (320 km) per day, often returning to the same building or barn to nest each year (swallows live for up to sixteen years). Their journey takes about four weeks and the males usually arrive first. The return journey (*22 September*) takes two weeks longer, because of head-winds.

-9.2C Lednathie 1890 ❄

23.0C Rumleigh 1991 ☀

14 APRIL

4C min / 11.9C max

'While the tea was hoisting out of the gun room, a thick haze that had prevailed all the morning just then cleared away, and we saw the land (the Lizard) not more than four leagues distant. The cutter hailed to inform their Chief they saw the *Albert* (custom house schooner) to the southward ...'

Thus records the lawyer William Hickey, aboard the East Indiamen *Plassey*, re-entering the English Channel on its return journey today in 1770. Hickey watches tea (some sixty chests) and cognac being transferred – probably to the most infamous of the Cornish family smuggling firms, the Carter brothers of Prussia Cove (John Carter is known as the 'King of Prussia') near St Michael's Mount (♥). Between 1700 and 1850, smuggling is one of Britain's biggest industries. And despite the popular image – moonlit coves on clear summer nights – in reality smugglers preferred bad weather: wind, rough seas, fog, haze and mist.

'There is twilight and soft clouds and daffodils – and a great weariness. Spring! *Excellentissime* – Spring? We are annually lured by false hopes. Spring! *Che coglioneria*! Another illusion for the undoing of mankind.'
Joseph Conrad, author, 1898

'This horrible wind always makes me bilious and savage. People and things all look disfigured and hideous under it.'
Matthew Arnold, poet, London, 1862

'Owing, I presume, to the unsettled weather, I awoke with a feeling that my skin was drawn over my face as tight as a drum.'
George and Weedon Grossmith, Diary of a Nobody, 1888

Until the iceberg in the North Atlantic sinks the 'unsinkable' *Titanic* tonight in 1912 – with the loss of 1517 – it's routine for ships to brave the sea's fury without enough lifeboats. In fact, the *Titanic* exceeds the lifeboat standard of the day, with twenty, even though this is barely enough for half the number aboard. (The law, dating from 1894, requires only sixteen lifeboats for ships over 10,000 tons.) The fact is that safety at sea hasn't caught up with the surge in passenger shipping since the 1820s. No ships in the busy North Atlantic carry enough lifeboats for passengers and crew (*25 January*) because, in the event of a serious accident, it's assumed that help will be readily available from other vessels. After the *Titanic* disaster, of course, passenger liners don't just carry enough lifeboats for all; they loudly trumpet the fact in their advertising.

-10.6C Achnagolchan 1966 ❄

130.3mm Snowdon 1926 🌢

'The north wind, which is the great enemy of the budding flowers and trees, blowing during nearly the whole spring.'
Matthew Paris, historian, Chronica Majora, 1255

24.4C Jersey-St Louis 1943 ☀

15 APRIL

4C min / 12.2C max

WORDSWORTH'S DAFFODILS / BLUEBELLS

A blustery walk in Grasmere (☂) this afternoon in 1802 inspires one of William Wordsworth's most famous poems (see over). 'When we were in the woods beyond Gowbarrow Park,' writes Wordsworth's sister Dorothy in her diary, 'we saw a few daffodils close to the water side. We fancied that the lake had floated the seeds ashore and that the little colony had so sprung up. But as we went along there were more and yet more . . . I never saw daffodils so beautiful. They grew among the mossy stones about and about them, some rested their heads upon these stones as on a pillow for weariness and the rest tossed and reeled and danced and seemed as if they verily laughed with the wind that blew upon them over the lake, they looked so gay ever glancing ever changing.'

BLUEBELLS FLOWER TODAY
In an average spring, bluebells first flower today (source: Woodland Trust). The bluebell is one of our native flowers most threatened by global warming. Its crucial competitive advantage is its ability to bloom early on the woodland floor, growing rapidly to produce a new bulb before spring leaves on the trees cast it in shade. Milder winters, which encourage other species to do this also, mean . . . goodbye, bluebell woods.

Two years later Wordsworth, with whom Dorothy lives, writes 'Daffodils' (*20 April*). Published in 1815, it's by no means the first time that William uses his sister's notes (*30 September*). Dorothy's journals are a source for others too, notably their friend Samuel Taylor Coleridge. Her notes only come to light in 1931, when the children's writer and artist Beatrix Potter buys the Wordsworths' Lake District home, Dove Cottage, and finds a bundle of old papers in the barn. Their discovery, just as critics are busily reappraising women's role in literature, prompts a surge of interest in her.

-10.0C Newton Rigg 1892

26.1C Cambridge Botanic Gardens 1934

I wandered lonely as a Cloud
That floats on high o'er Vales and Hills,
When all at once I saw a crowd,
A host of golden daffodils;
Beside the lake, beneath the trees,
Fluttering and dancing in the breeze.

Continuous as the stars that shine
And twinkle on the milky way,
They stretched in never-ending line
Along the margin of a bay:
Ten thousand saw I at a glance,
Tossing their heads in sprightly dance.

The waves beside them danced, but they
Out-did the sparkling waves in glee:
A poet could not but be gay,
In such a jocund company:
I gazed — and gazed — but little thought
What wealth the show to me had brought:

For oft when on my couch I lie
In vacant or in pensive mood,
They flash upon that inward eye
Which is the bliss of solitude,
And then my heart with pleasure fills,
And dances with the Daffodils.

16 APRIL

4C min / 12.3C max

RAIN AND CABS

'When we got outside the Drill Hall it was raining so hard that the roads resembled canals, and I need hardly say we had great difficulty in getting a cabman to take us to Holloway. After waiting a bit, a man said he would drive us, anyhow, as far as "The Angel," at Islington, and we could easily get another cab from there. It was a tedious journey; the rain was beating against the windows and trickling down the inside of the cab.

When we arrived at "The Angel" the horse seemed tired out. Carrie got out and ran into a doorway, and when I came to pay, to my absolute horror I remembered I had no money, nor had Carrie. I explained to the cabman how we were situated. Never in my life have I ever been so insulted; the cabman, who was a rough bully and to my thinking not sober, called me every name he could lay his tongue to, and positively seized me by the beard, which he pulled till the tears came into my eyes. I took the number of a policeman (who witnessed the assault) for not taking the man in charge. The policeman said he couldn't interfere, that he had seen no assault, and that people should not ride in cabs without money.

We had to walk home in the pouring rain, nearly two miles, and when I got in I put down the conversation I had with the cabman, word for word, as I intend writing to the Telegraph for the purpose of proposing that cabs should be driven only by men under Government control, to prevent civilians being subjected to the disgraceful insult and outrage that I had had to endure.'
George and Weedon Grossmith, Diary of a Nobody, *1888*

'Extraordinary storms of violent rain and the most vivid rainbows I ever saw.'
John Gielgud, actor, Glasgow, 1935

'This day in the afternoon, stepping with the Duke of York into St James' Park, it rained, and I was forced to lend the Duke my cloak, which he wore throughout the park.'
Samuel Pepys, diarist, 1688

-11.7C Braemar 1892 ❄

'What a Good Friday we're having! Rain, wind, cold, skating on all the ponds, icicles hanging from the eaves and George Bernard the shepherd blowing his nail.'
George Bernard Shaw, dramatist, Surrey, 1897

29.4C Camden Square 1949

17 APRIL

3.9C min / 12.2C max

LAST FROST / THE CANTERBURY TALES

Today, in central Britain, is the average day for the last frost of the year. Obviously this doesn't mean that a frost is especially likely – just that, from the dates of last frosts in recent years, this is the average. It's a key moment in any gardener's calendar, since plants in the British Isles are divided into those that are 'frost-hardy' and those that are not. Only after the last frost can gardeners in the south traditionally regard it as safe to put out frost-sensitive bedding plants.

'I stayed on deck till the sun had set. The water was fairly dark blue with rather high white-crested waves as far as one could see. The coast had already disappeared from sight. The sky was one vast light blue, without a single little cloud. And the sunset cast a streak of glittering light on the water. It was indeed a grand and majestic sight.'
Vincent Van Gogh, artist, Ramsgate, Kent, 1876

'The happiest, brightest, most beautiful Easter I have ever spent. I woke early and looked out. As I had hoped the day was cloudless, a glorious morning. My first thought was "Christ is Risen".'
Reverend Francis Kilvert, diarist, Clyro, Wales, 1870

FROST HARDY	FROST SENSITIVE
Geraniums	Dahlias
Phlox	Potatoes
Delphiniums	Petunias
Euphorbia	Pelargoniums
Violas	Nicotiana (tobacco plant)
Peonies	Camelias
Aquilegias	Apple (and other fruit) blossom

Of course, today is only an averaged date – last frosts routinely occur much later, so no wise gardener puts out sensitive plants yet. The traditional guide is Chelsea Flower Show Week (*23 June*) in the south, a fortnight later for anyone in more elevated or colder places.

Geoffrey Chaucer's pilgrims depart from Southwark for Canterbury today, in 1387 – according to the work's nineteenth-century editor, Walter William Skeat – with a cheery weather reference:

When in April the sweet showers fall
That pierce March's drought to the root and all
And bathed every vein in liquor that has power
To generate therein and sire the flower;
When Zephyr also has with his sweet breath,
Filled again, in every holt and heath,
The tender shoots and leaves, and the young sun
His half-course in the sign of the Ram has run . . .

-10.0C Braemar 1922 ❄

27.8C Whitstable 1949 ☼

18 APRIL

3.9C min / 12.5C max

BRITAIN'S GRASS ROOTS

In typical weather, now is the moment of the 'spring flush' of grass growth; when, winter over, grass – suddenly conspicuously greener – starts growing crazily. It's a crucial moment, because – be in no doubt – grass is what Great Britain's greatness rests upon. Grass is the root of the wool industry (*7 May*), by which Britain becomes a great trading nation, by which the Industrial Revolution is underwritten, and by which we build the largest empire in modern history. It's not hard to make the case that it's only for the sake of grass that, for five centuries, our navy rules the waves.

> 'The grass grows so quick that I feel like a barber faced with a never-ending queue (& not a chinaman's either, to be trimmed with one snip)'. *J. R. R. Tolkien, academic and writer, 1944*

The grass of the British Isles is like no other grass. 'The fineness and almost perpetual greenness of our turf cannot be found in France or in Holland,' notes the diplomat and essayist Sir William Temple in 1685 (thus dismissing, at a stroke, our two supreme trading adversaries). He's right, of course. In a general sense, grass is what makes our land so green and pleasant; why the overwhelming sensation, on returning to Britain from abroad, is one of *greenness*. Even setting aside the wool trade, British culture has always been grass-rooted: from the roast beef of olde Englande, to the dairy herds that make the best cream, butter and hard cheese in the world. What other country has fresh milk in every local shop? Or a velvet sward so perfect, we've had to invent things to do with it (football around 1100, bowls by the thirteenth century, cricket from 1300, golf by the 1400s, rugby around 1823, croquet in the 1830s, lawn tennis in 1874 and hockey by 1849)?

Grass is what makes Britain beautiful – an observation not wasted on eighteenth-century landscape gardeners such as William Kent and 'Capability' Brown as they set about 'tweaking' our verdant landscape, with lakes and clumps of trees, into scenes of romantically idealized pastoral beauty. 'England's greatest contribution, perhaps,' observed the architectural historian Christopher Hussey in 1948, 'to the visual arts of the world.' And directly descended from these pastures green is that centre-piece of every English garden, that shorn turf so 'delicious' – Henry James's words – 'to the sentient bootsole.' What could we mean, of course, but the lawn? (*21 March*)

-10.0C Lagganlia 1986

26.1C South Farnborough 1945

19 APRIL

THE FIRST SALMON 'The rainbow last night was a good sign, and I woke feeling that it was going to be a massacre. But the window was wet and the slate sky icy. Still, it might clear at noon. We started at the Mill Pool straight away ... And so we cast slowly down, and the east wind blew that rain through everything ... and the more we cast the more it grew upon us that hope was dead. From 8.30 till 1.15 we wandered on the banks, like lost souls staying for waftage. It didn't clear at noon. We tried the Crooked Pot; we tried the Ardgalleys; we tried up above the snipe marsh to the bitter end ... The east and watery wind blew Macdonald's casts on the bank, to my secret joy ... I only had a cap and that was a cold poultice. I wore it back to front, in the vain effort to keep the rain from running down my neck. I paid out a sticking line with slippery, frozen fingers: the horrible and slightly rasping stickiness of cold wet deer's fat ... It became really impossible to go on. I couldn't feel the line. I forgot whether it was hail then or sleet. I asked Macdonald if he would mind my using his spinning rod ... I had four casts and caught the bottom. At the sixth cast, I caught it again, but it was a little different. It just seemed to move. I struck ... It was a salmon. Oh, God, it was a salmon.

I shouted to Macdonald, who came, thinking I was snagged. I said: "I think I have a fish." He looked at me as if I was mad, then at the line to see if he was. He said: "You have. Yes, certainly you have." Then he began to become hysterical.

I remember Macdonald saying: "Well, if we can only get this one on the bank, we can call it a guid day." The important thing was the weather. Just for that twenty minutes the wind veered west and the sun shone. It woke the salmon up and they began to move up once more. But before moving they felt lively and took. I am sure that I hooked this fish during the only three or four minutes when it would have been possible to take fish by any means. The sun went in again, the wind went east, the rain came down: there was a silver cock salmon on the bank.' *T. H. White,* England Have My Bones, *1936, on catching his first salmon in 1934.*

-8.1C Inverdruie 1969 ❄

25.6C Cambridge 1893 ☀

HEINEKEN
REFRESHES THE
POETS

On a breezy afternoon in the Lake District in 1802, Dorothy Wordsworth notes in her journal how the daffodils toss and reel and dance their little heads off – a scenario that so inspires her brother William that he writes his poem 'Daffodils' (*15 April*). Precisely 180 years and five days later, Dutch brewing giant Heineken arrives at the same spot in the Lake District (♥), with their advertising agency, to film an action replay of the classic scene for their much-loved campaign, *Heineken refreshes the parts other beers cannot reach*. There's a hitch, however. No wind. More awkwardly, no daffs. The producer has been promised that this is *the* moment. The place will be carpeted. But, after an exceptionally cold winter, with several record-breaking lows, and despite a mild March and April, there are – well, none.

With an expensive film crew lolling about, there is no time to reason why. So popular are the Heineken ads – in which someone becomes 'refreshed' in an unusual or unexpected way – that the script for today has been written by a member of the public following a competition. Lack of daffs cannot be allowed to hold the show up. London is called. Six thousand are ordered, in two juggernauts. Finally, everything is ready: a top-hatted Wordsworth takes his seat amongst the sea of yellow blooms, and filming begins:

MUSIC: *Romantic throughout.*

SOUND EFFECTS: *Nib on paper, writing.*

VOICE: 'I walked about a bit on my own . . .'

SOUND EFFECTS: *Nib on paper, crossing out.*

SOUND EFFECTS: *Nib on paper, writing.*

VOICE: 'I strolled around without anyone else . . .'

SOUND EFFECTS: *Nib on paper, crossing out.*

SOUND EFFECTS: *Can opening and drinking.*

SOUND EFFECTS: *Nib on paper, writing.*

VOICE: 'I wandered lonely as a cloud that floats on high o'er vales and hills, When all at once I saw a crowd, a host of golden daffodils . . .' (FADING)

VOICE-OVER: 'Only Heineken can do this, because it refreshes the poets other beers cannot reach.'

135.4mm Honister Pass

-7.8C Alwen 1969

28.9C Cambridge 1893

21 APRIL

4.5C min / 13.1C max

A storm on the Yorkshire coast tonight in 1850 is noteworthy for only one reason: it is accurately forecast by a Victorian inventor by means of his 'jury of philosophical counsellors' – twelve leeches. Analysing the instincts of animals to predict the weather isn't new. Even the movements of leeches when wind and rain are imminent is documented. Inspired by the observations of Edward Jenner (inventor of vaccination) and William Cowper (poet), Dr George Merryweather, a Whitby GP, begins experimenting with leeches in 1849.

The result is one of the more elaborate appliances to grace the sober history of meteorological instrumentation. Merryweather unveils his 'Atmospheric Electromagnetic Telegraph, conducted by Animal Instinct' –

'3.15pm: I am now sitting on a bench, facing the sun in Regent's Park. They are mowing the lawns everywhere and the smell of cut grass, plants and warm earth is delicious. Nothing is so beautiful as England in April.' *Sylvia Plath, poet, London, 1960*

'Tempest Prognosticator' for short – in 1851. Twelve pint bottles, each containing a leech and an inch of water, sit on a circular stand. In the neck of each bottle, a piece of whalebone is set loosely in a brass tube, and connected to a hammer on a stand above. The principle is simple: when a storm approaches, the sensitive leeches wriggle up their jars, releasing the hammers that ring the bell. The more the bell rings, the more likely the storm. What is astonishing, however, is that this exotic instrument – in appearance, a cross between a toy merry-go-round and a Hindu temple – actually works. Merryweather spends a year listening out for the bell to toll, to prove 'that no storm can escape me, without my possessing a previous knowledge of its approach'. And sure enough, his storm predictions are often accurate. The apparatus is displayed at the Great Exhibition of 1851 (*1 December*). Lloyd's of London commission tests which bear out its claims to accuracy, and Merryweather nurses hopes that 'our Whitby pigmy temples would be distributed over the world', or at least 'used by the Government and the Shipping Interests of the country in stations all round the coasts'.

-10.0C Dalwhinnie 1981

It's not to be, of course. The need for a storm warning system is undoubtedly vital, but the scientists at the government's Board of Trade, where Robert Fitzroy (*1 August, 25 October*) is applying his empirical procedures, prefer mercury barometers and storm glasses to leeches. Not a single Tempest Prognosticator is sold.

28.1C Ross-on-Wye 1893

22 APRIL

4.7C min / 13.1C max

'FULL CONDITIONS' / INVENTION OF ICE CLIMBING

Following rain, snow and Atlantic storms, the rocks are 'heavily iced' for today's Glencoe meet of the Scottish Mountaineering Club in 1906 (☻). 'In Scotland . . . we cannot afford to wait for good weather, which may possibly never come. We must just take the rough with the smooth, and take our buffeting and our pelting as if we enjoyed it, which, indeed, we do, at least in retrospect,' says club stalwart Harold Raeburn in his account of today's ascent of one of the steepest cliffs of Ben Nevis.

BRITISH ICE – FROM NAILED BOOTS TO THE INVENTION OF THE CRAMPON
The abrasive, irritating, gifted, almost fearless Englishman Oscar Eckenstein (German father), a wandering, bearded scientist with a Sherlock Holmes pipe, is remembered – when he is remembered – for two things. First, his remarkable fear of kittens. Second, following regular climbing trips in the 1880s to the Lakes and Wales, his redesign into its modern form of the ten-point crampon, in 1908, and the short ice axe – the two crucial technical tools of modern ice climbing. Both are immediately adopted by the French and Swiss in the Alps.

Following the mid-nineteenth-century British 'invention' of mountaineering, Raeburn, along with Hugh Munro (9 February), Professor Norman Collie and W. W. Naismith, now develop their own, uniquely Scottish, approach: insisting that snow, ice, and blizzard – 'full conditions' – are essential ingredients for a 'guid' day on the hill.

'Ice climbing' may sound a little hopeful in Britain in these mild, modern times. But, in colder days, the Scottish mountaineering tradition crucially contributes to this now highly technical activity. Without the great snow and ice faces of the Alps Scottish lovers of ice-climbing have to be more creative, making the most of the shaded gullies, with their frozen waterfalls, which cut the big massifs. Of course, such 'old-style' winter climbing – five to eight hours in a sunless gully in the face of avalanches and spindrift, with numb fingers and toes – is not to all tastes. But Raeburn takes up the cause with irrepressible enthusiasm: his classic ascents include Crowberry Gully in 1893 (in 'full conditions', of course), Green Gully on Ben Nevis in 1906 in a blizzard, and Observatory Ridge in 1920. For a century from 1880, especially in spring, Scotland offers some of the finest pitches of technical ice climbing in the world, and it becomes a training ground for many of the great Himalayan expeditions. And after Eckenstein invents the crampon, modern ice climbing is born.

-10.7C Kingussie 1899

182.1mm Seathwaite 1970

26.2C Jersey 1984

23 APRIL

4.8C min / 13C max

BATTLE OF BARNET / SILVERSTONE

'In the morning, right early there was such a thick mist that neither of them might see the other perfectly,' writes John Warkworth, in his *Chronicle*, of the Battle of Barnet (♥) today in 1471. 'The Earl of Oxford's men had upon their lord's livery a star with streams, which was much like King Edward's livery, a sun with streams. And the mist was so thick that a man might not properly judge one thing from another; so the Earl of Warwick's men shot and fought against the Earl of Oxford's men, thinking and supposing that they had been King Edward's men. And at once the Earl of Oxford and his men cried "Treason! Treason!" and fled away from the field.'

Such is the chaos of the Battle of Barnet. Perhaps the uneasy alliances and orchestrated betrayals – hallmarks of the thirty-year struggle for the English crown known as the Wars of the Roses – are in the soldiers' minds as they try to work out who's killing whom in the fog. In the murk and confusion, the slaughter is heavy. It's the beginning of the end for the House of Lancaster. The all-powerful Earl of Warwick, 'plucker down and setter up of kings', dies with a knife through his eye. His naked body is dragged to London, where King Edward is received as a triumphant hero.

A fortnight of rain turns the 2000 Silverstone Formula 1 Grand Prix (🏁) into a Young Farmers' Tractor Pull. The 200 acres (80 hectares) of farmland set aside for parking become a quagmire. Police close roads leading to the notoriously inaccessible track, and blanket fog exacerbates the problem because helicopters can't fly. Mikka Hakkinen, one of the drivers, has to jump on a passing motorbike to reach the circuit in time. Most spectators are not so fortunate. Typically – this is April – by the time of the start, the sun is shining. Over an empty grandstand. Accelerating away from the grid, Michael Schumacher puts two wheels on the verge – no drier than the car park – and spins, blowing his chance of victory. David Coulthard goes on to win. In the ensuing press furore, the organizers are accused of 'buffoonery' and 'classic British amateurism'. A plan for a £45 million redevelopment to improve the circuit and its infrastructure is hastily assembled. It could have been worse. In 1981, the country from the Pennines to Salisbury Plain is blanketed with 8 inches (20 cm) of snow: the most severe late April weather of the twentieth century.

-10.6C Dalwhinnie 1981

26.1C Cambridge 1893

SNOW ON THE BLACK HILL

'Around eleven that morning, Amos had looked towards the west and said, "I don't like the look of them clouds. Better get the ewes off the hill."

It was late in the lambing season and the ewes and early lambs were on the mountain. For ten days the weather had been lovely. The thrushes were nesting, and the birches in the dingle were dusted with green. No one had expected any more snow.

"No," Amos repeated. "I don't like the look of it."

He had a chill on his chest, and his legs and back were stiff. Mary fetched his boots and gaiters and noticed, all of a sudden, that he was old. He bent down to tie his laces. Something cricked in his spine, and he sank back into the chair.

"I'll go," said Benjamin.

"Quickly now!" his father said. "Before it comes to snow."

Benjamin whistled for the dog and walked over the fields to Cock-a-loftie. From there he took the steeper path up the escarpment. He reached the rim, and a raven flew off a thornbush, croaking. Then the cloud came down and the sheep, when he could see them, were like little packs of vapour – and then it began to snow. The snow fell in thick woolly flakes. The wind got up and blew drifts across the track. He saw something dark close by: it was the dog shaking the snow off his back. Icy trickles ran down his neck, and he realized his cap was gone. His hands were in his pockets but he could not feel them. His feet felt so heavy it was hardly worth bothering to take the next step – and, just then, the snow changed colour. The snow was not white any more, but a creamy golden rose. It was not cold any more. The tussocks of reed were not sharp, but soft and downy. And all he wanted now was to lie down in this nice, warm comfortable snow, and sleep . . .

It was white when he woke, and it took him some time to realize that the whiteness was not snow, but bed-linen . . . The sharp spring sunlight streamed through the window.' *Bruce Chatwin,* On the Black Hill, *1982*

'April cheats me of all its advertised joys: the wind is howling at all hours, and the rain is raining at most of them. A bloodier April England never saw.' *Vita Sackville-West, writer, Sissinghurst, Kent, 1932*

-12.8C Garforth 1908

27.8C Cambridge 1893

25 APRIL

4.9C min / 13C max

1908 SPRING SNOWSTORM

By the standards even of winter storms, the blizzard that blows this Saturday in 1908 is extreme – and this is spring. Snow begins to fall on Thursday, so heavily that Newmarket races have to be abandoned mid-programme. By Friday morning, nearly 9 inches (22 cm) of snow lie across East Anglia. Cloudless skies and sunshine bring a rapid thaw and it all looks like a fleeting aberration. But clouds regather, and southern Britain wakes today to the worst snowstorm for twenty-seven years (since the Great Blizzard of 1881, *18 January*). The usually sheltered western Home Counties bear the brunt of it, with more than 2 feet (60 cm) lying in north Hampshire and Berkshire (a confused swallow is seen at Newbury Bridge). Oxford receives well over a foot (42 cm), a figure not exceeded since. There's even a foot of snow on Alderney in the Channel Islands. Temperatures sink to -12.8C at Garforth in West Yorkshire – it remains the latest date in the year for temperatures below -10C anywhere in the UK.

> 'Thank God, we have safely crossed the English Channel. Yet we have not done so without making a heavy contribution in vomiting. I, however, had the worst time of it.'
>
> *Leopold Mozart, accompanying his eight-year-old son, Wolfgang Amadeus, 1764*

April is the month with the most surprises up her sleeve. By now spring is usually well under way. Blackthorn and wild cherries are in flower. Oaks and ashes are coming into leaf around now, but recent, warmer Aprils (notably 2007) tend to favour oaks (which respond to temperature) over ashes (which respond to day length). But snow this late is by no means rare. In 1919, even later than this, wet snow brings down many telephone and power cables around London and the Chilterns, thawing rapidly and causing floods. In 1950, deep snow damages trains and confuses cuckoos in the south-east, especially Kent, and in 1981, a cricket match is snowed off ('OWZAT! APRIL'S GONE CRAZY!' says the *Daily Mail*). But the spring snowstorm of 1908 remains April's finest joker. With typical April caprice, next day the sun shines brightly, a rapid thaw sets in and severe flooding follows. By the following Wednesday there's no trace of snow, temperatures are back to a spring-like 17C, and everyone's left wondering if they've imagined the last few days.

-12.2C Corstopine 1908

27.8C Cambridge 1893

26 APRIL

WINDSCREEN WIPER INVENTED

It's snowing so hard, today in 1908 (25 April), as Mr Gladstone Adams motors back to Newcastle after visiting London to watch the FA Cup Final, that he has to stop continually to clear snow from his windscreen. In a bad mood anyway (Newcastle United lose 3–1 to Wolverhampton Wanderers), he eventually becomes so exasperated that he folds down the windscreen altogether – and then gets so cold, he vows to do something about it. When he gets home, he invents the windscreen wiper.

Cars, at this juncture, are still basic affairs: windscreens only appear in 1903 (made of ordinary glass, so inflicting terrible injuries in accidents). Henry Ford's Model T has just gone into production in the USA and Rolls-Royce has just opened its first factory at Crewe. Cars lack indicators, hard tops, heaters and electric starters. Indeed, cars are still enough of a novelty that on arrival in London Mr Adams's Daracq-Caron car is put on display in a shop window in Oxford Street.

Adams registers a patent for his invention in 1911, but numerous others later claim the idea. Prince Henry of Russia fits a hand-operated rubber wiper to his Benz car when he sets off from Hamburg to London in 1911, while in America Mary Anderson files a patent for a manually operated rubber wiper blade as early as 1905. Adams has great difficulty in interesting anyone in his invention, and wipers don't become standard equipment in America until 1916. 'Automatic' wipers arrive after the First World War, frequently vacuum-operated, so that they slow down or speed up with engine speed. The major British importer reports selling six in 1920, but in 1924, when companies like Sunbeam and Daimler start fitting them, they finally catch on.

'MY FIRST AUTOMATIC WINDSCREEN WIPER' – BY THE EDITOR OF THE *AUTOCAR*, 1923

'As we left Penryn (📍) it began to rain. Have I mentioned my Folberth automatic windscreen wiper? I think not. Yet it was of more solid value than any other of the gadgets that helped to earn for us our White Knighthood. For hours and hours on end it clicked patiently back and forth across the screen, utterly defeating the rain and rendering the steering . . . a joy instead of the purgatory it would otherwise have been. It never ceased clicking and cleaning from Penryn to Penzance.'

-8.3C Kincraig 1957
25.8C Inverdruie 1984

CULLODEN /
DUCKWORTH-
LEWIS METHOD

The freezing sleet, slicing across Drumossie Moor, could easily have rendered the Duke of Cumberland's muskets useless. It's 1746, a century before the all-weather percussion cap, and the flintlock 'Brown Bess' is a primitive weapon even under ideal conditions. In the rain and wind, Cumberland's fusiliers must perform seven separate actions: rip open a cartridge and pour powder down the barrel; drop in the lead ball and paper wadding; ram these with a rod; tap the remaining powder into the 'priming pan' and close it. Only then can the flint hammer be cocked and the musket aimed and fired. In conditions like today's, the damp priming powder often doesn't ignite. Or it ignites in the priming pan without setting off the main charge (a 'flash in the pan'). But Cumberland's men are well drilled – and well prepared. Each musketeer has a waterproof leather wallet to keep his cartridges dry. Moreover, they fight with the wind, which carries the freezing rain, at their backs.

The Highlanders haven't a chance. Despite their fearsome Highland Charge (Shakespeare's 'strange screams of death' in *Macbeth*), their broadswords are no match against Cumberland's storm of lead. This – the last pitched battle on mainland British soil – is Culloden (⚔). The Highlanders will never charge again. The battle, and the Jacobite rebellion, is over in sixty minutes. The Gaelic way of life is gone for ever.

Heavy rain stops play at a one-day cricket match at Cardiff today in 1997, with Glamorgan still needing 65 to beat Warwickshire. When further play becomes out of the question, a new formula is applied to calculate *what would have happened had the game been played to its natural conclusion.* It's the first time the Duckworth-Lewis method has ever been used in Britain. Frank Duckworth, a retired statistical consultant, and Tony Lewis, a university mathematics lecturer, base their formula on the analysis of hundreds of previous one-day games. After a good deal of harrumphing from senior cricket fans, the International Cricket Council accepts the formula – even though no one understands it. Today, the Duckworth-Lewis calculation decides Glamorgan would have won by 17 runs.

-6.1C Glenlivet 1956

25.6C New Malden 1916

28 APRIL

4.9C min / 13C max

FIRST BRITISH
NIGHT FLIGHT

Defying turbulent winds, the pioneering French aviator Louis Paulhan lands his Farman aeroplane in a field on the outskirts of Manchester just before dawn today in 1910. In doing so, according to the *Daily Mail*, he claims a victory that is 'not the greatest of the century, but of all centuries'.

'The dawn widened into an exquisite spring day. Soft, wool-like puffs of sound came from the thrushes' throats in the trees. The uneasy year, tortured by its spring of adolescence, broke into bud-spots in hedge, copse, spinney, and byre.'
Stella Gibbons, Cold Comfort Farm, 1932

'It being cold even for April, there was a fire in the drawing-room; Mr Finsworth had up some fine port, although I question whether it is a good thing to take on top of beer ... we sat around in easy chairs, and Teddy and I waxed rather eloquent over the old school days, which had the effect of sending all the others to sleep.' *George and Weedon Grossmith*, Diary of a Nobody, 1888

25.5C Kinlochewe 1984
-7.8C Glenlivet 1956

When the *Daily Mail* offers a £10,000 prize – one of a series put up during its championing of early aviation – to the first person to fly the 185 miles (296 km) between London and Manchester within twenty-four hours, it's 1906. The money seems safe: Europe's foremost aviator can only fly just over 650 feet (200 metres). *Punch* offers the same sum to the first person to swim the Atlantic. By 1910, however, Claude Grahame-White, an enterprising Mayfair car dealer, believes it can be done. His first attempt, on a misty, wintry April day a week earlier, fails when the westerly wind funnelling down the Trent valley (♀) grows too strong, damaging his Farman plane. Four days later, Grahame-White is ready for another attempt. But this time, Paulhan crosses the Channel to compete with him.

The London and North Western Railway agrees to whitewash its sleepers at every junction to show the way. The public, most of whom have never seen an aeroplane before, gather in their thousands. Paulhan audaciously sets off twelve hours before anyone expects, forcing Grahame-White to make the first-ever British night flight in an attempt to overhaul him. As dawn approaches he's back in the Trent valley, where the headwind is once again gathering force. Then his engine begins to lose power, and he is forced to pull out. Paulhan tries to avoid the turbulence by climbing: 'My machine rose viciously and then dropped so quickly that I was almost torn from my seat.' By the time Paulhan nears Manchester, according to *The Times*, 'no-one cared whether the aviator who approached was Frenchman or Englishman. It was enough that he was a hero of the air.'

29 APRIL

4.7C min / 13.1C max

BERTIE WOOSTER GETS THAT APRIL FEELING

"'How's the weather, Jeeves?"

"Exceptionally clement, sir."

. . .

After breakfast I lit a cigarette and went to the open window to inspect the day. It certainly was one of the best and brightest.

"Jeeves," I said.

"Sir?" said Jeeves. He had been clearing away the breakfast things, but at the sound of the young master's voice cheesed it courteously.

"You were absolutely right about the weather. It is a juicy morning."

"Decidedly, sir."

"Spring and all that."

"Yes, sir."

"In the spring, Jeeves, a livelier iris gleams upon the burnished dove."

"So I have been informed, sir."

"Right ho! Then bring me my whangee, my yellowest shoes, and the old green Homburg. I'm going into the Park to do pastoral dances."

I don't know if you know that sort of feeling you get on these days round about the end of April and the beginning of May, when the sky's a light blue, with cotton-wool clouds, and there's a bit of a breeze blowing from the west? Kind of uplifted feeling. Romantic, if you know what I mean. I'm not much of a ladies' man, but on this particular morning it seemed to me that what I really wanted was some charming girl to buzz up and ask me to save her from assassins or something. So that it was a bit of an anti-climax when I merely ran into young Bingo Little, looking perfectly foul in a crimson satin tie decorated with horseshoes.'

P. G. Wodehouse, The Inimitable Jeeves, *1923*

'The 30 h.p. two-seater made short work of the run to Godalming ... The hedges, fresh with the glories of spring, flashed past; the smell of the country came sweet and fragrant on the air. There was a gentle warmth, a balminess in the day that made it good to be alive, and once or twice he [Drummond] sang under his breath through sheer lightheartedness of spirit. Surrounded by the peaceful beauty of the fields, with an occasional village half hidden by great trees from under which the tiny houses peeped out, it seemed impossible that crime could exist – laughable.' *Sapper (H. C. McNeile),* Bulldog Drummond, *1920*

-9.9C Eskdalemuir 1973
25.8C Creebridge 1993

'Rain. The cuckold for the first time this year.'
Lady Eleanor Butler, one of the 'Ladies of Llangollen', 1789

30 APRIL

4.9C min / 13.6C max

Spring is well and truly in the air: you are most likely to hear your first cuckoo (on averages taken from the last five years) today, and also to see the white 'snow' of hawthorn blossom in the hedgerows – traditionally known as 'Maythorn' because it heralds the arrival of May. Not that this necessarily means spring weather, of course. In 1115, apart from heavy snow, according to Simeon of Durham, 'nearly all the bridges throughout England were broken by ice'. And in a violent storm in 1859, four hundred drown when the clipper *Pomona* is wrecked in Blackwater Band off County Wexford (✛). It's almost the last gasp for the beautiful clippers, the fastest working sailing ships ever built – steam has taken over.

'A chill afternoon, all the cherry trees lurid in the cold yellow light of a hail storm.' *Virginia Woolf applies Impressionist techniques to describing Kensington Gardens, 1939*

'We crossed to England ... It was a bad arrival: France in sunshine, Dover cliffs veiled in rain.' *Christopher Isherwood, Anglo-American poet and novelist, 1948*

'Spring comes late to Camusfeàrna(♀). More than one year I have motored up ... early in April to become immobilized in snowdrifts on the passes twenty miles from it ... By mid-April there is still no tinge of green bud on the bare birches and rowans nor green underfoot, though there is often, as when I first came ... a spell of soft still weather and clear skies. The colours then are predominantly pale blues, russet browns, and purples, each with the clarity of fine enamel; pale blue of sea and sky, the russet of dead bracken and fern, deep purple-brown of unbudded birch, and the paler violets of the Skye hills and the peaks of Rhum. The landscape is lit by three whites – the pearl white of the birch trunks, the dazzle of the shell-sand beaches, and the soft filtered white of the high snows. The primroses are beginning to flower about the burn and among the island banks, though all the high hills are snow-covered and the lambs are as yet unborn. It is a time that has brought me, in all too few years, the deep contentment of knowing that the true spring and summer are still before me.' *Gavin Maxwell*, Ring of Bright Water, *1960*

-7.9C Kindrogan 1973 ❄

'And since tis a bad day
Rise up rise up, my merry men,
And use it as you may.
I have accordingly been busy.'
Sir Walter Scott, 1829

25.0C Kensington Palace 1952 ☀

MAY

1 MAY

5.1C min / 13.7C max

**MAYDAY /
CALAN MAI**

Today, Mayday, has a better chance of being fine than the first day of any other month. That's because in May anticyclones (highs), which bring fine, settled weather, are more likely than in any other month. Mayday is a moment in the year pregnant with weather-linked, pagan ritual significance. Maia, the goddess of May, is celebrated for spring growth and replenishment. Maypoles are fertility symbols. In these mild times, Mayday heralds less the start of the third month of spring than the start of summer proper.

'Yesterday I climbed to Camelot on a golden day. The orchards are in flower and we could see the Bristol Channel and Glastonbury too, and King Alfred's tower and all below. I found myself weeping. I shall go up there at night and in all weather, but what a good way to see it first.'
John Steinbeck, writer, at Bruton, Somerset where he rents a cottage in 1959 to explore Arthurian sites

For the Iron Age Celts, the night of 31 April–1 May is the second most important night in the year (after 31 October–1 November, the Celtic New Year). It's a livestock festival, heralding the beginning of the 'lighted half', the pastoral season when cattle are led from lowland enclosures up to summer pastures on the hill, where, signalled by the hawthorn flowering, rich grass now grows. At the heart of the festival (Beltane in Ireland and Scotland, Calan Mai in Wales, Roodmas in England) is the marriage of two ancient and anthropomorphic folk figures: the Green Man, symbol of growth, and the May Queen, who emerges from her winter retreat and represents fertility. Great bonfires are lit on hilltops. Summer begins.

'The first of May brought a burst of summer weather. All the trees and hedges came into full leaf overnight; and from behind the latter, in the evenings, cries could be heard of: "Nay, doan't 'ee, Jem", and "Nay, niver do that, soul", from the village maidens who were being seduced.' *Stella Gibbons,* Cold Comfort Farm, *1932*

'Suddenly summer arrived, perfect hot weather, lovely cloudless skies, leaves rushing out from winter into summer with hardly any spring.'
Richard Crossman, Labour Minister of Housing and Local Government, Scunthorpe, 1966

27.4C Lossiemouth 1990

114.3mm Loch Quoich 1944

-8.9C Braemar 1927

2 MAY

5.1C min / 14C max

THE TIME OF
THE BLACK
FROST

'This morning there fell a great deal of snow, so that there is now a considerable storm on the ground, with frost,' Edinburgh's *Caledonian Mercury* reports today in 1740. The winter of 1739–40 (known as the 'Hard Winter', the 'Great Frost' or, in Ireland (*13 February*), the 'Time of the Black Frost') has become a yardstick for severity, unequalled until 1963 (*6 February*). Below-average temperatures begin in August 1739 and last a year. During the coldest months, thousands die from hunger, cold and related diseases (including scurvy and 'accident hypothermia' – shivering until you have a heart attack). In Scotland, the frost continues until the end of May.

'I brave it, and go out to buy the papers, intrigued to read the detailed dissection of the election results. And nothing out there has changed, except the weather. It's glorious and will undoubtedly be taken by those who believe in such things as a good omen.' *Deborah Bull, journalist, the day after Labour under Tony Blair wins the General Election, 1997*

'Still bitterly cold. Wearing thickest winter clothes. Everything nature-wise most backward. Snowdrops only just dead; the wind-flower out, and primroses; bluebells sprouting. Motored to Bath ... in a snow blizzard.' *James Lees-Milne, architectural historian and diarist, 1979*

One reason why this winter features so highly in the pantheon of 'great winters' is because, for almost the first time, there are accurate newspaper reports and systematic observations to tell us precisely how cold it is. The most valuable record is that of a Rutland squire called Thomas Barker, of Lyndon Hall (), one of a new breed of amateur scientists to emerge in the eighteenth century (*10 April*). Barker records his first weather observation in 1733, aged eleven, and continues a journal, almost without interruption, for the next sixty-two years, recording atmospheric pressure, temperature and rainfall. It's a work of permanent scientific value. From it we know about not only the winter but also its aftermath: 'So cold & dry a season as follow'd the frost, did not repair the injuries ... but occasion'd fresh damage.'

28.3C Rothes 1990
-7.8C Braemar 1938

Rain affects the first of the MCC's new limited-over, knock-out cricket tournaments today in 1963. Billed as cricket's FA Cup, sponsored by Gillette, the new one-day matches are intended to boost the flagging popularity of county cricket. The publicity makes much of the thrill of seeing a game from start to finish in one day. Unfortunately, due to the rain, today's first game (between Lancashire and Leicestershire) runs into a second day (*27 April*).

RADIOACTIVE
RAIN

Spring rain brings an unexpected bounty to the hills of southern Scotland, Cumbria, North Wales and Northern Ireland today in 1986 – in the form of caesium-137, iodine-131, strontium-90 and other radioactive fall-out in the form of aerosols. A week ago an explosion in reactor no. 4 of the Chernobyl power plant, near Prypiat in the Ukraine, causes the most dreaded of all civilian nuclear disasters – meltdown. It's the world's worst-ever nuclear accident. Residents of the surrounding area see a radioactive cloud hanging over them after an explosion that visibly glows. The initial concern, once the media hears of the crisis (Europe is only alerted twenty-four hours after the blast, and then not by Russian sources), is focused on the areas immediately surrounding Prypiat. But as weather systems begin to circulate the fall-out, the question arises: who else is going to be affected?

'These lodgings are COLD. As I sit I have a rug over my knees but my hands are growing colder as I write: I am smoking my last cigarette. In consequence of which I feel savage.' *Philip Larkin, poet, Wellington, Shropshire, 1945*

'For eleven hours Caroline and I drove around the constituency, in cold weather which turned to hail and snow. I sat on the roof of the car in a blanket with rubber overtrousers, wearing a woolly cap and anorak. It was freezing.' *Tony Benn, Bristol, Election Day – won by the Conservatives under Margaret Thatcher – 1979*

Following heavy rain in the British Isles, it's mushrooms and milk that are the immediate worries. Tests show that lower-lying dairy areas are not seriously affected, but upland areas, where more rain falls (as moving air is forced over hills it cools, and the moisture it contains condenses), are of great concern. Restrictions on moving sheep are imposed in June, covering nearly nine thousand farms and 4 million animals; all sheep due to be sent to market for slaughter have to be tested with special monitors. Caesium is the greatest worry. Although sheep excrete it rapidly when moved to lower pastures, the element has a half-life of thirty years.

By 2006 restrictions are still in place, though they now apply to fewer than four hundred farms: 359 in Wales, nine in the Lake District and ten in Scotland. Northern Ireland is finally cleared in 2000. However, although the restricted area has now decreased by 95 per cent (to around 466 square miles/750 square km) at least two hundred thousand sheep will almost certainly never be allowed to enter the food chain.

-7.2C Dundeugh 1967

28.6C Barbourne 1990

4 MAY

5.3C min / 14.5C max

FEN BLOW /
DAWN CHORUS

Today in 1955 wind howling across the pancake-flat landscape of the Fens (♀) whips up a dust storm reminiscent of Oklahoma in the 1930s. This is a 'Fen Blow', a phenomenon that occurs in the region after dry spells, when strong winds suck up the topsoil into a blinding, choking storm. Today's gale, one of the most notorious, gusts up to 65 mph (104 kph) and lasts forty-eight hours. Drainage dykes and plough furrows fill with drifts of fine dust.

Fen Blows tend to occur between mid-March and mid-May – when fields are ploughed and sown but the seeds haven't yet germinated. Significant ones happen in March 1968, May 1972, May 1975 and March 2004. In 2002, the March March march (a long, flat walk from the town of March, held in March, with no point other than its name) ends abruptly when walkers are engulfed by a brown sand cloud that stings the eyes and skin. But though disconcerting for walkers, for farmers these storms can be disastrous: any wind over 25 mph (40 kph) will lift dry topsoil, and up to 2 inches (5 cm) of soil, including seeds and fertilizer, can be scattered, while young crops are damaged by the sand-blasting effect of the windborne topsoil. Almost certainly, the Fen Blow is a problem of the farmers' own making. 'Deflation', or lowering of the landscape through erosion by wind, is increasing because so many hedges have been removed in the Fens.

> 'I reckon the Spring is at least a fortnight later than last year for on Shakespeare's birthday, April 21st, it being the tercentenary, Ilbert [a fellow undergraduate at Balliol] crowned the bust of Shakespeare with bluebells and put it in his window, and they are not plentiful yet.'
> *Gerard Manley Hopkins, poet, 1866*

The dawn chorus of birdsong – our oldest early-morning alarm call – peaks now because most of the migrant birds have arrived. As the last traces of winter bow out, male songbirds sing to attract a mate and protect their territory. You have to be up with the lark to hear it, of course. At 4 a.m. the roll call begins, often in this order: blackbird, robin, wren, tawny owl, warblers (including blackcaps and chiffchaffs), song thrush and blue tit. There's also a dusk chorus, but as it tends to be less windy early in the morning, the clarion call to love and war is sweetest at dawn. As soon as it's light enough for birds to look for food – about 4.45 a.m. – the chorus ends.

-9.4C Lynford 1941
28.3C Cheltenham 1990

WET WEATHER
AND *THE
WEALTH OF
NATIONS*

Today's snowstorm in 1767 comes as no surprise in southern England. It's merely the latest example of the baffling, freak weather that has characterized the 1760s, heaping further misery on an embattled population. The winters of 1763, 1766, 1767 and 1768 are all exceptionally cold, long and stormy. A succession of disastrously wet and chilly summers offers little respite.

'There was no weather in this city: there was rain, of course, and sometimes the sun shone, but there was no weather. You couldn't tell which direction the wind was coming from, or even if there was any wind at all. The branches of the plane tree jostled outside the window every time a bus went past; London air was kept in a continuous slow swirl by the passage of traffic through the streets. But the cycle of Atlantic depressions and ridges of high pressure, with the wind swinging round, southwest to northeast, northwest to south, seemed to pass clean over the top of London's head. I listened out of habit to the shipping forecasts, but they were like news from abroad.'
Jonathan Raban, Coasting, *1986*

'Such a run of wet seasons a century or two ago would, I am persuaded, have occasioned a famine,' Gilbert White reflects in 1768. Successive harvest failures may no longer mean mass starvation in England, but there's still sufficient scarcity, in the autumns of 1766 and 1767, to prompt rioting as well as attacks on farmers and middlemen in the grain trade, accused of exploiting the dearth for profit. These protests eventually result in Thomas Pownall's 1773 Corn Law. This legislation revises the regulations about how much grain can be exported at what price, and bans its export altogether when home prices grow worryingly high.

This paternalistic act, which stifles the market economy, inspires heated debated in the Commons. The influential 'liberal' conservative (Whig) MP Edmund Burke insists that an unrestricted, free market system is the most effective long-term means of marrying grain supply with demand. It's an idea that is crystallized three years later when Adam Smith publishes *The Wealth of Nations*, perhaps the most famous book ever written about economics, in which he rails against the restrictive practices of the grain trade. It lays the intellectual foundation for the great era of free trade and economic expansion that dominates the heyday of the British Empire.

-7.7C Kinbrace 1981

122.9mm Llechwedd Quarry

28.0C Jersey 1995

6 MAY

5.7C min / 14.8C max

THE FOUR-MINUTE MILE / MOMENT OF SPRING

It's windy today at Iffley Road, Oxford (♥), in 1954 as Roger Bannister prepares to try and break one of athletics' greatest landmarks. Many believe that 'the four-minute mile' (even the phrase 'had a beauty in it, a symmetry,' says Norris McWhirter) is insuperable. In a strong wind, Bannister knows it is: 'A wind of gale force was blowing which would slow me up by a second a lap. In order to succeed I must run not merely a four-minute mile, but the equivalent of a 3 min. 56 sec. mile in calm weather . . . I spent the afternoon watching from the window the swaying of the leaves. "The wind's hopeless," said Joe Binks [a former athlete] on the way down to the track. At 5.15 there was a shower of rain. The wind blew strongly, but now came in gusts, as if uncertain. As Brasher, Chataway and I . . . lined up for the start I glanced at the flag again. It fluttered more gently now . . . Yes, the wind was dropping slightly. This was the moment when I made my decision. The attempt was on.' Bannister (below) runs 3 minutes 59.4 seconds: the 'Miracle Mile' is broadcast live by BBC Radio, igniting a wave of national pride. *Forbes* magazine, in 2005, declares it 'the greatest athletic achievement of all time'.

'A good many people at Mrs Yorke's and the women quite superbly dressed. The weather being cold, they wore heavy stuffs, and the floors were almost impassable from rolls of brocade.' *Matthew Arnold, poet and critic, London, 1877*

'The usual people fainting in the heat (Royal Weather).' *Chips Channon, politician and diarist, at the Silver Jubilee of George V, 1935*

One May, I witnessed that actual, fulminating moment [the very beginning of spring] here, and felt like a witch doctor whose spell had worked. There'd been two days of cold, heavy rain and I'd come to the valley to try and get beyond it. I sat on a log by the side of the river and watched the cloud begin to lift. Small bands of swifts and martins appeared, drifting in from the south. Then – it seemed to happen in the space of a few seconds – the wind veered round to the south-east. It was like an oxygen mask being clamped to the face, so sudden that I looked at my watch for the time. It went down in my diary: "6 May. Spring quickening, 4pm." *Richard Mabey, naturalist and writer, Chess Valley, Chilterns*

-5.6C Camps Reservoir 1980 ❄

28.6mm Cheltenham 1995 ☀

WOOL: CLOUDS
AND RAIN
CONVERTED
FOR EXPORT

Even in the chilliest spring, grass in the fields is now growing furiously – at its greenest, at its most nutritious. British grazing is the very best in the world (*18 April*), and for two thousand years one particular product has capitalized on this grazing more perfectly than any other – wool.

Wool is the British weather, commodified. It's the ethereal – clouds, rain and sunshine – made material. That is, made tradable, exportable and taxable. Shortwool sheep thrive on our high, windswept moors and mountains. Longwool types thrive on the vast lowland grazings of the Midlands and East Anglia. Consequently, wool's relationship with England's history is as long and entwined as the yarn it spins. Wool is how the Romans keep warm, far from their Mediterranean sunshine (*4 October*). By the eighth century, the Emperor Charlemagne is insisting on woollen cloaks from the north of England. England is a wool exporter long before the Norman Conquest. By the twelfth century, Henry of Huntingdon is writing of England's 'most precious wool'. By the end of the thirteenth century, the wool tax is the main source of royal revenue. By the fourteenth century the Lord Chancellor, in the House of Lords, sits on the Woolsack – as he still does: a vast red leather cube of legendary hardness and discomfort, stuffed with pure English wool – a symbol and constant reminder of the relationship between this one item and our national prosperity.

The great monasteries that now form such grand and picturesque ruins? Those moated, half-timbered medieval manors in East Anglia? The magnificent, honey-coloured Cotswold churches? All due to profits from wool. Almost everyone benefited. In the countryside, the humble tenant farmer gradually elevated himself into that conservative heart of the shires, the 'yeoman farmer'; in the towns, wool created the merchant class.

Wool is Britain's first world-beating export industry: a truly international commodity, it turns the country into an outward-looking trading nation. In the sixteenth century a pastoral revival, whereby arable land reverts to grassland, lays the foundations for England to become the first industrial state; our enormous wool exports supply the manufacturing capital that underwrites the Industrial Revolution – and make Britain the world's leading nation, politically and economically, for the next two centuries.

-8.3C Garforth 1917

29.0C Waddon 1976

6.1C min / 14.8C max

THE WETTEST PLACE IN ENGLAND

So much rain falls today in 1884 on a tiny village in the Lake District that it still holds the record as May's wettest-ever day – nearly 7 inches (172.2 mm) arrives in twenty-four hours. The event is received with weary resignation by the 129 inhabitants of the hamlet of Seathwaite in the Borrowdale valley (♥), for this is consistently the wettest inhabited place in England. The ribbon of nobbly, white-washed Lakeland farm cottages – harled in a vain attempt to keep the damp at bay – also holds the record for April's wettest day (in 1970 – it's even wetter than this). Indeed, Seathwaite can proudly claim eight of the ten wettest days in England in the last hundred years. The record for the highest annual rainfall *anywhere* in Britain (inhabited or uninhabited) – 257 inches (6527 mm) in 1954 – is held by the aptly named Sprinkling Tarn. Which sits in the hills directly above – you've guessed it – Seathwaite.

'The bells had begun to peal and, after the night's storm, London was having that perfect, hot, English summer's day which, one sometimes feels, is to be found only in the imaginations of the lyric poets.'
Mollie Panter-Downes, VE Day, 1945

Set in a deep valley surrounded by high mountains (the hamlet is on the main walkers' route to Scafell Pike, England's highest mountain), Seathwaite bears the consequence of its topography: it's situated in the lee of the first high ground encountered by west-moving Atlantic air-streams, heavy with moisture from their ocean crossing. Forced up over the high ground, the air cools as it rises, the water vapour condenses and – whoosh! Rain, rain and more rain, in steady, unremitting torrents, sometimes for months on end. The average annual rainfall is 140 inches (3556 mm) or nearly five times more than London (the average annual rainfall across England is 33 inches – 838 mm). Seathwaite's rivals for the British crown are Blaenau Ffestiniog in Snowdonia and Fort William in the Scottish Highlands – both for exactly the same reason.

However, life in Seathwaite is not lived – so to speak – under a cloud. Locals (today, just ten) enjoy the benefits of unrestricted water consumption, the soothing sound of running water where the 'becks' (or streams) drain the fells into the River Derwent, and a thriving trout farm. They can also rejoice in the knowledge that their climate is only a quarter as wet as the wettest places in the world, Mount Wai'ale'all in Hawaii, or Cherrapunji in India. Sunburn is also seldom a problem.

172.2mm Seathwaite 1884

-7.8C Lynford 1941

28.0C Plumpton 1976

9 MAY

6C min / 14.8C max

RETURN OF
THE WEATHER

'Violent rainstorms in many districts but most of the time it has been sunny and very warm.' It may not sound like it, but there's something extraordinary about today's weather bulletin in 1945 on the BBC Home Service. As the rider to the report reveals: 'We give you Britain's weather on the day it happened. That's something you haven't had since the war began.'

For five years, weather forecasts have been censored. The geographical position of the British Isles, on the edge of the Atlantic, means Allied military intelligence have advance warning of any eastbound depressions – potentially a key tactical advantage (*4 June, 6 June*). Accordingly, for five years barometers, thermometers and books about medieval weather lore have been dusted down and re-consulted – especially by fruit growers and farmers, for whom the secrecy has been a major problem.

Today's bulletin ends with the evening forecast: 'Sporadic rain over the whole country with thunder in places ... There is also news today of a long-lost friend – the large depression. It's turned up again between Ireland and the Azores.' Perhaps it was better not to know.

> 'I mighty weary to bed, after having my hair of my head cut shorter, even close to my skull, for coolness, it being mighty hot weather.'
> *Samuel Pepys, diarist, 1666*

> 'A mainly sunny day threatening showers that did not happen. A friend and I walked from Eisteddfa Gurig to the top of Plynlimon and ate our lunch in the lee of the cairn piled up at great cost of energy by the people of the Bronze Age about whose life up there we know so little ... In the clear air the views were magnificent ... days of clarity all around the compass are rare indeed.'
> *William Condry, journalist and conservationist, 1997*

The snow that falls in Derbyshire today in 1853 is so fine that it drifts under doors and through keyholes, filling homes like white sand. In 1978, the largest daytime temperature range ever is recorded at Tummel Bridge, west of Pitlochry (♀): a hypothermic morning (-7C) turns into a golden afternoon (22C). And in 2006, early-morning commuters along the east coast are bemused to find their cars covered in yellow dust. The explanation? Following a mass pollination of birch trees in Denmark, strong winds have carried millions of tiny pollen grains all the way across the North Sea.

-7.2C Lynford 1941 ❄

27.8C Southampton 1976 ☀

10 MAY

6C min / 15.1C max

BRITISH ICEBERGS / THE BLACKTHORN WINTER

'An island of ice has actually been stranded upon the island of Fowla, the most western of the Shetland islands,' *The Times* reports, today in 1818. 'This iceberg is said to extend full six miles in length, and, of course, is an object of terror to the natives.' Well it might – Britain's most remote permanently inhabited island (♥) measures less than five square miles itself. Two months later, *The Times* reports the probable origin of the ice: 'Four hundred and fifty square miles of ice has recently detached itself from the eastern coast of Greenland and the neighbouring regions of the Pole.'

> 'A filthy day, very wet and stormy. But I wore my fur-lined boots and took an extra jersey. Now I can see how people get eccentric.'
> *Barbara Pym, novelist, 1939*

Though a berg of this size is exceptional, Britain's northernmost islanders are familiar enough with great lumps of ice drifting towards them across the North Atlantic. During the coldest phases of the so-called 'Little Ice Age' (*3 March*) – especially 1690–1728 – the drifting ice even brings Inuit and Sjo-Same hunters, in their kayaks, to the Orkneys and Western Isles. In their seal-skin suits, shy of the shore and invariably alone, these nomads from Greenland and Norway cause great alarm. Accounts of 'seal people' and 'mermaids' exist from 1676 to the mid-1800s, and early in the eighteenth century, an 'Indian man' paddles his canoe up the River Don to Aberdeen: 'He died soon after and could give no account of himself,' the writer Francis Douglas later records. The belief, which becomes part of the folklore of the north and west coasts of Scotland, is that these curious beings are half-human, half-seal.

In 1955, a record week-long cold snap starts today. Though rare now, frosts in mid-May were once regular enough to have their own name: the 'Black Thorn Winter' or the 'Festival of Ice Saints' (named after Saints Mamertus, Pancras, Gervais and Boniface whose feast days fall now). Similar cold snaps in 1935 (with snow as far south as Cornwall), 1922 (-4C in Oxfordshire), 1902, 1891 and 1855 all cause heartache to farmers and fruit growers. And in 1698, during the coldest May on record, 'there fell snow . . . which if the ground had been cold and hard as in winter . . . it would have been ten to twelve inches deep,' according to the church annals in Guestwick, Norfolk (✆).

27.2C Langar 1959
-8.3C Lynford 1941

'THE WATER-SPLASH' / BOB DYLAN

'Pools of water glistened on the turf and players were soon slithering around like tyros on an ice-rink,' *The Times* reports of today's rugby league Challenge Cup Final in 1968, between Leeds and Wakefield Trinity. Dubbed the 'Watersplash', and with the match played partly in a thunderstorm, the Wembley pitch resembles a paddy field. The denouement of arguably the most famous final of all comes when Don Fox, the Wakefield prop forward and 'man of the match', has a straightforward conversion to win the game with the final kick. He misses. From 15 metres, in front of the right-hand post, he somehow toe-pokes the ball wide. Soaked and distraught, Fox sinks to his knees and buries his head in his hands, creating an immortal image of British sporting pathos. Eddie Waring, the commentator, says: 'He's missed it . . . Eee, poor lad.'

'All the night the air was very clear, and the stars over all the heaven were brightly shining. And the tree fruits . . . were sorely nipt.' *Anglo-Saxon Chronicle, 1010*

'Severe drying exhausting drought. Cloudless days. The country all dust.' *Gilbert White, naturalist and writer, Selborne, Hampshire, 1785*

'May is usually the worst and coldest month in the year but this beats them all and out-herods Herod. A black bitter wind violent and piercing drove from the East with showers of snow. The mountains and Clyro Hill and Cusop Hill were quite white with snow. The hawthorn bushes are white with may and snow at the same time.' *Reverend Francis Kilvert, diarist, Clyro, Wales, 1872*

In 1966, under Tupperware skies, Bob Dylan waits at the Aust ferry terminal to cross the River Severn (🌀). Dylan is heading to Cardiff for the next gig of his unhappy 'Judas Tour': hardcore folkies believe the skinny singer has betrayed their movement by playing an electric guitar. With the bridge under construction in the distance, the location may lack the glamour of Golden Gate Park, but Barry Feinstein snaps away regardless. Dylan's moody glower isn't because he's boo-ed off at last night's gig, Barry later says: 'It was the British weather that made him look so gloomy.' The photo becomes a rock icon.

-9.4C Lynford 1941

28.3C Long Sutton 1945

12 MAY

6.8C min / 16C max

THE GREAT FAMINE

Heavy rain today, in 1315, heralds the start of the worst, most sustained and least anticipated spell of bad weather in the Middle Ages. 'The inundation of rain consumed nearly all the seed, so that now was seen the fulfilment of the prophecy of Isaiah, and in several places hay was so hidden under water that it could neither be cut nor gathered. Sheep also perished in flocks,' according to the *Vita Edwardi Secundi*. The rain just goes on and on . . . and on.

> 'Today dawned blissful, not a breath of wind, warm and the sun out, a great silence ... Heat shimmers off the shingle, weeks of soaking rain have left the Ness a hopeful green.'
> *Derek Jarman, film-maker, Prospect Cottage, Dungeness, Kent, 1991*

The trouble is, everyone's been taking the weather for granted. For a century it's been mild, warm and stable with almost no extreme events (*20 February*). The population has exploded, but the economy hasn't kept pace. The British Isles are ravaged by wars – in Scotland, the Welsh Borders and Ireland – under a weak king, Edward II. According to medieval belief, the king makes the weather: harmony in weather means harmony in the state. For Edward – struggling with his father's legacy of debt and unfinished wars – the weather now heaps misery on his already disastrous reign. The only missing ingredient for catastrophe is a bad harvest.

Modern study of tree growth rings in ancient oaks confirms what the chronicles say. In 1315: 'God sent a dearth on earth, and made it full hard.' The harvest is a disaster. 'The Great Famine' – the greatest of the millennium – follows. It lasts in England from 1315 to 1318; 10–15 per cent of the population die from starvation or disease. After that, things go from bad to worse for poor old Edward. With the institutional authority of the Church and the government undermined, a failed campaign in Scotland (*2 July*) and mounting debt, the King is exposed. His queen and the barons rebel. In 1326, he's deposed. His attempt to flee England is foiled by the weather (contrary winds keep his ship in the Bristol Channel), culminating in his capture and that moment every schoolboy remembers: a red-hot poker up the bum in Berkeley Castle (🌭). Who'd believe what heavy rain can lead to?

-6.9C Grizedale 1974 ❄

30.6C Camden Square 1945 ☀

13 MAY

6.9C min / 16C max

PICNICS 'We rode to the top of Helvellyn; on arriving at the top we had luncheon and made sherry cobbler with sherry and frozen snow.' So records Charles Wood, companion to the sixteen-year-old Prince of Wales on a tour of the Lake District, today in 1857. A picnic at 950 metres (3,117 ft), in mid-May, in snow, may seem a peculiar expedition for the future King. But this is the mid-ninteenth century, and Britain's gone picnicking mad.

The picnic – described by Georgina Battiscombe as 'the Englishman's grand gesture, his final defiance flung in the face of fate' – is a by-product of the Romantic Movement and its obsession with wild places. Prior to this, eating in a wet or wasp-haunted field is entirely unconnected to pleasure. It's Dorothy and William Wordsworth, Coleridge, Walter Scott and Byron, all inveterate picnickers, who unite the two ideas. Undeterred by weather, they forge into the great outdoors, with 'cold mutton in our pockets', in all seasons. In June 1802, Dorothy Wordsworth writes, of a fishing picnic: 'I sat out at the foot of the lake until my head ached with cold.'

By mid-century, grimly Spartan 'rustic dinners on the cool green ground' have evolved into fashionable events: a social meal to which each guest contributes a share of the food, taken outside by choice. Queen Victoria enjoys frequent picnics at Balmoral. Tennyson describes one in his poem 'Audley Court', and picnics have become a standard episode in the novels of Surtees, Dickens and Jerome K. Jerome.

By 1861, when Mrs Beeton publishes her *Book of Household Management*, things have moved on a little. Her recommended picnic 'Bill of Fare' includes: 'A joint of cold roast beef, a joint of cold boiled beef, 2 ribs of lamb, 2 shoulders of lamb, 4 roast fowls, 2 roast ducks, 1 ham, 1 tongue, 2 veal-and-ham pies, 2 pigeon pies, 6 medium-sized lobsters, 1 piece of collared calf's head, 18 lettuces, 6 baskets of salad, 6 cucumbers. Stewed fruit well sweetened; 3 or 4 dozen plain pastry biscuits, 2 dozen fruit turnovers, 4 dozen cheesecakes, 2 cold cabinet puddings in moulds, 2 blancmanges in moulds, a few jam puffs, 1 large cold plum-pudding (this must be good) . . .'

-6.1C Dalwhinnie 1935 ❄

28.9C Southampton 1998 ☼

Have café culture and the rural gastro pub done for the picnic? It seems not. Picnickers are still to be found on Helvellyn, in all weather, on any day of the year (*30 October*).

14 MAY

6.9C min / 15.8C max

EXPLORING THE CLOUDS IN A VINTAGE GLIDER

It's thundery today in 1963 where Peter Scott, the fifty-one-year-old naturalist, painter and Olympic sailor (and son of Scott of the Antarctic) is gliding near Durham (♀). Thunder clouds make for interesting gliding: useful for gaining height, they contain dangerously violent air currents – entering a storm cloud is only for the very brave or very foolish. This early in the year a storm cloud is unlikely to be anywhere near as fearsome as later in the season (July and August). Even so, Scott's account of his successful attempt to achieve his 5000-metre 'Diamond Height' badge, taken from his autobiography *The Eye of the Wind* (illustration below), gives a good idea of what goes on inside one of these monsters:

> 'Cloudy, no sun, hail, rain, snow all day, very cold, dirty, N: W, east.'
> *A Catholic clergyman, tutor of one Charles Jermyn Bond of Edmundsbury, 1740*

> 'A very cold morning – hail and showers all day ... William tired himself with seeking an epithet for the cuckoo.'
> *Dorothy Wordsworth, 1802*

'Laboriously, I crept ... up to 4,000 feet and circled into the bottom of a little grey patch of cumulus.

Then a remarkable thing happened. The lift suddenly and startlingly increased. From 100 feet per minute my rate of climb had increased to more than 1,500 feet per minute. At about 9,000 feet I put on my oxygen mask ... still the altimeter needle went round half as fast again as the second hand of a watch. But at 13,500 it became lighter and turbulent and the lift fell off, so I turned to the north-west and flew straight and level.

From a bright white mist I flew with absolute suddenness into almost twilight darkness, the glider dropped like a downgoing lift and a hideous roar broke out as hail bombarded my perspex canopy. A moment later lightning began to flash, but so deafening was the roar of the hail that I heard no thunder. I was now back in lift again, and going up fast though not quite so fast as before ...

-6.7C Braemar 1915 ❄

♀

Ice crystals were penetrating into my cockpit from the cracks round the canopy. Already there was a thin coating over my knees. The inside of the perspex was entirely frosted over, and so were the faces of the instruments ...

By now I had a serious problem on my hands ...'

30.0C Bromley 1943 ☀

SEVEREST-EVER HAILSTORM

The severest hailstorm ever to strike Britain, according to the Tornado and Storm Research Organization (TORRO), strikes Hitchin and Offley (☉) in Hertfordshire today in 1697. On the TORRO Hail Scale (a sort of Richter Scale for hailstorms) this storm is graded H8 out of a possible 10. Several British hailstorms have been graded H7, but this is the only one to make the magic H8 – though how a storm can be confidently graded three centuries after the event is a little hard to understand. The hailstones are up to 4 inches (110mm) in diameter – about the size of tennis balls.

Hail, snow's harder cousin, forms when super-cooled water droplets collide with particles of dust inside Cumulo-nimbus thunderclouds (*14 May, 1 July*) forming ice pellets. Curiously, hail most often occurs in warm, muggy conditions when the air has a high moisture content and rises rapidly in convection clouds (see *Definitions*). Equally curiously, hailstones tend to be larger in summer, gaining in size as they circulate within the strong currents until they are too heavy for the up-currents of air to support their weight, when they fall as hail. The damage hail does depends on amount, size of stone, intensity and duration. A hailstorm does damage (*6 August, 5 September*) when it reaches or exceeds TORRO intensity H3.

Such storms occur, on average, about fifteen days a year, and mainly within a belt from Lancashire and Greater Manchester to the Thames valley and East Anglia: 77 per cent of these storms occur between May and August, peaking in June, and invariably in showers.

'Epsom Races, and actually snow at the "Derby".' *Colonel Peter Hawker, 1839*

'Wonderful morning of coldness and sun – frost, only little birds chirping, and my Regency chair standing, silent against the white door, its black glistening, its brass gleaming with the remains of the old lacquer.' *Denton Welch, writer and painter, 1946*

'It was a mild and lovely evening. The rays of the sun looked heavy, as they frequently do towards the approach of a summer sunset, and lay between the tunnels of green leaves like long rods of gold. There were no clouds in the blue sky, whose colour was beginning to deepen with the advance of night, and the face of the whole country-side was softened by the shadows which were slowly growing in the depths of the woods and hedgerows.' *Stella Gibbons, Cold Comfort Farm, 1932*

-9.4C Fort Augustus 1941 ❄

28.2C Lee on Solent 1998 ☼

16 MAY

6.6C min / 15.5C max

'Rain spreading south on a cold front' is the gist of the BBC TV forecast this morning in 2005. But it's not *what* the forecast says that so shocks us, as *how* they tell us. The weather's had a make-over – the first for twenty years. Gone are the cuddly symbols, those models of elegant expression that have aroused such limitless affection over the years. In their place is a new digital, 'fly-by' computer graphic swooping around a brown – yes, *brown* – British Isles. Viewers are stupefied. Why must we now take travel sickness pills to watch the forecast? Where's our green and pleasant land gone? Why is Scotland so small? Amateur meteorologists are the most aggrieved: where are the pressure charts? The isobars? The fronts? Yet more proof – as if we needed it – of BBC 'dumbing down'. Change is always disconcerting, we are haughtily informed; soon we will love the new look as much as the old.

'Although it was ... the day I had scheduled for our departure, a full-blown gale was raging. The rain was sheeting down, and the wind buffeted against the damp cliffs of the creek. In the open Atlantic, outside, the surface of the sea was torn to smoke as the squalls rushed across it.' *Tim Severin*, The Brendan Voyage, *waiting on the Dingle Peninsula Ireland, to depart in an ox-hide curragh across the Atlantic, in a reconstruction of St Brendan's sixth-century voyage, 1976*

It's all a far cry from the first live TV weather forecast from Lime Grove Studios in 1954 (*11 January*). Met Office forecaster George Cowling has an extravagant five minutes – an eternity, on air – to draw his own chart on to a map hanging on an easel, using wax crayons. His brief: to show 'how the weather expected tomorrow is conditioned by the weather experienced today'. Every day, you see, isn't just a random one-off. Cowling tells his listeners that the wind will make it a good day for drying washing, earning himself a sound ticking-off from his employers and taking forecasting into a new popular era.

-6.5C Dalwhinnie 1995 ❄

The BBC forecast remains the shortest, most watched programme on television. Eight million tune in nightly, even though, according to a recent psychologist's study, 70 percent of viewers can't remember a thing that is said.

27.8C Camden Square 1925 ☀

17 MAY

6.7C min / 15.5C max

THE HELM,
CUMBRIA'S
VERY OWN
WIND

A gale-force wind howls in the eastern side of the Lake District today in 1939 – yet, strangely, in close neighbouring villages it's still. This is the Helm, Britain's best-known named wind. Not as famous as the Mistral, Sirocco or Chinook, perhaps, the Helm can blow at any time of year but is commonest in spring. With its accompanying cloud (the 'Helm Bar'), it roars off Cross Fell (♥) at the top (the helm) of the Pennines at up to 90 mph (145 kph), fast enough to take a roof off a barn or pin customers of the Hartside Café into their cars. Often it blows for days, maddening locals.

'When the Helm is on,' writes the Reverend John Watson in 1847, 'in a few minutes the wind is blowing so violently as to break down trees, overthrow stacks . . . blow a person from his horse, or overturn a horse and cart . . . its sound is peculiar . . . it has been compared to the noise made by the sea in a violent storm.' The deafening roar is loudest down chimneys. It's caused by air from the east rising over the Pennines, cooling and rushing down the western slopes, and it blows whenever there is an easterly breeze, stopping as soon as the direction changes. It's not, however, Britain's only named wind (*15 July*).

Snow ruins the celebrations marking King George V's Silver Jubilee today in 1935, particularly in Yorkshire. 'Small villages in the dales were 2–3 feet deep in snow and villagers had to dig themselves out of their homes,' reports *The Times*. Cars have to be abandoned and trains are derailed on frozen points. Devon and Cornwall resemble scenes from a Christmas card as England records its lowest-ever May temperature: -8.6C in the Rickmansworth frost hollow (*10 December*) – a disaster for fruit and vegetable growers from South Wales to Kent. Snow is not uncommon this late in the year. It arrives in 1891, while 1955 brings the worst May snowstorm for sixty years to Birmingham, the Cotswolds and the Chilterns – the last time there's been substantial May snow around the London area. In 1923, *The Times* reports that Scotland Yard is complaining of fog impairing the control of traffic and the detection of crime (*19 December*).

28.9C Huddersfield 1952

-8.6C Rickmansworth 1935

6.9C min / 15.6C max

MAY BANK HOLIDAY / THE YEAR WITH NO SPRING

A huge fall of snow marks the Whitsunday Bank Holiday weekend in 1891. Tens of thousands of pleasure-seekers, lured away from home by a mini-tropical heatwave the preceding week, are bitterly disappointed. 'May has turned traitor,' reports *The Times*. The reputation of 1891 (*10 March*), which has become known as 'the year with no spring', is confirmed as fruit is damaged by frost across southern England, 6 inches (15 cm) of snow falls in East Anglia and the observatory on top of Ben Nevis records -10C at night.

'Rain in Oxford too. The undergraduates look pale grey and soiled. I was thinking of the way we talk about the grubby glum appearance of people and clothing in Communist countries but honest to God, there are few sadder sights than British folk, forever under these grey skies.' *Martha Gellhorn, American war correspondent, 1986*

'To Yarmouth for a Festival of Labour Rally. It was so cold on the pier that the band was playing with literally nobody in sight.' *Tony Benn, politician and diarist, 1963*

William Gladstone only instigates Bank Holidays twenty years earlier, but Whitsun's reputation for atrocious weather is already – well, snowballing. Partly, this is because the holiday is tied to Easter (Whitsun being a Christian festival, celebrated seven weeks after), so it can fall on any date between 11 May and 14 June. The Whit weekend of 1891 is the worst, however. Needless to say, no sooner is everyone back at work than the skies clear. *The Times* wonders if the motto of the stoical British holidaymaker will be *Meminisee jurabit*: 'Perhaps one day even these things will be pleasant to recall.'

'Then came the fine weather. On the 18th May – even though it was not a Sunday – they heard the peal of church bells on the far side of the hill. Amos harnessed the pony and they drove down to Rhulen, where Union Jacks were fluttering from every window to celebrate the Relief of Mafeking. A brass band was playing and a parade of schoolchildren passed down Broad Street with pictures of the Queen and Baden-Powell. Even the dogs wore patriotic ribbons tied to their collars. As the procession passed, she nudged him in the ribs, and he smiled. '"Be the winter as makes me mad." He appeared to be pleading with her. "Some winters seem as they'll never end."' *Bruce Chatwin,* On the Black Hill, *1982*

-6.1C Shap 1996 ❄

30.0C Camden Square 1952 ☀

19 MAY

6.6C min / 15.7C max

THE HALIFAX STORM / A NATION CAUGHT IN THE RAIN

At 4 p.m. today, in 1989, a two-hour cloudburst delivers 193.2 mm – 7.6 inches or nearly four months' rain for a typical lowland area – on Walshaw Dean Lodge, Hebden Bridge. It's become known as the 'Halifax Storm'. Seathwaite's 172 mm in 1884 (*8 May*) retains the *official* title for May's record rainfall, as the Halifax total is too localised for radar to confirm it in full. Still, is it any wonder that, throughout history, the British have been the greatest innovators in waterproofing?

A BRIEF HISTORY OF MODERN BRITISH WATERPROOFING
1824, Charles Mackintosh introduces his rubberized raincoat – the first genuinely waterproof waterproof • 1843, Charles Goodyear invents vulcanized rubber – 'Wellington' boots follow • 1852, Samuel Fox introduces his light-weight, collapsible umbrella (*5 December*) • 1853, John Emary patents his waterproof wool, re-naming his company 'Aquascutum' (Latin for 'water shield'). • 1894, John Barbour starts selling oilskins in South Shields to sailors, dockers and fishermen, soon branching into the waxed cotton adopted by the country set • 1901, Thomas Burberry submits his close-woven, almost water-proof 'gabardine' worsted raincoat to the War Office. With epaulettes and D-rings, it becomes, on the Western Front, known as the 'trench coat' (*22 January*) • 1954, following the invention of proofed nylon, Noel Bibby intro-duces the Peter Storm outdoor clothing line, based on a seamless (so non-leaking) French prototype: the monk's habit or cowl. *La cagoule* is born • 1949, Leslie Cohen crosses the trenchcoat with the plastic bag to create the Pakamac. Even the Queen wears it, and within 15 years, 60,000 are selling a week • 1961, re-interpreting Inuit (Eskimo) style, the hooded, caribou-skin *anoraq* is repackaged as a short, hooded jacket lined with fake fur. The anorak is appropriated by Mods – then trainspotters.

-6.7C Stronvar 1903
28.3C Glenbranter 1948

193.2mm Walshaw Dean Lodge

Heavy snow sees today's annual Ten Tors adventure training weekend on Dartmoor (✆) abandoned in 1996. Organized by the Army for up to 2400 fourteen- to twenty-year-olds, the weekend event routinely delivers extreme weather. In 1998, temperatures reach 26C. In 2005, rain and winds lead to mass retirements. While in 2007, the event is abandoned again when a fourteen-year-old girl is swept away by a brook – expanded from 3 to 15 feet (1 to 5 metres) in the heavy rain.

20 MAY

7.1C min / 15.8C max

DUFFER'S FORTNIGHT / WEMBLEY WEATHER

'I saw the first Mayfly of the season, a green drake, and walked processionally beside him, down the bank. The Mayfly doesn't generally happen till June nowadays: but we call him by his old name still. He was a genuine Mayfly before the adoption of the new calendar, and country memories are long,' T. H. White writes today in 1934. The first appearance of the mayfly, best known of the British up-winged insects, is a call to trout anglers. And the sunnier the spring, the earlier the *Ephemera danica* appears.

'Two scholars of Wadham ... standing neare the head of a boat, were presently with a stroke of thunder or lightning both struck off out of the boat into the water, the one of them stark dead.'
Philosophical Transactions of the Royal Society, 1666

'The mournful foghorn sounded all through the night. The moon faded, dawn came up under a milk-white sky, calm and very warm.'
Derek Jarman, film-maker, Prospect Cottage, Dungeness, Kent, 1991

The arrival of the mayfly is linked to several aspects of the weather – rainfall, water temperature and hours of sunshine – though entomologists and meteorologists can't agree on precisely how. Suffice to say, between mid-May and the end of the first week of June the mayfly emerges after two years underwater in larval form and spends a day, max. two, of rapture in the spring sunshine. As rivers and lakes come alive, trout, who find them irresistible, abandon all caution. Thus the period of the mayfly hatch is known as 'Duffer's fortnight' – if you can't catch a trout now, then you never will.

'Heat exhaustion' means 250 spectators receive medical attention at Wembley during today's FA Cup Final in 1989. With the recent Hillsborough Disaster (Liverpool's semi-final against Nottingham Forest, at which 95 fans die) haunting the match, Liverpool and Everton battle it out in 'tropical' temperatures touching 30C. In extra time, the Reds (Liverpool) win. The FA Cup Final nearly always delivers the 'proverbial Wembley weather': as Nick Hornby, in the stands for a final, writes in *Fever Pitch* (1992), 'It's as if you are in a cinema watching a film about another more exotic country.' There's a 70 per cent chance of sunny periods, and an 80 per cent chance of dry weather, on Cup Final day. The FA Cup Final has never been postponed on the day in all its history, and in the last twenty-five years only 1992 and 1994 have been wet.

-6.7C Glenlivet 1956

29.5C Barbourne 1992

21 MAY

7.1C min / 16.1C max

LONGEST BRITISH TORNADO / CHARLES LINDBERGH

Britain's longest recorded tornado cuts a swathe from Berkshire to the north Norfolk coast this Sunday afternoon in 1950. Eyewitness accounts do not immediately conjure images of the Hollywood blockbuster *Twister*. The tornado is observed to pick up a cat (its four legs spread-eagled, cartoon-style) and pluck some poultry clean of feathers. In a series of kangaroo hops, the phenomenen leaves a trail of overturned hen coops, deroofed sheds, slates, sheets of corrugated iron and assorted tree branches wherever it touches down. It lasts four hours, and covers more than 100 miles (160 km).

EYEWITNESS ACCOUNT OF A SOLAR ECLIPSE, 1724

'Though it was very cloudy, yet now and then we had gleams of sun-shine, rather more than I could perceive at any other place around us,' begins the antiquary Dr William Stukeley, a friend of Isaac Newton, in his account of today's 5.30 pm eclipse. With the thin clouds 'doing the office of [smoaked] glasses,' he records how 'a most gloomy night with full career came upon us. At this instant we lost sight of the sun ... not the least trace of it to be found, no more than if really absent ...

Now I perceived us involved in total darkness ... like a great dark mantle ... thrown over us ... and the horses we held in our hands were very sensible of it, and crowded close to us, startling with great surprise. As much as I could see of the men's faces that stood by me, had a horrible aspect ... immediately after, the whole appearance of the earth and sky was entirely black. Of all things I ever saw in my life, or can by imagination fancy, it was a sight the most tremendous.'

'Spiralled down thru' hole in clouds, and soon encountered 2 hours of heavy fog,' records the pioneering American aviator Charles Lindbergh this morning in 1927 as he approaches the west coast of Ireland twenty-six hours after leaving Long Island. It's the first non-stop, solo flight across the Atlantic (*15 June*). The sight of fishing boats alerts him that land must be near. Swooping as low over the boats as he can, the lost, exhausted and disorientated Lindbergh closes his throttle and roars: '"Which way is Ireland?" ... of course the attempt was useless, and I continued on my course.' Seven hours later, he touches down at Le Bourget, Paris.

-3.9C Grantown-On-Spey 1969

30.6C Camden Square 1922

22 MAY
7.4C min / 16.6C max

THE DERBY AND GOING

So wild is the weather – flurries of snow and driving sleet – on the Epsom Downs (☂) today in 1867 that there are ten false starts at the Derby. Riding round Tattenham Corner, Hermit, an unfancied horse, finds advantage in the now sodden turf, cocks his head and charges from the back of the field to win at 66–1. The race makes and breaks the fortunes of two young aristocrats, but, more significantly, the date of the Derby is moved, as a direct result, to later in the calendar – when the weather has proved to be equally bad (*31 May*).

When a squall hits America's Cup challenger *Shamrock II* off the Isle of Wight, today in 1901, the yacht is demasted and Britain nearly loses a second monarch inside six months (Queen Victoria dies on 22 January). Edward VII is a guest of tea magnate Sir Thomas Lipton.

HOW WEATHER AFFECTS GOING

Going, or the state of the turf at a racecourse, is fundamental to racing. And weather is fundamental to the going. There are seven grades of surface: heavy, soft, good to soft, good, good to firm, firm and hard (though the latter is seldom used). Clerks at racecourses assess the going before and during meetings, usually with a sharp-pointed walking stick. It is a subjective assessment, and in 2007 the Jockey Club is trialling a scientific device called a 'Going Stick'. This stick measures both lateral movement and downward resistance in the soil, and should ensure greater consistency in going reports at Britain's fifty-nine racecourses.

As most horses have a preference for one type of surface, going is as vital a part of the punting equation as trainer, jockey, distance and weight carried. Things to look out for are hooves (a horse with 'soup-plate' feet is likely to run well on soft ground) and 'action' (horses with snappy, 'daisy-cutting' strides tend to race well on firm turf). A great horse, however, can win on any surface. Desert Orchid was thought to favour good ground when he lined up for the 1989 Cheltenham Gold Cup in rain and mud. At long odds, 'Dessie' clinched a memorable victory, delighting punters and elevating himself into the equine hall of fame.

-4.4C Braemar 1894

32.8C Camden Square 1922

23 MAY

7.7C min / 16.9C max

CHELSEA FLOWER SHOW

'So here we are at Chelsea in the middle of a drought, where the rain flumes off the tents and the hats and the noses and the broad-leaved *gunnera*,' writes Caitlin Moran in *The Times* today in 2006. 'People walk past the "drought-tolerant" gardens and laugh. People walk past the Wetlands garden and laugh. The modish "New Zealand" garden sends rivulets of black volcanic New Zealand mud out into the walkways. All of Chelsea is a water feature this year.'

The world's greatest garden festival has, in the near century it's been held at Chelsea, attracted almost every kind of weather disaster. In 1935 exhibitors frantically try to save prize plants with heaters as snow carpets the country (*17 May*). In 1971 there's severe flooding, and the roof of the commercial stand collapses due to weight of water. In 2006 a 350-feet-deep (100-metre-deep) borehole is urgently sunk to save wilting plants; it becomes operational just as pelting rain arrives. Then, in 2007, the warmest April on record sets off a premature burst in plant growth. As flowers start blooming around a month ahead of schedule, distraught nurseries rush their charges into darkness, cool them in giant refrigeration units, and even, in one instance, put straitjackets around individual buds to prevent them opening.

'Chelsea Week' – the week followed by the May Bank Holiday – is a traditional weather landmark for gardeners: it's regarded as safe to put out frost-sensitive bedding plants after the show (*17 April*). Otherwise, today in 1932 sees catastrophic flooding (in one of the wettest years of the twentieth century), especially around Derby and Nottingham.

'It must be a fine sight to see London without smoke. [The miners' strike] does not affect us here, with unlimited firewood. We have hot days & cold nights, but crops and fruit promise well so far, as there are thunder showers between times.'
Beatrix Potter, writer and artist, Lake District, 1921

'The weather deteriorated. We lost sight of the lighthouse on Slyne Head in the low clouds. Rain showers began sweeping regularly across us; and it grew colder. By evening we were in the grip of our first gale, and driving faster and faster out to sea ... For twenty-four hours Brendan ran before the gale ... Then, after I had calculated that Brendan had been driven about 100 miles off the coast, the wind eased and swung into the west, blowing Brendan back toward land and safety.' *Tim Severin,* **The Brendan Voyage,** *off Ireland, beginning a trans-Atlantic journey in an ox-hide curragh in the wake of St Brendan, 1976*

-3.2C Braemar 1997 ❄

31.7C Isle Of Grain 1922 ☀

24 MAY

8.1C min / 17.1C max

SUMMER OF RAVE / SEBASTIAN'S 'CROCK OF GOLD'

'By ten in the morning the sun's breaking through, the temperature's rising, and our rush has dwindled . . . We sprawl on the grass,' records dance culture chronicler Simon Reynolds of this perfect May Bank Holiday morning in 1992. If '67 is the Summer of Love, and '88 is the Second Summer of Love, then '92 is the apogee of Acid House – and the last, the biggest and the most notorious of the free, illegal, Ecstasy-fuelled raves is today, at Castle Morton Common (♀). The weather's *so* gorgeous that London's press posse turn up – clinching Castle Morton's place in dance history.

It's the scene's last stand, as the tabloids stoke Middle England's indignation with tales of '40,000 drug-crazed New Age hippies . . . terrorising locals and desecrating the tranquillity of England's green and pleasant land'. The numbers are closer to twenty-five-thousand, but the Conservative government is seeking re-election and needs to look tough. Provisions are drafted into the Criminal Justice Bill banning gatherings of more than a hundred at which amplified music 'wholly or predominantly characterised by a succession of repetitive beats' is played.

It's the end of the free party. From here the movement turns mainstream, urban and highly commercial: superclubs, superstar DJs, massively branded labels, clothing, drinks and drugs. 'Muddy fields and hastily erected marquees were replaced by steel and chrome, and £30 entrance fees . . . terra techno turned into slinky house. Shiny clubs, shiny drugs, shiny people, and shiny music. It did not feel bad anymore. It had become respectable,' reports the *Observer*.

'It was about eleven when Sebastian, without warning, turned the car into a cart track and stopped. It was hot enough now to make us seek the shade. On a sheep-cropped knoll under a clump of elms we ate the strawberries and drank the wine – as Sebastian promised, they were delicious together – and we lit fat, Turkish cigarettes and lay on our backs, Sebastian's eyes on the leaves above him, mine on his profile, while the blue-grey smoke rose, untroubled by any wind, to the blue-green shadows of foliage, and the sweet scent of the tobacco merged with the sweet summer scents around us and the fumes of the sweet, golden wine seemed to lift us a finger's breadth above the turf and hold us suspended. "Just the place to bury a crock of gold," said Sebastian. "I should like to bury something precious in every place where I've been happy and then, when I was old and ugly and miserable, I could come back and dig it up and remember."'
Evelyn Waugh, Brideshead Revisited, *1945*

-3.5C Altnaharra 1997 ❄

32.2C Camden Square 1922 ☀

SCOTTISH POTATO FAMINE

Today in 1846 a spell of hot weather begins that will suck the watery landscape of north-west Scotland so dry that 'men swore they had seen salmon swimming in red dust'. The crofting communities are worried. Following a mild winter and a cold spring, the success of the potato crop is now in doubt. Because of the weather, most communities in this corner of Britain no longer grow grain. Instead they have adopted the potato as their staple. Potatoes grow in virtually any soil, and flourish when the weather is cool and moist. The same heavy rainfall and high winds that destroy grain prove ideal for potatoes. The greenfly which spreads the common potato disease, known as 'curl', rarely moves when the wind is above 8 mph (13 kph), while heavy rainfall washes plants clean of the parasite. So marked is the success of this tuber that by 1840 the Highlands is effectively a potato economy.

'It was fine to drive the moors in the numb night, to drive fast in a motor-car made for it ... It was fine to see the sun rising cold and yellow on the near side, whilst the night, long before sunrise, retreated on the off: to feel the dew and the slight mist: to overtake another early bird in a fast motor, and to race him by common consent to Middlehampton, all before breakfast in the quickening sun.' *T.H. White, author, 1934*

The sun is gladly shining,
The stream sings merrily,
And I only am pining,
And all is dark to me.
Emily Brontë, 25 May 1839

Strangely, the Western Highlands avoid the worst of the 1845 blight caused by the new fungus *Phytophthora infestans*, which leads to famine in Ireland *(8 September)*. In fact, barrels of potatoes are even exported, at high prices. But the mild winter allows the fungus spores to survive: healthy tubers are planted next to slightly infected ones, and by today in 1846 the first characteristically pitted potatoes are being dug up in the Western Isles. By the end of September, as the Reverend Norman Macleod says, 'the most momentous calamity in the condition of the Highlands that has occurred for a century' is apparent. For a decade, the climatic changes brought about by the end of the 'Little Ice Age' *(3 March)* provide ideal conditions for the blight to thrive. The Highland potato crop fails, on and off, until 1857. The crisis, exacerbated by a wave of clearances in favour of sheep farming in the late 1840s, creates an upheaval in Highland society which is still apparent today.

-7.2C Braemar 1914 ❄

31.7C Farnham 1953 ☀

7.8C min / 16.8C max

TEMPEST AT THE GLOBE

Rain falls with quiet determination at the official opening (and first performance) of *The Tempest* at the new Globe Theatre in London in 2000. The 'groundlings' in the open-air courtyard huddle under their white cagoules like 'sodden Klu Klux Klan men', the *Daily Mail* reports, while the cast milk the play's numerous references to weather. 'Is the storm overblown?' (Audience: 'No!') Vanessa Redgrave, as Prospero, changes the words in the final soliloquy from 'bare island' to 'in this wet island by your spell.' (Cheers and applause.)

Getting wet is just an occupational hazard at the Globe – today as much as in Elizabethan times. Curiously, though, there are few recorded incidents of plays being interrupted by weather in the sixteenth and seventeenth centuries. Possibly, because the galleried seats and stage are partially covered, it's only the peasantry in the pit who get drenched – and who cares about them? Other irksome disturbances – background noise, smells, pickpockets, the occasional riot – prove more noteworthy than rain. Overcast, bleak and wintry weather does blight the opening of John Webster's masterpiece, *The White Devil*, in 1612 at the Red Bull Theatre. The playwright blames the play's failure on 'so dull a time of winter, presented in so open and black a theatre'.

'About the first hour of the day, the thunder and lightnings approaching nearer, one clap more terrible than the rest, as though it would bear the heavens down on the earth, struck the hearts and ears of those who heard it dumb by its sudden crash. With that crash a thunderbolt fell on the bed room of the queen, where she was then staying with her sons and family, and threw the bed to the ground, crushing it to powder, and shook the whole house. In the adjacent forest of Windsor, it threw down or split asunder thirty-five oak trees.' *Matthew Paris, historian, Chronica Majora, 1251*

'Among all the severall judgments on this nacion: god this spring when rye was earing and eared, sent much terrible frosts, that the eare was frozen and so dyed, and cometh to nothing: young ashes that also leavd were nippy, and blackt, and those shootes died, as if the lord would continue our want, and penury, wee continuing our sins.' *Ralph Josselin, Puritan minister and diarist, Essex, 1648*

-4.1C Grantown-On-Spey 1990 ❄

29.3C Norwood 1880 ☀

27 MAY

8C min / 16.9C max

BARON VON WARREN It's an idyllic summer night in 1940 as, at 8.30 p.m., a plane piloted by Captain Warren leaves Dishforth, Yorkshire (♥) for a German airfield in Holland. As the Whitley bomber and its crew cross the North Sea, lightning starts flashing around it. The aircraft rocks and buckets in the electrical thunderstorm, so to get out of the weather Warren asks his navigator for a new course. It's dark when, at their estimated time of arrival over the Dutch coast, they search for a pinpoint. At last they spot the Rhine estuary. As anti-aircraft fire hoses up at them they follow the river, turn to starboard as planned and search for the German airfield. Suddenly the second pilot calls: 'I've got it!' Bombs away! 'Give me a course for base,' says Warren. At first light, as their estimated time for arriving home approaches, they drop down through the cloud. Below them is a city, with the sea beyond. It all looks vaguely, unsettlingly familiar – rather like the west coast of England, in fact. The dreadful truth strikes Warren. It's Liverpool. 'According to my calculations, we can only have bombed something inside England. Christ, what are we to do?'

THE DRENCHED LIEUTENANT'S WOMAN As filming starts, today in 1980, of John Fowles's *The French Lieutenant's Woman*, so does the rain. It rains, and it rains – and does so throughout the summer in Lyme Regis, Dorset, where most of the action is set. And when the rain brings one bonus – the rough sea needed for the opening scene – United Artists forbids Meryl Streep to stand, exposed, on the slippery sea wall known as the Cobb (she's just too expensive to sacrifice to the British weather, apparently). In the end, an art director pulls on her black cape and lashes himself to the quayside as the waves crash around him. The film is eventually nominated for five Oscars.

It turns out that their magnetic compass has been completely thrown out of true by the storm. They have followed the Thames estuary, not the Rhine, dropping their full bombload directly across the runway of the Fighter Command base at Bassingbourn, Cambridgeshire. Happily, no one's hurt. Less happily, nor is the runway. From this misadventure, Bomber Command learns a salutary truth: that a full bombload, delivered directly on to an airfield does . . . virtually no damage at all. Captain Warren is demoted and thereafter known as Baron von Warren (two Spitfires later drop Iron Crosses on RAF Dishforth), closing an incident that exposes just how much work Bomber Command has to do before it's an effective fighting force.

150.4mm Ben Nevis 1902
−6.1C Lednathie 1899

28.9C Tern Hill 1944

28 MAY

8C min / 17.1C max

DUNKIRK · There's little wind, a bit of low cloud, and the odd thundery shower over the Channel today in 1940. In short, the weather's perfect: the Luftwaffe can't see what's going on under their noses. Which is fortunate, given that what's going on has been called 'the most extensive and difficult combined operation in naval history'.

The ideal conditions begin the day before yesterday, as the remains of the British Expeditionary Force find themselves trapped against the coast of northern France by the advancing German army. The Admiralty's call goes out over the BBC for 'small ships' (sailing boats of 30–100 feet/9–30 metres) that can get into the shallow waters to ferry troops out to Royal Navy destroyers.

> 'We are almost freezing here in the midst of beautiful verdure with a profusion of blossoms and flowers: but I keep good fires, and seem to feel warm weather while I look through the window, for the way to insure summer in England, is to have it framed and glazed in a comfortable room.'
> Horace Walpole, writer, 1774

On the beaches, the advantages of the weather are less obvious. 'Not a breath of air was blowing to dissipate the appalling odour that arose from the dead bodies that had been lying on the sand, in some cases for several days. We might have been walking through a slaughterhouse on a hot day,' writes a soldier later. Tomorrow, brighter, clearer conditions bring German air attacks, though fortunately mist and cloud return, remaining until the 30th. After this, the weather deteriorates (from a military perspective). Wind begins to generate heavy surf, and by the last day of the month embarkation is difficult, almost more than the exhausted troops can manage, especially as the Germans attack again. Despite this, 68,000 British soldiers are evacuated on the 31st, the highest one-day total. The weather 'improves' alarmingly over the next few days, but fog at night allows the evacuation to continue.

By 4 June, although 100,000 Allied prisoners of war have been taken, 338,000 British and French troops are safely back across the Channel. Churchill later admits to the House of Commons that when Operation Dynamo is launched he expects only 20,000–30,000 to be saved. But in the general elation, he sternly warns: 'Wars are not won by evacuations.' Ironically, the phrase 'Dunkirk spirit' is often invoked to mean rallying round in the face of bad weather. In fact, Dunkirk is a triumphant success chiefly because the weather is so kind.

-5.0C Alwen 1961

28.3C Kensington Palace 1944

LOUTH FLOODS / A BATSMAN'S SUMMER

It's a Sunday in 1920, and around 2 p.m., above the east Lincolnshire town of Louth (♥), the heavens darken forbiddingly and then open. As water pours from the sky the River Lud rises rapidly – 6 feet (2 metres) in ten minutes at one stage. Suddenly, there is a deafening roar. Without warning, apparently from nowhere, a torrent of water over 650 feet (200 metres) wide sweeps through the town, pouring into houses. Horrific situations develop in minutes: one man watches from his top-floor window as his neighbour drowns; a mother who scrambles on to a dresser with her three children is engulfed by the water and watches as one by one they drown.

In minutes, twenty-two are drowned. All six bridges are destroyed, fifty houses collapse and a further seven hundred are damaged. The flood subsides almost as fast as it arrives, and still no one is certain of the cause. Though the rainfall is heavy, the river should have been able to carry it. The most likely cause of the disaster is that a bridge upriver of Louth becomes dammed by junk and debris, until, when the weight of water becomes too great, it collapses, releasing the torrent. Did a dumped iron bedstead – the litter equivalent of today's shopping trolley – cause the most lethal flood of the twentieth century?

The year 1947 renews its bid to be the most interesting, weather-wise, of the twentieth century. Following the long, snowy winter (*4 March*), the warmest extended summer on record begins today when the mercury hits 31.7C in Lincolnshire. The glorious weather continues into September. This golden summer is remembered as the zenith of county cricket. Scarcely a match is rain-affected, and the sun-baked wickets provide ideal conditions for batting. Middlesex's Denis Compton thrills a war-weary public and tops the figures with 3816 runs and 18 centuries – records for a first-class cricket season that will probably never be broken. The commentator, John Arlott, describes the endless days of sunshine as 'an enjoyment that seemed like magic'.

-4.4C Rickmansworth 1936
32.8C Horsham 1944

3O MAY

8.3C min / 17.5C max

FIRST
HELICOPTER
MOUNTAIN
RESCUE

There are no gales, no floods and no blizzard today in 1947. In fact, it's windless and sunny in mid-Wales: perfect conditions, you might think, for the RAF's first helicopter mountain rescue trial. Well, not exactly. Helicopters are still novelty machines – hardly beyond the experimental stage – and the exercise is to investigate 'the practical use of a helicopter in mountain search and rescue'. RAF Mountain Rescue (not to be confused with civilian mountain rescue – *14 March*) is a specialized unit formed during the Second World War to locate aircraft crash sites in remote areas, rescue casualties, recover bodies, and make safe anything explosive or dangerous.

On this occasion, a 'crash site' is duly earmarked: a grassy slope high in the Cambrian mountains north of Llandovery (☀). Three RAF teams are involved, some acting as casualties, some as searchers, some as stretcher-bearers lugging sledge stretchers. Numerous official observers and press photographers are also present. And then there is the sun – getting hotter all the time. From the start, the long shadow of cock-up looms low over the day. One convoy's vehicle breaks down. A jeep gets stuck in a bog. One party, wearing ordinary blue RAF uniforms and without food, is soon too exhausted to continue. The exercise is abandoned, by which time so many vehicles are broken down that personnel can't be evacuated. The moral? Even the military – even RAF Mountain Rescue – can mess up a simple walk in the Welsh hills on a sunny day.

And the helicopters? Everyone's very impressed, especially with the way they handle confined valleys – as well they might be: most have never even seen a helicopter before. From such inauspicious beginnings, the helicopter soon comes to symbolize mercy missions (*31 January, 2 February, 27 March, 22 November*), though its first real test, in proper mountain weather, doesn't take place until the Glen Doll tragedy in twelve years' time (*1 January*).

-3.9C Tarfside 1961 ❄

162.8mm Seathwaite 1865 🌢

In Selkirk (☂), in 2003, intense rain causes a flash flood with uncanny similarities to that in Boscastle in 2004 (*16 August*): gusting winds and hail up to ground-floor windows block critical drainage culverts, resulting in water levels rising to car windscreen height.

31.7C Kensington Palace 1947 ☀

31 MAY

8.1C min / 17.7C max

GREAT DERBY DAY DISASTER

Towering black clouds amass in the sunshine, with just the odd rumble of distant thunder, before the Big Race on Epsom Downs (♥) this afternoon in 1911. But it's only when the racing is over at 4.59 p.m. that the first fork of lightning splits the sultry sky. The show starts with 'peals coming in sharp, decisive cracks, closely resembling cannonading, whilst the lightning flashes were of dazzling intensity,' reports local amateur meteorologist Spencer Russell. Soon the lightning is flashing at thirty strokes a minute, as a stupendous electrical storm hits its stride. In fifty minutes, 2.4 inches (62 mm) of rain falls, along with 'hailstones as big as walnuts'. It's 'an inferno of water, mud, thunder, lightning, and hail,' reports a local paper, as horses plunge, children scream and everyone huddles under cover.

A number of people sheltering against a wall are struck by lightning – two die. A horse pulling a cart of race-goers is struck and killed, along with one of the passengers. Eight more are struck sheltering in a marquee, while another man, leaning on metal rails along the side of the course, is struck and badly injured. Three hay-ricks are set ablaze. In all, this afternoon, seventeen people and four horses are killed across the south-east in one of the most violent electrical storms ever to hit England. Of course, for the Derby it's not *that* unusual (see above).

A SAFE BET – AWFUL WEATHER AT THE DERBY

18 May 1820	Driving rain and gales blow away tents. Appositely named Sailor, who likes soft going, wins
27 May 1830	Heavy rain and hail. Thirteen false starts
15 May 1839	Bitter east wind; snowstorm during race
20 May 1863	'While last year it was iced champagne, claret cup, and silk overcoats, now it ought to be hot brandy and water, foot baths, and flannels' – Charles Dickens.
22 May 1867	Wet snow, ten false starts – so unpleasant that it contributes to moving the race date to two weeks later in the calendar (*22 May*)
May 1891	So wet during one race that jockeys are all a couple of pounds overweight by the end
May 1924	Torrential rain. New motorized buses and
May 1925	charabancs become mired in mud and have
May 1926	to be towed out
31 May 1979	1.16 inches (30 mm) of rain in ninety minutes.

-5.1C Carnwath 1975 ❄

32.2C Camden Square 1947 ☀

JUNE

1 JUNE

8.4C min / 17.8C max

JUTLAND – THE
BATTLE NEVER
FOUGHT

Summer officially starts today, though June is seldom especially warm. The hottest day of the year falls this month only about a quarter of the time (about the same as August). June is the only month showing no sign of warming in the last three centuries; on average, it's hardly warmer than September, and hasn't been warmer than July since 1970. It was last the warmest month in 1960, 1966 and 1970, and the driest month in 1925 (when it also records the highest monthly sunshine record of the century). Still . . . long days, short nights, flowers, lightning, hay fever – summer's here.

As the mist and fog clear over the North Sea this morning in 1916, a question forms, one that will re-form again and again: who's won the Battle of Jutland (✠)? This 'Clash of the Dreadnoughts', the largest battleship conflict of all time, is the only naval engagement of its kind during the First World War. It's the one at which Admiral Beatty says, 'There appears to be something wrong with our bloody ships today,' and at which Admiral Sir John Jellicoe (commander of the Grand Fleet) is, according to First Sea Lord Winston Churchill, 'the only man on either side who could have lost the war in an afternoon'.

Argument and controversy will rage for ever over the performance of the British fleet. A crucial point is whether Admiral Evan-Davis's battlecruiser squadron can or cannot see a signal hoisted by Admiral Beatty. The signal, consisting of three pennants (see over), ordered: 'Alter course to South South East and pursue the enemy.' Did he see them and choose to ignore them? Or were they obscured by smoke and fog? Whatever the reason, the admiral fails to respond and pin down the German Grand Fleet.

Revisionist historians believe it may be the ships themselves which initiate the deterioration of weather conditions, leading to vital signals being obscured. First, the immense quantities of soot and smoke from the coal-fired steam boilers and heavy shells of 250 vessels in a relatively small sea area (130 square km/50 square miles) would provide just the conditions that generate fog – as London's chimneys did before the Clean Air Acts took effect (*6 December*). Secondly, cold water pushed to the surface by whirling propellers criss-

31.7C Mildenhall 1947

1 JUNE

crossing the water at up to 25 knots (29 mph) would immediately initiate condensation in the warm air above the water's surface. Are Beatty's signals simply indiscernible through the murk?

Whatever the answer, Jutland is so indecisive that it's sometimes called 'the battle that was never fought'. Both sides claim victory – the Germans because they have sunk considerably more ships, the British because the German Grand Fleet flees to port and doesn't emerge for the rest of the war. Neither side really wins, and more than 8500 seamen die.

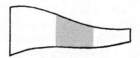

'Alter course to South South East and pursue the enemy.'

2 JUNE

8.8C min / 17.9C max

In thick fog on the Mull of Kintyre (🎈), an RAF Chinook helicopter flies into a hillside at 150 mph (240 kph) today in 1994. Four crew and twenty-five passengers – comprising virtually the entire senior Northern Ireland intelligence team from MI5, the Royal Ulster Constabulary and the British army – all die. It's the worst helicopter crash in peacetime Britain and a massive setback for the anti-IRA campaign. There are no witnesses. Pilot negligence is blamed, a verdict bitterly contested by the pilots' families as a cover-up for technical deficiencies in avionics equipment that, it is alleged, the Ministry of Defence know all about. Successive RAF and parliamentary inquiries over eight years all fail to reach a comprehensive conclusion.

'The weather may be called delightful at present; sun shining, small breeze blowing, ground green as leeks. My windows "look to the Forth," . . . and I get a view nightly of the Sunset. It is very grand to witness the great red fiery disc, sinking like a giant to sleep, among his crimson curtains of cloud – with the Fife and Ochil hills for bedstead! I often look at him till I could almost break forth into tears, if it would serve, – or into some kind of poetical singing, if I could sing.' *Thomas Carlyle, satirist and historian, Edinburgh, 1822*

In 1975 snow falls heavily today – the latest date in the year that such a widespread fall has occurred since the nineteenth century. Six inches (10 cm) blanket the cricket ground at Buxton (⛄), delaying the county match between Derbyshire and Lancashire. The following day, in bright sunshine, the uncovered pitch is sodden and virtually unplayable. When it starts to dry out, the unpredictable bounce causes huge problems for the Derbyshire batsmen who are bowled out for 42 and 87. Dickie Bird, one of the umpires, recalls one batsman handing him his false teeth for safekeeping while he takes guard: 'Look after these . . . I won't be long,' he says.

In grey drizzle in 1953, Queen Elizabeth II is crowned at Westminster Abbey. The weather amuses millions across the Commonwealth, watching on television, but does nothing to diffuse the blazing patriotic fervour of the crowds lining the procession route back to Buckingham Palace. The vast Queen Salote of Tonga defies the by now pelting rain in an open-topped, horse-drawn carriage. She is accompanied by the diminutive Sultan of Kelantan. 'Her lunch,' declares the writer and composer Noël Coward.

33.3C Hunstanton 1947

3 JUNE

8.7C min / 18.1C max

PASSING OUT
PARADE

In 25C blazing sunshine today in 1957, a record thirty-five guardsmen faint while rehearsing Trooping the Colour for the Queen's official birthday. Twenty are charged with disciplinary offences.

'Astounding and enchanting change in the weather, which becomes warm. I carry chair, writing-materials, rug, and cushion into the garden, but am called in to have a look at the Pantry Sink, please, as it seems to have blocked. Attempted return to garden frustrated by arrival of note from the village concerning Garden Fête arrangements, which requires immediate answer, necessity for speaking to the butcher on the telephone, and sudden realisation that Laundry List hasn't yet been made out, and the Van will be here at eleven. When it does come. I have to speak about the tablecloths, which leads – do not know how – to long conversation about the Derby. Shortly after this, Mrs S. arrives from the village, to collect jumble for Garden Fête. After lunch, sky clouds over, and Mademoiselle and Vicky kindly help me to carry chair, writing-materials, rug, and cushion into the house again.' *E.M. Delafield, novelist, 1930*

Of course, standing to attention as straight as . . . well, as a guardsman, for ninety minutes in the midday sun, wearing 26 oz (0.74 kg) of beaver wool crowned with a black bearskin, the wonder is that more of the Queen's men don't keel over. They avoid fainting by not standing on their heels and by wiggling their toes and flexing their muscles – invisibly, of course. Mind games are used, too, such as counting backwards from 500 in threes. If someone begins to go, guardsmen either side try to prop him until he recovers. Barley sugars are often secreted behind their white 'buff belts', which sergeant-majors are adept at unwrapping (single-handed and wearing gloves) and slipping into the guardsman's mouth while apparently adjusting his chin-strap. The common error among novice guardsmen, especially if it's windy, is to over-tighten the leather band that fits the bearskin to the head: on hot days the head swells, the sweat-soaked band shrinks and . . . TIMBER! Should the worst happen, a falling guardsman gains respect by breaking his jaw – proving that he was guardsman-straight to the last. Yee-ees, sah!

The charging of the fainting men prompts questions in Parliament. Labour MPs enquire of the Secretary of State (John Hare) what crime the men have actually committed. Hare replies that they have been charged under Section 69 of the Army Act with 'falling out on parade'. (Laughter.) Hare adds that the award of punishments was 'contrary to standing orders, and he was taking steps to see that there is no recurrence'. (Loud cheers.)

34.4C Waddington 1947

4 JUNE

8.8C min / 18.4C max

THE MOST IMPORTANT FORECAST EVER?

Strong winds, low cloud, big seas in the English Channel and a cold front stretching from Scotland to the western Atlantic: it's a forecasters' nightmare today in 1944. Unfortunately for Group Captain J. M. Stagg, an accurate forecast is precisely what he *must* deliver. General Eisenhower, Supreme Commander of the Allied Expeditionary Force, is poised to launch Operation Overlord, the most ambitious seaborne invasion in history, otherwise known as D-Day *(6 June)*. But to go, or not to go? Meteorologists from the British Met Office, the Naval Meteorological Service and the US Army Air Forces, overseen by Stagg, have studiously researched the best moment. But now the weather has gone disastrously awry.

Reluctantly Eisenhower makes his decision. He postpones his plan to go tomorrow, the 5th. The forecast is, at best, marginal; at worst, a storm. The consensus among the meteorologists is that the outlook won't change – a grave situation, indeed. Then a Navy vessel, situated south of Iceland, reports sustained, rising pressure. Stagg believes that a ridge of high pressure is developing behind the cold front, and moving east towards the Channel. If he's right, there could be a window of fair weather on the 6th – just long enough to land forces and secure the beach-heads in the critical first few hours of the campaign. Crucially, German forecasters are unlikely to have spotted this.

The weather is so volatile that the commanders-in-chief and their staff are not convinced. Stagg, a tall, conscientious, blue-eyed Scotsman, explains that his confidence in his forecast – perhaps the most important ever – is high. In the library at Southwick House HQ, as the rain drives against the windows, the atmosphere is tense. Eisenhower paces the floor, asking his commanders if there is any reason why the invasion should not now be launched. Finally he makes up his mind: 'OK, we'll go.'

'Very fine day, and being the King's birthday, the town was in a bustle and hurry ... Poor soldiers sweating and stinking ... City Buckeens on hired horses and with borrowed boots and spurs; young misses slipping away from their mammas to meet their lovers; old maids taking snuff, and talking of old times; pickpockets waiting for a lob, and old bawds and whores for a cull; handkerchiefs in constant employ, wiping dust, sweat and dander from the face and head.'
Sir Vere Hunt, High Sheriff of Limerick, gambler and opportunist, Dublin, 1813

'Went for short wet walk in Ken. Gardens & saw the poor damp soldiers of all colours in their tents, looking most Un-victorious happy & glorious.'
Violet Bonham-Carter, political hostess, 1946. The soldiers were preparing for the victory celebrations in London.

31.1C Glasgow Airport 1939

5 JUNE

9C min / 18.2C max

DEATH OF KITCHENER

A storm off the Orkneys today in 1916 kills a national hero. Lord Kitchener, Secretary of State for War (and owner of the remarkable drooping moustache and pointing finger in the 'YOUR COUNTRY NEEDS YOU!' recruitment posters), is on his way to Russia – a diplomatic mission to raise support for the war effort. Aboard the armoured cruiser HMS *Hampshire*, he leaves the naval base at Scapa Flow (⚓) bound for the Russian port of Archangel. Mountainous seas and a Force Nine north-easter, however, mean that the ship heads north in the lee of the Orkneys for shelter. This channel has not been cleared by minesweepers, though merchant shipping has been using it for months. No sooner have they left, however, than the wind shifts to the north-west. The waves are so high that the two escorting destroyers cannot keep up, and turn back. The *Hampshire* strikes a mine within sight of the Brough of Birsay (✣) and sinks in minutes: 643 – all but 12 crew – are lost.

'For me, ... a nice day is ... when there is a mild breeze. This brings into life all the sounds in my environment ... because it suddenly gives a sense of space and distance ... The misunderstanding between me and the sighted arises when it is a mild day, even warm, with a slight breeze but overcast. To the sighted, this would not be a nice day, because the sky is not blue.' *Professor John Hull*, Touching the Rock: An Experience of Blindness, *1988*

'I hear people talking about the horrible weather. Horrible? This is all we've got. It's like saying horrible life.' *Julie Christie, actress, Daily Mail, 2000*

A collective national shock-wave, almost panic, greets the news of Kitchener's death – not unlike that following Princess Diana's. In the absence of a body, conspiracy theories multiply. These vary from claims that he's been assassinated by the government, to sabotage by Irish Republicans, to an elaborate hoax by a failed film-maker called Frank Power, alleging that Kitchener is still alive and well.

Did Kitchener's death change history? With Russia on the brink, could his advice to Tsar Nicholas have prevented the Russian Revolution? Certainly, the moment is a tipping point. As Alexander McAdie (1863–1943), a Harvard meteorologist, puts it: 'If the storm centre had passed over the Orkneys a few hours earlier, the eastern channel would have been selected. If the fury of the northwest wind had been less, a rescue would have been effected. But it was not to be. The destiny of Russia, perhaps the fate of Europe itself, hung upon a forecast of weather made that June afternoon in Scapa Flow.'

31.7C Camden Square 1933

6 JUNE

9.1C min / 18.4C max

D-DAY The navy needs winds below 12 mph (19 kph). The gliders must have no fog. The army needs no rain, so that the ground is firm. All of this has to fall in a period when the tide is low, at dawn, and the moon is nearly full. And the weather has to hold for thirty-six-hours to land the forces and secure the beaches. That's all that is required to launch Operation Overlord, involving some 3 million troops, 4000 ships and 12,000 planes. The weather on the 5th, the original launch date, is dreadful and the decision to postpone *(4 June)* has averted disaster. But is it good enough today? Bad weather gives the Allies one advantage: the Germans are off guard. The storm prevents their naval patrols and reconnaissance planes from surveying the Channel. Field-Marshal Rommel, the German commander, returns briefly to Berlin to see Hitler and celebrate his wife's birthday.

The weather is far from perfect. Before dawn, the airborne divisions run into chaos parachuting through thick cloud. Preliminary RAF bombing raids miss their targets. Waves up to 5 feet (1.5 metres) high leave soldiers seasick, with sodden battle gear. On Juno Beach, the landing is severely disrupted by strong winds. On Omaha Beach, soldiers plunge into neck-deep water and artillery is lost; many amphibious Sherman tanks sink. But the sun does intermittently shine, visibility is good and the wind dies off during the day.

'The day came and we all sailed early and in great spirits; but throughout the day the wind increased from the westward, the sky darkened, and at nightfall we all turned back in some alarm – it's not easy to stop machinery of this magnitude. The meteorological officer was pestered by everybody, which seemed to have the right effect, because the next day the "ridge of high pressure" in the Atlantic took a turn to the northward and we steamed up the Channel in bright weather ... It was just like the Pied Piper: from every port poured hosts of landing craft of every shape and size ... A weird armada, but we had all confidence in it.' *Lieutenant R.G.C. Macnab, aboard HMS* Glasgow

With hindsight, we know that the decision to go is dead right. President Truman later awards the chief meteorologist, Group Captain Stagg, the US Legion of Merit. For the rest of June, the weather is foul: at the time of the next 'moon and tide' window, the Channel is blasted by the worst weather in twenty years. It's impossible to say how long the invasion could have been put off; as Eisenhower writes to Stagg: 'Thank the Gods of war we went when we did.'

150.0mm Southery 1963

33.3C Camden Square1950

7 JUNE

9.4C min / 18.3C max

BOB MARLEY It's either drizzling or raining, or about to drizzle, or about to rain, all day at Crystal Palace in 1980 where Bob Marley's playing his last UK gig. It's raining as the Wailers come on and start playing 'Sun is Shining'. It's raining as Bob comes on-stage, playing his guitar. And then, at the exact moment he takes up position, centre-stage, the rain stops, the clouds seem to part and brilliant sunshine blazes down. The great man begins: 'Sun is shining, the weather is sweet . . .' Man . . . it's religious.

Less divine is the weather for the Queen's Silver Jubilee celebrations in 1977. The thousands who camp out overnight for a better view of the royal procession through London do so in pouring rain. Those celebrating around the beacon bonfires lit across the country make as merry as blustery winds allow. Those on Scafell and Skiddaw in the Lake District are ankle-deep in – yes, snow. It's no better for the anti-monarchists: 'We were going to march to Buckingham Palace to proclaim the republic,' complains one 'Stuff-the-Jubilee' rally organizer to the *Guardian*. 'But unfortunately it was too cold and only five turned up.'

DRIVING WITH THE ROOF DOWN

'Now, at last, we are in early summer, [I am] reminded of how a quite small difference in the ambient temperature – from say [14–17C] – makes the whole difference to open-top motoring. And a time of year when, at dusk, those traces of ground mist seem to make twin SUs run so sweetly. All my great drives in classics seem to have been in early June, when the hours of daylight are so long and, in the far north, blend into the dawn. I remember as a student trying to get to Glen Brittle in the Cuillin Hills in one day from Hampstead, being determined to climb Sligachan over the weekend and return. I was in the [Jaguar] SS, and took the last ferry across to Skye in broad daylight at 10 pm. That last stretch I was so exhausted that I never once changed gear nor, I am ashamed to say, did I drive much on the left. My black box would have recorded eighteen hours without, in those days, a single mile of motorway.

But it is in conditions such as these that you grow as one with your mount. Every beat of the exhaust and creak of the bodywork is part of the romantic experience: something you will never capture in a Volvo wagon.'

Alan Clark, politician

33.1C St. James's Park 1996

8 JUNE

9.3C min / 18.1C max

Eyes streaming? Nose running? Constantly sneezing? If so, it's because the worst part of the year for hay fever sufferers begins around now, continuing through to mid-August. Hay fever may be caused by several tree or weed pollens, but in the 'right' weather conditions (warm, with dry ground), it's the pollinating grasses – rye, cocksfoot, timothy and meadow fescue – peaking about now, that set the largest number of people (15–20% of the population) snivelling. Possibly because summers start earlier now than 20 years ago, hay fever is on the rise.

The wettest part of a record wet summer begins today in 1768: thirty-six out of the next forty-four days are thundery and wet.

'A pouring wet day. Nevertheless my Father and I started in the rain for the Vale of Arrow, he riding the Vicarage pony sheltered by two mackintoshes and an umbrella and I on foot with an umbrella only. We plodded on doggedly through the wet for 6 miles, casting wistful glances at all quarters of the heavens to catch any gleam of hope. Hope however seemed to be none. The rain fell pitilessly. We reached the Harbour more like drowned rats than clergymen of the Established Church.' *Reverend Francis Kilvert, Herefordshire, 1872*

'A wonderful June morning. The buttercups were up to my knees. There was a breath of wind just stirring the tops of the elms, and the great green clouds of leaves were sort of soft and rich like silk. And it was nine in the morning and I was eight years old, and all around me it was early summer, with great tangled hedges where the wild roses were still in bloom, and bits of soft white cloud drifting overhead, and in the distance the low hills and the dim blue masses of the woods round Upper Binfield. And I didn't give a damn for any of it. All I was thinking of was the green pool and the carp and the gang with their hooks and lines and bread paste.' *Georg*

32.2C Cromer 1915

203.2mm Camelford 1957

9 JUNE

9.1C min / 18.3C max

MIDGE DAY Bad news for Scotland and Scotland-lovers. In average weather, today is 'midge day', the day when rising temperatures and day length prompt the first hatching of *Culicoides impunctatus* – the Highland Biting Midge. Of the thirty-seven species of midge recorded north of the border, this is the one responsible for 90 per cent of human misery there. There is good news, however. Because of the late, chilly spring of 2006, and the record heat of the second half of the summer, midge numbers are down in the summer of 2007. In Fort William (📍) – a notorious stronghold, because of boggy, acidic topsoil in which the midge larvae overwinter – they are down by around half.

If overall midge numbers depend on the character of the season, minute-by-minute misery depends on specific weather conditions. As a rule of thumb, midges love warm, damp days – dull and overcast is ideal. What they loathe is bright sunshine, dryness, cold, heat or wind. Indeed, wind of more than about 5 mph (8 kph) and they can't cope at all, which is why they tend to disappear above 1500 feet (500 metres). When it's windy they lurk in heather, waiting for the weather to change.

While today is the start of the midge year, it's the second hatching, in mid-July or early August, that produces the most vicious bloodsuckers – pregnant females. After fertilization a female requires a 'blood meal' to develop the eggs. Having found her target, she drinks for up to four minutes if undisturbed, releasing throughout (this is the scary part) a powerful pheromone which attracts other females. Hence the mass attacks, the clouds of airborne piranhas at the height of the tourist season in July and August, that render a simple action like lacing your boots unendurable. More itchy facts? An area of boggy peat just 6 feet (2 metres) square contains up to half a million insects, and most insect repellents have little or no effect against midges.

Although they occur throughout the British Isles, it's Scotland with which they are associated as indelibly as tartan or whisky – albeit less heavily featured in the brochures. The midge is reckoned to cost the Scottish tourist industry £286 million a year in lost revenue (recently, 49 per cent of tourists said they would not return to Scotland at the same time of year because of them). Scots should start praying – climate change promises to deliver perfect midge weather.

32.8C Cranwell 1940

10 JUNE

9.4C min / 18.2C max

SKEGNESS IS SO BRACING / DROOPING THE COLOUR

On a day of continuous bright sunshine with a high of 22C, the poster artist John Hassall visits Skegness (🌂) for the first and only time in 1936. Twenty-eight years earlier, in 1908, Hassall created one of the most famous holiday posters ever – the Jolly Fisherman skipping to the slogan 'Skegness is so bracing' – for the Great Northern Railway Company, to advertise the special three-shilling excursion to the seaside from King's Cross. Today, he receives the 'freedom of the seafront' for his part in Skegness's fame. 'The reality of Skegness has eclipsed all my anticipations,' he says. 'It is even more bracing and attractive than I had been led to expect!'

In 1948, sunshine blazes down on a quarter of a million people gathered at Horse Guards Parade for Trooping the Colour on King George VI's official birthday. They are baffled when the loudspeakers announce that the ceremony is postponed due to 'unfavourable meteorological conditions'. Meanwhile, Major General John Marriott of the Scots Guards, in charge of the parade, has retreated to his office to pray, in vain, for a thunderstorm. Earlier that morning, after conferring with the King, Marriott has taken the decision to postpone the parade. Rain was then falling in sheets, and the forecast offered little encouragement. 'Hardly had I done so [ordered the postponement],' Marriott later recalls, 'when a blazing sun appears and the rest of the day is perfect. Another call from Buckingham Palace. Can I cancel my cancellation?' Too late. 'Have we now reached the stage where no one in authority dare say "carry on" if a meteorologist says it is going to rain?' demands Hugh Linstead MP in *The Times* the next day.

In 1982, on the other hand, the parade goes ahead in the rain – and the Household Cavalry have their boots filled to the brim. In 2001, seven hundred guardsmen splash through inches of water in perfect time, watched by the Queen from under a see-through umbrella and a sodden Prince Philip in his dripping bearskin – 'Drooping the Colour,' says the *News of the World.*

In 1960, today dawns 'cold, blustery and wet' as Francis Chichester makes final preparations in Plymouth on *Gypsy Moth III* for the start of the inaugural single-handed transatlantic yacht race. Instructions read simply: 'Leave the Melampus Buoy to starboard and thence by any route to the Ambrose Light Vessel, New York.'

31.7C Maldon 1970

11 JUNE

9.3C min / 18.3C max

DUBLIN DELUGE / PERSHORE DOWNPOUR

A day of raging cloudbursts. In a storm in Dublin today in 1963, a rain gauge records 7.2 inches (184 mm), more than 3.1 inches (80 mm) of which falls in one hour, 2–3 p.m. To give some idea how much rain this is, in an average June Dublin expects total rainfall of 2–2.4 inches (50–60 mm). Statistically, Dublin can expect 0.6 inch (15 mm) of rain in an hour once every five years, 0.8 inch (20 mm) once a decade, and 1.4 inches (35 mm) once every fifty years. That's how unusual 3.1 inches in an hour is. What's more amazing is that the 'Dublin Deluge' is trounced by a storm in Pershore, Worcestershire (☻), also today, in 1970: 2.6 inches (67 mm) of rain falls in twenty-five minutes.

'As wet as ever a day came.' *Sir Vere Hunt, New Birmingham, Ireland, 1813*

To put *these* figures in context, the record British rainfall for an hour is 3.6 inches (92 mm), the record for June is 9.6 inches (242.8 mm) (*28 June*) and the record for any month ever is 11 inches (279.4 mm) (*18 July*). So no prizes – but today's still mighty wet.

MAKING THE MOST OF A SUMMER'S DAY, BY BRIDGET JONES
8st 13 (v.g., too hot to eat), cigarettes 0 (v.g., too hot to smoke), calories 759 (entirely ice-cream).

Another wasted Sunday. It seems the entire summer is doomed to be spent watching the cricket with the curtains drawn. Feel strange sense of unease with the summer and not just because of the drawn curtains on Sundays and mini-break ban. Realize, as the long hot days freakishly repeat themselves, one after the other, that whatever I am doing I really think I ought to be doing something else ... The more the sun shines the more obvious it seems that others are making fuller, better use of it elsewhere: possibly at some giant softball game to which everyone is invited except me; possibly alone with their lover in a rustic glade by waterfalls where Bambis graze, or at some large public celebratory event, probably including the Queen Mother and one or more of the football tenors, to mark the exquisite summer which I am failing to get the best out of. Maybe it is our climatic past that is to blame. Maybe we do not yet have the mentality to deal with a sun and cloudless blue sky, which is anything other than a freak incident. The instinct to panic, run out of the office, take most of your clothes off and lie panting on the fire escape is still too strong.

31.1C Cambridge 1900

12 JUNE

9.7C min / 18.5C max

THREE FINE DAYS AND A THUNDER-STORM

After three days of good weather, the old definition of a British summer rings true in 2006. As the temperature nudges over 31C at Heathrow today – a remarkable figure for the first half of June – clouds gather, thunderstorms begin to rumble and dogs dive under sofas across the Midlands and the Pennines.

'It was shining blue weather, with a constantly changing prospect of brown hills and far green meadows, and a continual sound of larks and curlews and falling streams. But I had no mind for the summer, and little for Hislop's conversation, for as the fateful fifteenth of June drew near I was over-weighted with the hopeless difficulties of my enterprise.' *John Buchan, The Thirty-Nine Steps, 1915*

'It rained very hard ... and so he [Ralph Greatorex, maker of mathematical instruments] was forced to stay longer than I desired.' *Samuel Pepys, diarist, 1661*

'Since the Coronation, London has been unbearable. Bad weather, teeming crowds, traffic at a standstill, general frustration.' *Noël Coward, writer and composer, 1953*

Thunderstorms occur when dry, warm weather (brought by southerly winds) encounters moist, cooler air off the Atlantic. Water vapour cools quickly as it rises, forming fluffy, white cumulus clouds. These build and build into the distinctive cumulo-nimbus tower or thunder clouds, often anvil-topped (*8 April, 1 July, 21 August*). A thunderclap is the sound of air violently expanding as it's heated by a flash of lightning; the rumble is caused by the noise passing through layers of the atmosphere at different temperatures. Thunderstorms tend to be accompanied by brief, heavy and localized downpours.

The distance between you and a thunderstorm can be roughly measured by dividing the lag in seconds between the flash and the thunder, by three (for km) or five (for miles). It is rare to hear a thunder clap more than 12½ miles (20 km) away.

The phrase 'three fine days and a thunderstorm' has been attributed to Charles II and George III amongst others, but it's probably just an old country saying. As with many weather adages, it has some truth: 90 per cent of thunderstorms occur in summer (the west of Ireland is an exception), generally between 1 p.m. and 6 p.m. They often signal a change in the weather. South-east England has the most thundery days in the British Isles. In June 1970, thunder is reported on twenty days when an equatorial weather pattern – fine nights, warm mornings and afternoon thunderstorms – is established.

30.2C Scarborough 1897

31C Heathrow 2006

13 JUNE

9.6C min / 18.6C max

LONDON'S RAINIEST DAY Starting today, in 1903, the occupants of the Victorian terraces of Camden Square in north London endure 58½ hours of continuous rain – an all-time record for a British rainstorm. Either side of this, there's plenty more rain, too: it falls at the rate of 1 inch (25.4 mm) a day for six days, notwithstanding one dry day. 'Thus on every acre of ground,' declares *The Times*, 'there fell 487 tons of water.' There is, needless to say, widespread flooding – at Hammersmith Broadway underground station the water rises to within a foot of the platforms – but the rain is highly localized to East Anglia and the south-east. Snow blankets the Scottish Highlands.

> 'June is the month when roses tumble over the walls, the tall spikes of delphiniums tower above the jungle of the borders, at the mercy of the gales that nearly always turn up ... to humble our pride and challenge our foresight. The farmers take the "June drop" in their stride. Though it flattens the corn and brings violent rain to devastate the hay, it also thins out the apples for them. The poor gardener has no such compensations for his shattered hopes.' *Margery Fish, gardening writer, Somerset, 1965*

All in all, 1903's summer is a stinker across Britain, down there with the very worst. As so often (*18 July, 13 October*), the finger of blame is pointed at distant volcanic activity. The major eruption the previous year is Pelee in Martinique, which spews millions of tons of ash and soot into the atmosphere. This circulates, screening out the sun. The effect is widespread, lasts months, and London gets the worst of it. Incidentally, the Camden rain gauge is set up by George James Symons, the 'Father of British Rainfall', in the drought year of 1858 (*23 June*). Symons, subsequently employed by Admiral Fitzroy, founder of the Met Office (*1 August*), collects figures going back to 1766 – which is why Britain has the longest set of rainfall data in the world, and why his former house in Camden Square bears a blue plaque.

Finally we have the drama, in 1967, of a cricket pavilion in Melksham, Wiltshire (☛), lifting off the ground with fifty people inside it, which inspires the founding, seven years later, of a British tornado research unit. Does that sound as optimistic as a Society for the Study of Live British Volcanoes? Or a Scottish Glacier Institute? Or, perhaps, the plot of an Ealing comedy? Nevertheless, as the delightfully named British Tornado and Storm Research Organization (TORRO) is regularly called upon to confirm, the British Isles has more reported tornadoes for its land area than anywhere else in the world.

28.3C Earls Colne 1948

14 JUNE

9.4C min / 18.4C max

A DRYING PITCH AND A SIZZLING SUN

Following heavy rain in 1980, three County Championship cricket matches scheduled to start today are delayed, then cancelled, without a ball bowled. The Sisyphean task of clearing water from the pitches at Swansea, Bristol and Bath during persistent and heavy rain is abandoned. This summer is so wet, and the clubs suffer so badly, that the Test and County Cricket Board rules that from 1981 all pitches must be fully covered in bad weather. There is uproar. Though this only brings the county game into line with international cricket (*31 July*), people know the change is fundamental. The Editor of *Wisden* laments the 'loss of a part of the very heritage of English cricket – a drying pitch and a sizzling sun. Some of the great feats of batsmanship have been performed under these conditions.' The initiative certainly rebalances the relationship between bat and ball – in favour of the batsman. Some say the slow decline in the quality of English spin bowling dates from this moment.

'It froze hard last night: I went out for a moment to look at my haymakers, and was starved [cold]. The contents of an English June are hay and ice, orange flowers and rheumatism. I am now cowering near the fire.' *Horace Walpole, writer, Twickenham, 1791*

'As in 1959, the Opposition are fighting a fine weather mood and a sense of complacency, yet I have to record that we can't say the electorate has never had it so good.' *Richard Crossman, Labour Cabinet Minister, fighting the 1970 election campaign. The Conservatives win*

'Day became night,' people on Wandsworth Common report today in 1914 when an epic thunderstorm pauses briefly over South London. The 'phenomenon quite outside the experience of any who lived through it' begins at midday when the sky turns liverish, then black, before the streets are illuminated with sheets of 'blood red or crimson lightning'. Some 2 inches (50mm) of rain (the average rainfall for the entire month) falls in forty-five minutes, flooding the underground and dozens of basement flats. In Kingston and Wimbledon, hailstones the size of 'plums' and 'walnuts' pelt suburban homes. Several church spires and seven people are struck by lightning: four sheltering under a tree on Wandsworth Common are killed. Meanwhile, a few miles away across the Thames, people doze in their deckchairs.

31.7C Colmonell 1896

Another apocalyptic thunderstorm strikes England in 1977, ending five days of brooding weather. Many comment on the extraordinary, illuminating power of the lightning at night. One strike knocks the computer at the Met Office HQ in Berkshire (☞) out of action.

15 JUNE

9.2C min / 18.7C max

ALCOCK AND BROWN A brisk 30 mph (50 kph) westerly in 1919 sweeps Captain John Alcock and Lieutenant Arthur Brown into the history books. At 8.40 a.m., they crash-land their Vickers Vimy biplane into a Connemara peat bog near Clifden in the west of Ireland (♥) (from the air it looks flat, green and just right for landing), having completed the first non-stop crossing of the Atlantic. Their journey has taken just sixteen hours – nearly four days faster than the fastest sea-crossing (on the Blue Riband-holding, 26-knot *Mauretania*) – with an average speed of 118 mph (139 kph) for 1900 miles (3040 km). During the flight, Brown clambers numerous times along the wing to clear snow from the air intakes of the engines. It's the dawn of a new civilian flying era, though it's another eight years before the next Atlantic crossing, solo, by Charles Lindbergh (*21 May*). As with all early transatlantic attempts, it is made from west to east, to capitalize on the prevailing south-westerly winds.

'The deluge fell ... near eight and forty hours without intermission ... we will affect to have a summer, and we have no title to any such thing ... we get sore throats and agues ... then they cry this is a bad summer – as if we ever had any other! The best sun we have is made of Newcastle coal.'
Horace Walpole, writer, 1768

'No rain had fallen for nearly a month, and our dry soil had become a hot dust ... A burning wind ... had brought quantities of noisome blight ... Bushes of garden Roses had their buds swarming with green-fly, and all green things, their leaves first coated with viscous stickiness, and then covered with adhering wind-blown dust, were in a pitiable state of dirt and suffocation. But ... in the night [rain] came down on the roof in a small thunder of steady downpour. It was pleasant to wake from time to time and hear the welcome sound, and to know that the clogged leaves were being washed clean, and that their pores were once more drawing in the breath of life, and that the thirsty roots were drinking their fill. And now ... how good it is to see the brilliant light of the blessed summer day, always brightest just after rain, and to see how every tree and plant is full of new life and abounding gladness; and to feel one's own thankfulness of heart, and that it is good to live, and all the more good to live in a garden.'
Gertrude Jekyll, gardener, 1900

30.7C Brampford Speke 1896

16 JUNE

9.7C min / 18.8C max

How do you make a day of ordinary weather extraordinary? Get James Joyce to describe it. Today is Bloomsday, the day in 1904 on which Joyce sets *Ulysses* in Dublin. Records show that the weather is . . . well, pedestrian: west-south-west wind, Force 4, cumulo-stratus clouds, patches of sunshine and rain approaching.

To Joyce, however, Bloomsday begins with 'warm sunshine marrying over the sea' as Stephen Dedalus departs the Martello Tower at Sandycove for the last time. Before long, 'A cloud began to cover the sun slowly, wholly, shadowing the bay in deeper green.' When Leopold Bloom walks to the butcher's shop, 'The sun was nearing the steeple of George's church. Be a warm day I fancy. Specially in these black clothes feel it more,' he notes.

The day turns sultry. When Bloom is on the way to a funeral at Glasnevin cemetery, he leans out of the carriage:

– The weather is changing, he said quietly.
– A pity it did not keep up fine, Martin Cunningham said.
– Wanted for the country, Mr Power said. There's the sun coming out again.
Mr Dedalus, peering through his glasses towards the veiled sun, hurled a mute curse at the sky.
– It's as uncertain as a child's bottom, he said.

At dusk, 'The summer evening had begun to fold the world in its mysterious embrace,' but later it breaks: 'The wind sitting in the west, biggish swollen clouds to be seen as the night increased and the weatherwise poring up at them and some sheet lightnings at first and after, past ten o'clock, one great stroke with a long thunder and in a brace of shakes all scamper pellmell within door for the smoking shower.'

Dedalus is terrified (as Joyce was) by thunderstorms: 'A black crack of noise in the street here, alack, bawled back. Loud on left Thor thundered: in anger awful the hammerhurler. Came now the storm that hist . . . and his heart shook within the cage of his breast as he tasted the rumour of that storm.' In the early hours, Bloom and Dedalus meander home, 'the temperature refreshing since it cleared up after the recent visitation of Jupiter Pluvius'.

31.0C Southampton 1893

17 JUNE

10C min / 19C max

WATERLOO WEATHER 'Had it not rained on the night of 17–18 June, the future of Europe would have been different,' observes the commentator in Victor Hugo's *Les Misérables*. The reference is to the monumental rainstorm that floods the battlefield at Waterloo before Wellington and Napoleon engage.

'It was a misty, heavy evening. There was a sense of blight in the air; the flowers were drooping in the garden, and the ground was parched and dewless. The western heaven, as we saw it over the quiet trees, was of a pale yellow hue, and the sun was setting faintly in a haze. Coming rain seemed near.' *Wilkie Collins,* The Woman in White, *1860*

'So miserably hot, that I was in as perfect a passion as ever I was in my life at the greatest affront or provocation. Then I sat an hour, till I was quite dry and cool enough to go swim; which I did, but with so much vexation, that I think I have given it over: for I was every moment disturbed by boats, rot them ... The only comfort I proposed here in hot weather is gone; for there is no jesting with those boats after 'tis dark ... Pox take the boats!' *Jonathan Swift, writer and satirist London, 1711*

'God, how it rains – A pigeon has come to live with me. I think it must be the ghost of the dove that flew out of the Ark.' *Vita Sackville-West, writer, Sevenoaks Weald, Kent, 1926*

Waiting for the ground to dry before launching the French offensive, Napoleon – who allegedly brags that beating the Allied armies would be 'nothing more than eating breakfast' – critically provides time for Marshal Blücher's Prussians to join Wellington's forces.

The same rain soaks the British Isles today. In fact, the whole of June is unsettled and wet, and temperatures throughout the summer are well below average. The decade of 1810–19 is the coldest since the 1690s (*7 November*). Poor harvests, exacerbated by the end of the grain trade with Europe during the Napoleonic wars, bring great hardship in Britain. Not long after Waterloo, Parliament passes the protectionist Corn Laws (*23 January, 5 May*), prohibiting the import of cheap wheat to keep the price of cereals and bread artificially high. When the nation should be celebrating Wellington's great victory over the French, grain prices are so high that widespread strikes and food riots break out.

Raunds – no, not a cut of beef, but a small market town in Northamptonshire (♥) – has the unusual distinction of recording the highest high today (27.8C) and the lowest low (-18.3C on 13 December) in the same year, 1920.

185.0mm Honister pass 1972

33.9C Little Massingham 1917

THE CONTRO-VERSIAL LIGHTNING 'ATTRACTOR'

During a storm today in 1764, the elegant steeple of St Bride's church in Fleet Street (🌑), described by the poet W. E. Henley as 'flight on flight of springing, soaring stone', is struck by lightning. 'One stone forced out of its place,' a contemporary records, 'broke through the roof of the building, another fell on top of an adjoining house; and many pieces of broken stone and shivers were scattered.'

'England, I charge thee, dress thyself in smiles for my sake! I will celebrate thee, O England! And cast a glory on thy name, if thou wilt remove for me thy veil of clouds.' *Mary Shelley, author, London, 1824*

The event has huge consequences. St Bride's is a site of antiquity. The first Christian church is built here in the sixth century and seven churches follow before Wren completes his tallest spire, a 234-foot (71-metre) 'madrigal in stone'. There is, then, public clamour when it loses 8 feet (nearly 2.5 metres) to a lightning strike. King George III consults the American polymath Benjamin Franklin, famous for his groundbreaking experiments conducting lightning. A decade earlier, Franklin revealed 'the sameness of the electrical matter with that of lightening'. By attaching 'upright Rods of Iron, made sharp as a Needle and gilt to prevent Rusting, and from the Foot of those Rods a Wire down the outside of the Building into the Ground,' Franklin surmises, 'would not these pointed Rods probably draw the Electrical Fire silently out of a Cloud before it came nigh enough to strike, and thereby secure us from that most sudden and terrible Mischief!' The destiny of tall buildings is to be changed for ever.

The concept of a lightning 'attractor' or rod, which renders the uncontrollable force of a storm harmless, is hugely controversial. It challenges a fundamental aspect of Protestant and Catholic theological meteorology – that storms are delivered by the 'Prince of the Power of the Air', Satan himself, and the only means to resist them is through prayer. Franklin is an arch-infidel, according to people like John Wesley. In the heated eighteenth-century struggle between science and religion, the lightning rod is an excellent weapon. And it works.

32.2C Ochtertyre 1893

Curiously, once the principle is accepted, it's the form that lightning conductors should take that becomes the subject of a national debate. The King has conductors with blunt ends fixed to his palace; Franklin (supported by the Royal Society)

☼ 18 JUNE

THE NEEDLES

I have never anywhere enjoyed weather so delightful; so warm and genial, and yet not oppressive, the sun a very little too warm while walking beneath it, but only too warm to assure us that it was warm enough. And, after all, there was an unconquered freshness in the atmosphere ... I suppose there is still latent in us Americans ... an adaptation to the English climate, which makes it like native soil and air to us.'
Nathaniel Hawthorne, American novelist and US Consul in Liverpool 1853–7, Leamington Spa, 1855

insists that pointed ones are more effective. As the devices begin to be installed after 1764 (atop St Paul's in 1770–1), the satirical press delight in the battle for the spires of the city – between 'blunt, honest George' and the 'sharp-witted colonist'.

The same storm of 1764 also reconfigures a famous coastal feature. The Needles, a row of jagged, chalk stacks, takes its name from the distinctive needle-shaped 120-foot (36.6-metre) pinnacle called Lot's Wife. In today's storm, it topples. The remaining stacks look more like chipped teeth than needles, but the name sticks.

Before

After

19 JUNE

10C min / 19.1C max

'Summer, June summer, with the green back on earth and the whole world unlocked and seething – like winter, it came suddenly and one knew it in bed, almost before waking up; with cuckoos and pigeons hollowing the woods since daylight and the chipping of tits in the pear-blossom. On the bedroom ceiling, seen first through sleep, was a pool of expanding sunlight – the lake's reflection thrown up through the trees by the rapidly climbing sun . . . 'Outdoors, one scarcely knew what had happened or remembered any other time. There had never been rain, or frost, or cloud; it had always been like this. The heat from the ground climbed up one's legs and smote one under the chin. The garden, dizzy with scent and bees, burned all over with hot white flowers, each one so blinding an incandescence that it hurt the eyes to look at them.' *Laurie Lee, poet,* Cider with Rosie, *1959. Lee moved to the Cotswold village of Slad (🍐) in 1917, aged three (14 December)*

'It was swelteringly hot, so we emptied ourselves into deckchairs in Hyde Park, and surveyed the scene round the Serpentine. You would have been appalled. The slightest change in the temperature seems to give City dwellers an excuse for filth. Londoners take off their clothes at the drop of a hat. Or a rise in the barometer. They were sprawled everywhere. The queens making NO concessions at all. Lying there in those terrible Vince Man Shop Briefs which are totally indecent, with their clothes piled beside them; the stained underwear on top.' *Kenneth Williams, actor, to Joe Orton, dramatist, 1967*

'Charles the Second declared a man could stay outdoors more days in the year in the climate of England than in any other. This was very like a king, with a palace at his back and changes of dry clothes.' *Robert Louis Stevenson,* Kidnapped, *1886*

'The weather is fair and warm; so that the public-houses on the road are pouring out their beer pretty fast, and are getting a good share of the wages of these thirsty souls. It is an exchange of beer for sweat; but the tax-eaters get, after all, the far greater part of the sweat; for, if it were not for the tax, the beer would sell for three-halfpence a pot, instead of fivepence.' *William Cobbett, radical journalist, MP and champion of rural society, St Albans, 1822*

32.8C Southampton 1893

20 JUNE

10.2C min / 18.9C max

LIGHTNING-STRUCK AND STEEPLE-LESS ASHTON

'The lightning fell on [the Reverend] Mr Pitcairn's right shoulder, made a hole in his coat . . . descended from thence down the lower parts of his body . . . and produced the sensation of a cord, tied close about his waist. A violent pain in his loins immediately followed; and from thence to his extremities there seemed to be a total stoppage of circulation, all sensation being lost . . . Besides shivering the glass of his watch, the lightning melted a little of the silver of it, and a small part also of half a crown in his pocket . . . From the middle of his thigh the lightning . . . went down the under side of it to the calf of his leg, and so to his shoe, which was split into several pieces . . . his face was blackened . . . his body was burned . . . and he lost in some measure the use of his legs for two or three days; . . . What is remarkable, Mr Pitcairn remembers very well to have seen the ball of fire for . . . a second or two, after he found himself struck.'

'On a very gloomy dismal day, just such a one as it ought to be, I went to see Westminster Abbey.' *Karl P. Moritz, Prussian clergyman visiting England, 1782*

In another part of this account of today in 1772, read before the Royal Society a year later, the two hapless clergymen involved describe the ball of fire (*12 July*) as 'the size of a sixpenny loaf, and surrounded with a dark smoke; that it burst . . . like the firing of many cannons at once; that the room was filled with the thickest smoke; and that they perceived a most disagreeable smell, resembling that of sulphur . . . Mr Wainhouse providentially received no hurt, except a slight scratch . . . from the broken glass . . . and a kind of stupefaction . . . and a continued noise in his ears.'

Curiously, it's by no means the first time that the parish of Steeple Ashton (♀), in Wiltshire, has been a lightning victim. In July 1670 the 93-foot (28-metre), 250-year-old church steeple, from which the village so proudly takes its name, is zapped. No sooner are repairs approaching completion than, in October, it happens again. This time the lightning, a plaque in the church explains, 'threw down the Steeple, and killed the two men Labouring thereon'. Evidently disheartened, the villagers this time do not rebuild, and Ashton remains steeple-less – perhaps wise, given that, in 1973, it happens *again*. 'MR BOLT', as the *Bath and Wiltshire Evening Chronicle* headlines, 'SURVIVES THUNDERBOLT'.

31.7C Camden Square 1936

21 JUNE

10.1C min / 18.9C max

LONGEST DAY / MIDSUMMER NIGHTMARE

Today (or tomorrow – the date oscillates) is the summer solstice, the longest day of the year. On a perfectly clear day, at the latitude of Birmingham, this means we receive our maximum possible daily dose of 16 hours and 50 minutes of the warming golden rays. Penzance in Cornwall receives less, Wick in northern Scotland almost 90 minutes more. Is this, then, Midsummer's Day? Actually, no. Because our terrestrial orbit round the sun is not circular, but slightly egg-shaped, our calendar is slightly out of synch with our astronomy. Midsummer's Day is 24 June.

The temptation, therefore, is to report tales of naked revellers and druids at Stonehenge missing sunrise because of pouring rain (as in 2006). We won't, however, because contemporary thinking suggests Stonehenge is less to do with the summer than with the winter solstice (the stones line up better). At somewhere like Cape Clear (☞), the 5000-year-old Stone Age tomb off southern Ireland, the stones line up a good deal more convincingly – again, pre-supposing the sun's rays get through.

'The seasons alter; hoary-headed frosts / Fall in the fresh lap of the crimson rose . . . The spring, the summer, the childing autumn, angry winter change / Their wonted liveries, and the mazed world / By their increase now knows not which is which' – Titania, Queen of the Fairies in Shakespeare's *Midsummer Night's Dream*, delivers her celebrated 'bad weather speech'. Although the play is set in Ancient Greece, Titania's words so clearly describe the disastrous run of British summers from 1594 to 1597 that it's generally agreed to be one of Shakespeare's few directly topical references. One of the specific features of the summer of 1596 – bad beyond Elizabethan experience – is the grain rotting in the fields before it ripens.

The food shortages spark riots in larger towns and protests in smaller ones, including Stratford where Shakespeare has recently bought a house. The authorities try to restrict maltsters thought to be squandering what little grain there is on making beer rather than bread. Others – including, in 1597, Shakespeare himself – are accused of hoarding to profit from soaring prices. The situation, prompting an urgent desire for social stability, leads to the Poor Law Act of 1601 whereby each parish becomes responsible for its own poor – the first coherent step towards a social security system.

112.0mm St. Albans 1936

31.7C Camden Square 1936

22 JUNE

10.3C min / 19.2C max

'WELLINGTON DEFEATED . . .' A fog story: on 18 June 1815, Wellington wins his historic victory over Napoleon at Waterloo. The news is received at Plymouth (🌦) and telegraphed to London by the system of semaphore signals in place at the time. Dense fog, however, obscures the second half of the transmission from the signal station on the tower of Winchester Cathedral (🌦). Thus, the message received in London is: 'WELLINGTON DEFEATED . . .' The capital is immediately plunged into gloom, until the fog clears and the whole of the message gets through: 'WELLINGTON DEFEATED THE FRENCH . . .', etc. Popular and charming as the tale is, it's apocryphal, of course. There are any number of reasons to discount it – the story is told of several victories during the Napoleonic Wars; no such signal system operates at Winchester Cathedral; the message would anyway have come via the shortest route from Waterloo, by Calais and Dover.

'The wind raves like ten demons at the window'. *William Hazlitt, essayist, 1823*

'Before leaving I touched an electric button by which I started a message which was telegraphed throughout the whole Empire: "From my heart I thank my beloved people. May God bless them!" At this time the sun burst out'. *Queen Victoria, on her Diamond Jubilee, 1897*

Besides, the story reported in today's 1815 *London Gazette* is almost as good: 'News of the Battle of Waterloo was rushed to London by Harry Percy, Wellington's only surviving unwounded ADC. He carried the despatch in a velvet handkerchief sachet an admirer had thrust into his hand as he hurried from the Duchess of Richmond's famous Brussels ball on the eve of battle. He had no sleep that night, nor the five nights following, and had to row himself ashore from the middle of the Channel [presumably because there's no wind]. His scarlet and gold tunic was still torn, dirty and blood-stained when he burst into a St James's ballroom, a captured French standard in each hand, and dropped to one knee before the Prince Regent. It was Shakespearian.'

This, June's mean wettest day in central England is, in 1799, the start of almost five months of continuous rain (with only eight rainless days between now and 17 November). In the 'perfect summer' of 1911 (*22 July*), poet Siegfried Sassoon turns down an invitation to attend the coronation of George V in favour of playing cricket in Kent, and the match is rained off. And in the great summer of '76, this is when the real heat kicks in.

33.9C Tottenham 1941 ☀

THE GREAT STINK / WELLINGTON'S NOOKY

Today in 1858 London is in the grip of a heatwave, the hottest for forty years. *The Times* reports the mean temperature of the Thames as 70F (21C) and 'favourable to evaporation'. Indeed, the Thames *is* evaporating. The result, as *The Times* also reports, is that it 'stinks most abominably'. Novelist and future Prime Minister Benjamin Disraeli's analysis is more pungent: 'a Stygian pool reeking with ineffable and unbearable horror' – the 'Great Stink' is on.

It's hard to emphasize just how bad the smell is. There's talk of moving the Law Courts to Oxford. The windows of the Houses of Parliament are draped with canvas sheets soaked in chloride of lime – some measure given that, as a witness later informs the hastily convened Select Committee on the River Thames, when this was tried last year 'the smell seemed rather more disagreeable than the smell of the river'. The heat-wave merely exacerbates a fundamental problem: the Thames, London's source of drinking water, is an open sewer. Not only has the capital's population doubled between 1800 and 1840, but an 1847 commission has abolished cesspits. The disastrous consequence is that 270,000 houses now sluice excrement directly into the river, with the discharge vastly increased by widespread adoption of the new, flushable 'water closet'. Evaporating in the heat, the Thames is now just concentrated raw sewage. As it oscillates to and fro – up with the tide, down with the ebb – directly under Parliament's nose, the result, finally, is action. Within a month, a Bill for the purification of the Thames is passed. The following year Joseph Bazalgette begins to effect his grand plan for a three-tier drainage scheme north and south of the river – still the heart of London's sewers today.

Meanwhile, on a night of pouring rain in 1814, the Duke of Wellington, back in London after a successful Iberian campaign, is hunting for sex. According to the courtesan Harriette Wilson, he arrives to find her ensconced with the Duke of Argyll. 'Come down, I say,' roars Wellington. 'Don't keep me here in the rain, you old blockhead.' Argyll, pretending to be the madam of the house, shrilly claims he can't recognize the great general. Wellington takes off his hat and holds his dripping face towards the light. 'You old idiot, do you know me now?' Argyll – as the madam – claims he does not. Wellington storms off, cursing 'all the duennas and old women that ever existed'.

32.2C Maldon 1976

24 JUNE

10.5C min / 19.1C max

MOSSDALE CAVING DISASTER

A run of dry weather means that the ground is hard as concrete when ten pot-holers, mainly from Leeds University, set out on a caving expedition in Wharfedale, North Yorkshire (♥), today in 1967. The Mossdale caverns are the potholer's equivalent of the North Face of the Eiger, a 6¼-mile (10 km) system graded V, or 'super severe'. Although no special equipment is required, much of the system involves crawling and squeezing along narrow passages that flood completely, after even small amounts of rain. The morning's forecast mentions the phrase 'thundery trough', but otherwise it hardly sounds ominous – and June has been so parched that thoughts of rain are the last thing on anyone's mind as they descend into the cave's dry mouth.

'There was not a breath of wind. It was the loveliest hour of the English year: seven o'clock on Midsummer Night.' *Stella Gibbons,* Cold Comfort Farm, *1932*

After three hours, four of the party have had enough. They surface and head for home, leaving no lookout above ground to signal the rapidly darkening sky. Rain starts to fall heavily, streaming off the hard, sun-baked earth crust. Without porous soil to absorb the flow, the underground streams swell rapidly. When one of the four pot-holers who surfaced early returns a short time later, she finds, to her horror, the beck full and the entrance to the cave already submerged.

By the time rescue teams arrive, access to the cave is already impossible. Trenches are dug to divert the flow, but for hours the system remains blocked. Only the following afternoon are the bodies of five men finally brought out from the Far Marathon Crawls, deep within the cave system. The sixth body is discovered the next day. After the inquest, the coroner orders the safe entrance to the cave to be sealed with concrete (it reopens in 1976). The accident, the worst in British caving history, remains the supreme lesson of the hazards of thunderstorms to cavers.

32.4C Gillingham 1976

25 JUNE

10.6C min / 19.4C max

'We were only half-way across the moor when a vicious jab of lightning almost blinded us and a great crash of thunder directly overhead seemed to rock the fellside. The first rain spattered on the dusty rocks and then the quick rat-tat-tat of hail and, within seconds, we were in the middle of a deluge. In its way it was invigorating and even entertaining for once you are wet through to the skin – this took about half a minute – there is no real discomfort. Within minutes the dried-up fellside was live with a hundred rivers and we splashed down and along them, knowing we could get no wetter, while the torrent poured down, hailstones and bathtubfuls of them, the lightning flashed and the thunder crashed and rolled. A good end to a good day.'
A. Harry Griffin, climber and journalist, Lake District, 1960

'Afternoon, practise forced landings in a hired farm field with Cowling. Then after tea in the Doc's old Morris to catch the last hours of lovely weather and swim from the river bank down by The Haycock. Lots of daddy-longlegs skittering noiselessly about our rooms, and a plague of earwigs is upon the land.'
Frank Tredrey, Pilot's Summer, *1939, a diary of a flying instructors' course at RAF Central Flying School, Wittering, 1935*

'If ... I had been sure on my own evidence that Gos [his goshawk] still lived, then no mist nor petty rain nor mean, unholy thunder would in the least particle have deterred a rejoicing patience. But in the worst summer within memory, to sit cramped day long without tobacco (perhaps the most maddening deprivation) in damp and cold and hunger unoccupied, in a place where one was only sure that hawks had been, with tools which one had never known to be successful, on an old quest which might from the start have been ridiculous, while all one's letters for help round Europe went not only without help but even without answer; it was bitter work ... Meanwhile it dewed, and stormed, and thundered.' *T. H. White,* The Goshawk, *1951*

'Rain-bow. Rock-like clouds. Sweet evening. Moonshine.'
The Reverend Gilbert White, 1771

'We ... had pictured ourselves sitting in the sunshine making tea by the cairn, while we ... lay on our backs and sleepily gazed at the clear blue sky through the heat haze. The reality was very different ... A wintry day in the summer is very like a wintry day in the winter.'
J. H. Bell, climbing Ben Nevis by the Tower Ridge, 1910

Sheffield and Hull floods 2007

33.5C East Bergholt 1976

26 JUNE

10.6C min / 19.5C max

WIMBLEDON
WEATHER:
'OUTBREAKS
OF CLIFF
RICHARD
LATER'

When the temperature inside Centre Court at Wimbledon hits 43C, today in 1976, four hundred tennis fans faint. This is the middle of the hottest tournament ever. The shade temperature reaches 32C each day, and, such is the risk of sunstroke, the All England Club threatens to remove any spectator without a shirt. Aptly, it is Bjorn Borg, 'the Iceman', who wins, on a baked, cappuccino-brown court. Wimbledon, of course, is inextricably linked with weather. But the weather most readily associated with the green grass of SW19 is not sunshine. In 129 years of the tournament, just five years have not been affected by rain.

Average June rainfall over south-west London is 45mm (1.8 inches), making rain at Wimbledon one of the surer things of a British summer. At the first Championships, in 1877, contested by twenty-two men ('requested not to play in short sleeves when ladies are present'), the final is postponed because of rain. At the first Wimbledon on the current Church Road site, in 1922, it rains on the opening day in front of King George V, and on every day thereafter. The courts become quagmires and the finals finish a week late.

More recently, rain (and hail) affect six of the first eight days in 1980. In 1991, only fifty-four of the 240 matches are completed in the first week, and play is scheduled on the middle Sunday for the first time. In 1996, Cliff Richard tries to lift the gloom on a grey Centre Court with a rendition of 'Singing in the Rain', prompting a cartoon captioned: 'Showers over Wimbledon may lead to outbreaks of Cliff Richard later'. In 2001, rain stops Tim Henman in full flight, denying him a place in the final (*6 July*). In 2002, a spectator has enough of the rain delays and streaks around the court: 'Perhaps he misinterpreted the umpire's call for "new balls"', a journalist wonders. In 2004, one of the wettest years of all, only three days are unaffected by rain. On the first day of the 2006 tournament – the tenth in a row to be disturbed by rain – 248 umbrellas are sold.

The weather also influences how the grass plays: when it's damp, players slip; on dry courts, the ball bounces higher and faster. In many ways, it's Wimbledon's weather-dependence that makes it the most exciting tennis tournament of them all. Shame about the retractable roof that arrives over Centre Court in 2009.

35.4C North Heath 1976

27 JUNE

10.7C min / 19.4C max

Glastonbury secures its reputation as a mud-fest today in 1997. There's been so much rain that oil drums and tents float on a brown sea across Worthy Farm, the eight-hundred-acre festival site in the Vale of Avalon (☉). Ninety thousand magnificently ill-equipped music fans are undeterred, however. Hundreds of tents are submerged (those not blown away in strong winds). There are rumours that the main Pyramid Stage is sinking. In the Dance Tent, thousands of twitching feet blend the earth into brown soup. The 'sewage gulper' is sent in to drain the sludge but somebody accidentally turns it on to blow instead of suck. Outside activities include mud-sliding, mud-surfing and naked mud-healing sessions. As cow pats have been churned into the mud which is ingested, eight festival-goers are hospitalized in isolation units after contracting E. coli. Despite this – or perhaps because of it – Radiohead's headline gig on Saturday night is rated one of Glastonbury's all-time greatest.

British summers hardly seem suitable for festivals at all, yet Glastonbury is the largest 'greenfield' music and performing arts festival in the world. Certainly it offers a unique experience: fans can see the best live music acts, while simultaneously catching diseases unheard of since the First World War. At the 1998 Glastonbury, trench foot is the prevailing condition as the rain once again converts the site into a quagmire. Fortunately, sufferers have left before gangrene sets in. Drainage is improved markedly for the following year. In 1999, two days of baking sunshine only end when Travis takes to the main stage on Saturday evening and starts singing 'Why Does It Always Rain on Me?'

In 1982, the highest rainfall for a single day in forty-five years is recorded on the Friday. In 2004, high winds prior to – and steady rain during – the festival ensure that the usual phantasmagoria of colour is reduced to green and blue cagoules. In 2005, 150,000 gather to see 385 live performances – and a month's worth of rain falls in a few hours. Lightning strikes the acoustic tent, after which the Undertones take the stage and sing 'Here Comes the Summer'. People swim to rescue their belongings and one bar sinks in the mud. It's generally agreed to be a vintage year. The chances of the Glastonbury Festival being rained on stand at, roughly, one in two: yet, 137,500 tickets to the 2007 festival sell out in under two hours.

35.5C Southampton 1976

28 JUNE

10.9C min / 19.9C max

The highest-ever June temperature – 35.6C (96.1F) – is recorded, today, in 1976, in Southampton. The sunniest part of mainland Britain is the stretch of coast between Sussex and the Isle of Wight. Here, in an average June, over eight hours of sunshine can be expected every day. Some Junes, the average has been as high as ten hours per day. Weymouth is, on average, the sunniest place on the mainland, and the sunniest island location is St Helier, Jersey (200 hours of sunshine in June 2001).

It is here, in this most consistently sunny strip from Margate to Portsmouth, that the great British bucket and spade tradition has its roots. Beach huts, sandcastles, lidos, Punch and Judy, piers, stripey windbreaks, miniature steam railways, Fawlty Towers-style hotels (actually based on the Gleneagles Hotel, Torquay), crazy golf, candy floss – the whole caboodle starts here, in the sunshine. The early railways facilitate mass travel, and Thomas Cook organizes the first 'seaside excursion' in 1843. The first Bank Holiday is declared in 1871. By the 1930s romantic railway posters are proclaiming that 'Summer Comes Soonest in the South' and 'The Sun Shines Most on the Southern Coast' before the advent of the jet age – and *real* sun – bring Costa Britannica to a cloudy end.

'Up at 9 shower bath, out in the garden for a few minutes & came in with an apologetic feeling, heat being intense. After, worked all day hard at the sketch of Last of England – then in the garden about 8 P.M. & came in with the same sensation as this morning. Lay on the sofa with shirt unbuttoned & vinegar compresses for about 1 hour, felt numb at the left extremeties (7 hours)! *Ford Madox Brown, artist, 1855*

'On a fine evening there are very few pleasures comparable to driving a light, open [Rolls-Royce Silver] Ghost on country roads. Some will get it from waiting for salmon to take, in dark peat pools, but I am too impatient, and can't stand the midges. In a Ghost you waft along, high enough to look over people's hedges, noiseless enough (as was the original intention) to leave horses unscared ... I drove for about forty minutes and on my return took a jug of iced lime juice and soda water to the music room where I played the piano, quite competently, until the light faded. A day filled with trivia, but douceur de vivre also! *Alan Clark, politician, Saltwood Castle, Kent, 1987*

242.8mm Bruton 1917
35.6C Southampton 1976

THE FLYING COACHMAN / NOCTILUCENT CLOUDS

We don't know the exact local conditions today in 1853 on Sir George Cayley's Brompton Hall estate, near Scarborough (♠). But we know he's picked the moment carefully, and the light breeze he requires is blowing across Brompton Dale, where he conducts all his glider trials. Like so many of the early flying experiments by this pioneering Yorkshireman, he gets it right. Everyone thinks the Wright Brothers are the first to fly in a heavier-than-air craft (i.e. not a balloon), but they aren't. Fifty years earlier, on this breezy day, in a glider Cayley's designed himself, the first-ever manned flight in a heavier-than-air craft takes place. Who is the pioneering pilot to claim this glorious honour? The answer is Cayley's coachman, one John Appleby. He gives notice to terminate his employment immediately on landing.

'Sit out in the white garden ... till 10 pm planning improvements ... A perfectly still, breathless evening, scented and warm.' *Vita Sackville-West, writer, Sissinghurst, Kent, 1953*

'Fine day, intermittent cloud and sun, but cold wind. Haven't yet worn summer clothes.' *James Lees-Milne, writer, 1981*

Cloudspotters should keep their eyes open at night for a collector's item that only appears at this time of summer: the noctilucent cloud. This mysterious creature is the only cloud lit by the sun *at night* – hence its name. It appears as silvery or milky blue, luminous ripples, so thin they are scarcely noticeable except against a dark, night sky. These are also the highest clouds, which is how they manage to reflect sunlight even long after sunset. Made of ice crystals, they appear now because this, the warmest time of year at ground level, is the coldest season at the top of the mesosphere, 30–50 miles up, on the fringes of space. Look for them in the northern sky after sunset.

35.6C Camden Square 1957

30 JUNE

10.7C min / 19.9C max

THE BLACK DEATH Bring out your dead! During a summer of rain, 'scarcely stopping by day or night', in 1348 there are fears for the harvest. But famine becomes a secondary concern when, today, the first English victim succumbs to one of the greatest pandemics in human history – the Black Death.

'To supper at the Camden Brasserie. It's a hot night and the shutters have been folded back so that the room opens directly on to Camden High Street. In France or New York this would excite no comment. In London, or in Camden Town at any rate, it draws the jeers of every passing drunk.' *Alan Bennett, writer, 1985*

'It is 90° around the house. (I can still only think in terms of Fahrenheit, yards, pints and ounces.) We opened all doors and windows to get a bit of air. Then I had the wheeze of changing into a nightshirt, pretending I was in the tropics ... The afternoon postman didn't turn a hair.' *Alec Guinness, actor, 1995*

The disease spreads quickly, decimating a malnourished population. There are two forms. The first and more virulent, pneumonic plague, attacks the lungs. Victims cough up blood and die quickly. The more common form, bubonic plague, includes noisome swellings (or buboes) in the groin and armpit. A sudden coldness is followed by prickling sensations and depression, over several days. Then death. For contemporary chroniclers, the Great Mortality heralds the end of the world. Within eighteen months, a third of the English are dead.

It's thought that the plague arrives in Europe following a monumental flood in Asia, which destroys the habitat of the black rat, driving the rodents west. The epidemic subsides in 1350 (though it doesn't die out for centuries – *10 November*). It's the persistence of the plague through the winters, however, that has surprised historians. Ordinarily, British winter temperatures should be cold enough to kill off the rat flea, which carries the bacterium. But frosts fail to materialize in the exceptionally mild winters of the late 1340s: 'From October until February ... there was not so much ice as would support the weight of a goose.' A report on the plague by the University of Paris in October 1348 also cites the 'unseasonable weather' as 'a particular cause of illness ... it is because the whole year was warm and wet that the air is pestilential.'

30.8C Knockarevan 1983

33.8C Barbourne 1995
127.0mm Princetown 1932

The consequences of the Black Death are profound. There are simply not enough people left to work the land. Wages and prices rise, the power of the Church declines (the priesthood is decimated), social mobility increases, feudalism begins to unravel and the authority of the Crown creaks.

JULY

1 JULY

11.2C min / 20.1C max

INSIDE THE
KING OF
CLOUDS

The arrival of July heralds high summer, the warmest month (usually) and the month with the hottest day (about 44 per cent of the time). July ties with August as the least windy month. A record frost comes to Norfolk today in 1960 (air temperature of -1C), while 1961 brings the highest temperature of the decade (33.9C). 'Weirdness' describes 1968: baking 33C heat in the south-east; almost total darkness from mid-morning in the Midlands and north-west; hailstones the size of tennis balls over Devon and at Cardiff airport. Then the hot North African airstream partly responsible for all this drops thousands of tons of red-orange Saharan dust as rain the following day over cars, pavements and washing in the south (*8 March*).

'I stopped ... to watch the wind's effect upon a field of unripe barley. Waves drove across it in all directions ... like hares frisking at play, chasing one another, retreating, changing course and double tracking. I could have watched for hours.'
James Lees-Milne, writer, Gloucester-shire, 1974

In 1957 the naturalist (and subsequently British Gliding Champion) Peter Scott, gliding over the Cotswolds, enters a cumulo-nimbus thundercloud to try for his Gold height badge. The air currents in a cu-nim are so fearsome that most pilots – even in Jumbos – do anything to avoid them. But glider pilots willing to brave the consequences can be carried up to 40,000 feet (12,200 metres). 'I would dearly have liked a reason for staying out in the friendly sunshine,' records Scott in his auto-biography, *The Eye of the Wind*. 'I ... plunged into the side ... trying to get into the darkest part, in the hope that it would be smoother ... a vain hope. There was a patter of rain, and later a patter of hail; and ... icing at the front of the canopy ... I was steaming up at 20 feet per second and bouncing about like a pea in a pod ... The altimeter crept up to 11,000, and the turbulence increased sharply ... Fascinated and seriously worried, I watched it go up 700 feet in about 30 seconds before, with a frightful bump, we flew into a violent down. A few moments later the air-speed shot up again, and once more with full air brakes out and 80 mph on the clock we climbed 700 feet; again ... a violent jerk and down; ... then suddenly we were out in the blessed evening sunshine. The panic was over. I tried to shut the brakes, but they were frozen open.' On landing, he finds that he has duly recorded a climb to Gold height.

34.8C Jersey 1952

2 JULY

11.1C min / 20.1C max

SWINGING
CRICKET BALLS

When temperatures reach 33C (91F) today during the 1976 heatwave, there's nearly a constitutional crisis in the Stewards' Enclosure at Henley Regatta (♥). Throwing 137 years of history to the wind, the Stewards take an unprecedented decision. Members and guests are permitted to remove their jackets inside the sardine-packed Stewards' Enclosure. It's the first time the dress protocol – which simply demands that people dress in accordance with established tradition – has been slackened. Ties and top buttons, naturally, must stay firmly fastened.

In similar heat in 1314, Edward II (*12 May*) makes a last, desperate attempt to restore his authority in Scotland. His markedly larger, better-equipped and very probably over-dressed army is cut apart by Robert the Bruce's soldiers in fierce hand-to-hand fighting at Bannockburn, near Stirling (✂).

'I am looking out upon a dark gray sea, with a keen north-east wind blowing it in shore. It is more like late autumn than mid-summer, and there is a howling in the air . . . The very Banshee of Midsummer is rattling the windows drearily while I write.'
Charles Dickens, author, Broadstairs, Kent, 1847

Meanwhile, heavy cloud and humid air play havoc with the ball at Edgbaston (☂) today in 1999. There's more swing than a 1970s dinner party as twenty-one wickets fall on the second day of the first Test between England and New Zealand. The Edgbaston wicket is renowned for jags and nips off the pitch, but the ball has never swung like this before. In stark contrast, day three is bright with a fresh breeze and the only thing swinging is Alex Tudor's bat. Batting conditions are so good that the 'Night Watchman' makes 99 not out, as England win comfortably by seven wickets.

There is no consensus among aerodynamic engineers as to why a cricket ball swings more in warm, humid conditions, but it does. The most widely accepted theory is that warm air holds more water vapour (saturated air at 20C has 3.6 times more water vapour than air at 0C). This affects the roughened side of the ball more than normal, reducing its critical speed and making it swing. Certainly, cricket grounds beside the sea like Hove and Swansea are notorious as batsmen's graveyards because of the moisture-laden sea breezes. Micro-climatic peculiarities like this at different county grounds are, of course, what give English cricket its charm.

148.1mm West Baldwin Reservoir 1968

35.7C Cheltenham 1976

3 JULY

11.3C min / 20.1C max

At 35.9C in Cheltenham, today is the hottest day of *that* summer, the summer of '76. It's the summer when, for ten long weeks, the sun never goes in, and we learn what it's like to be Mediterranean – when the roads melt, we share baths and day after day the news shows people queuing at standpipes and reservoirs reduced to cracked mud paving. For the first time in its 143-year history, the MCC at Lord's allows its members to remove their blazers – ties remain, obviously.

PHEW! WHAT A SCORCHER!
So when does the British summer's most overused headline first appear? It's often said to be in the *Sun* on 4 July 1976 – but it isn't. The answer seems to be, as with those other culture-defining headlines, 'FOG IN CHANNEL: CONTINENT CUT OFF' (*12 February*) and 'THE WRONG KIND OF SNOW' (*11 February*), less straightforward. Was it ever used? The *Sun*'s librarian can't find it – though of course it's been used thousands of times, by any number of papers, since.

The major heatwave begins during Wimbledon, in the last week of June (*26 June*). For over a fortnight temperatures exceed 32C (90F) somewhere every day. By the first week of August water is running out, and a drought bill is rushed through Parliament. Mains water in South Wales is switched off between 7 p.m. and 8 a.m. And still the sun shines. What's going on? A self-perpetuating high pressure area gets 'blocked' over the British Isles, like a boulder in a river, diverting the usual stream of Atlantic rains to the Med (which, satisfyingly, has a lousy summer). The hot weather continues (except in the north-west) throughout August until Prime Minister James Callaghan appoints a Minister for Drought (*24 August*). This – along with the looming August Bank Holiday – is the catalyst the heavens need. Thunderstorms finally arrive on the 29th, after which it rains and rains: all through September (the wettest since 1918) and October (the wettest on record).

The consequences? Apart from incessant heath and forest fires, the Notting Hill Carnival riots at the end of August are blamed on the heat (*7 July*). The financial cost is £500 million-worth of failed crops and £60 million-worth of subsidence-related insurance claims from home-owners living in clay areas (most of London). In 1977, a new National Water Authority is granted wide-ranging powers to co-ordinate national water management more effectively.

35.9C Cheltenham 1976

4 JULY

11.2C min / 20C max

We have today's heat in 1862 to thank for the story of *Alice's Adventures in Wonderland*. 'The beginning of *Alice* was told one summer afternoon when the sun was so burning that we had landed in the meadows down the river, deserting the boat to take refuge in the only bit of shade to be found, which was under a new-made hayrick. Here, from all three, came the old petition of "Tell us a story", and so began the ever delightful tale.' Thus recalls Alice Liddell – the eponymous Alice – later in life.

The question is: is this true? We know that the Oxford mathematics don the Reverend Charles Dodgson, aka Lewis Carroll, takes his three nieces Lorina (thirteen), Alice (ten) and Edith (eight) rowing on the Isis () from Oxford to Godstow today, as he records doing so in his diary. And in the seven-verse preface to the book, explaining how the story came about, he describes the 'golden afternoon' and 'dreamy weather'. The official weather records for the day, however, a little inconveniently show that it was cool and wet. Reputations are restored by the late Irish meteorologist H. B. Doherty, who has the idea of using the detailed coastal weather reports published by *The Times* to construct a weather map for the day. This shows that a front passes over Oxford in the early morning, with another moving in from the west in the late evening, both of which could have brought rain and cold. But in between, a transient ridge of high pressure could well have provided the balmy conditions that Dodgson and Liddell describe.

'Much too hot. I'm sitting legs apart, stayless, stockingless and floppy, in front of the window – it's boiling.'
Joyce Grenfell, writer and comedienne, London 1933

No record survives of the first BBC shipping forecast, read out at 10.30 a.m. in 1925 – all transmissions are live and tape recording lies far in the future. Although weather information has been available for eighteen months to vessels with radio-telephony equipment, the BBC broadcast is intended for small ships without such fancy kit. It is broadcast on long wave, the signal received most clearly at sea. By the 1980s, when almost all vessels have radios, the forecast is essentially redundant. By this time, however, this daily poem of weather and sea has become – especially to landlubbers safe in their beds – a heritage theme tune for an island nation. Dropping it is unthinkable.

34.1C North Heath 1976

5 JULY

11.4C min / 20.2C max

GUNTER'S ICES This muggy morning in 1822, fashionable Londoners read the following announcement in the *The Times*: 'Messrs. Gunter respectfully beg to inform the nobility and those who honour them with their commands that, having this day received one of their cargoes of ice by the *Platoff*, from the Greenland seas, they are able to supply their CREAM and FRUIT ICES, at their former prices.'

'Crossed in fog – watching the island grow on the radar screen, ourselves a blip ringed with concentric quarter mile circles. In mid-Sound, the boat broke into a sudden clear patch – astonishingly brilliant light and heat for a few moments, then plunged back into clammy grey.' *Christine Evans, poet, Bardsey Island, Lleyn Peninsula, Wales, 1997*

'After two dreary days alone at home I went to London ... It was dark with some drizzle after a stormy night, but I had sent £2 to the Poor Clares at Looe asking them to arrange good weather from 7 pm onwards and at 7 it cleared and remained fine throughout the night; a remarkable performance by those excellent women to whom I have sent another £3.' *Evelyn Waugh, author, 1956*

In normal years in the early nineteenth century, enough ice is obtained for fish dealers, confectioners and quaffers of champagne – from the winter harvest of lakes, ponds, purpose-built shallow reservoirs and Highland lochs – to last the summer. But following an unusually mild winter, ice-houses (instituted by Charles II in the late 1600s – sheets of ice were packed in salt, wrapped in strips of flannel, and stored underground until summer) are empty, pushing the price of ice cream up and forcing ships to make the arduous journey to Greenland to fetch ice. As the British Isles steadily warm up towards the end of the Little Ice Age (*3 March*), the international ice trade is growing.

So, too, is the British population's taste for ice cream. From their advent in the late seventeenth century until the early 1900s, Neapolitan ices are an exclusive pleasure of the well-to-do. The introduction of mechanical refrigerators in the 1890s popularizes the delicacy. Then the Wall brothers, a century after the arrival of the ice-bearing *Platoff*, have their titanic inspiration. They sell wrapped bricks of ice cream from an ice box on a tricycle. Alas, the summer of 1922 is notably cold.

On a very hot day in 1969, 250,000 gather in Hyde Park (♀) to see what is critically regarded as the worst-ever live Rolling Stones performance: in the heat, the fans simply take another layer off, but the thousands of white butterflies due to be released in tribute to deceased band member Brian Jones perish in an unventilated container.

34.4C Cromer 1959

6 JULY

11.3C min / 20.3C max

RAIN STOPS THE NEARLY MAN / DOG DAYS

On a cool evening in London SW19 in 2001, Tim Henman is on the brink of achieving his destiny – a Wimbledon final. Only the ageing Goran Ivanisevic stands in his way. Henman loses the first set, and scrapes through a tie-break to hold the second. Then, with a display of dazzling tennis, he thunders through the third 6–0 in just fifteen minutes, dismantling the Croat's game. As the fourth set begins, the first spot of rain falls.

Wimbledon, of course, wouldn't be Wimbledon without a little rain (*26 June*). The psychological stress of weather-interrupted matches is a well-documented quirk of the tournament. But still . . . not now. Not on Tim Henman, the 'Nearly Man' (he's lost in two previous semi-finals). Not when our boy is on the brink of becoming the first British men's finalist since 1938. With the match undoubtedly going Henman's way and the nation in a frenzy, at 6.18 p.m. rain stops play. It's a disaster. The spell is broken.

> 'The weather continued much the same way all the following morning; and the same loneliness, and the same melancholy, seemed to reign at Hartfield – but in the afternoon it cleared; the wind changed into a softer quarter; the clouds were carried off; the sun appeared; it was summer again. With all the eagerness which such a transition gives, Emma resolved to be out of doors as soon as possible. Never had the exquisite sight, smell, sensation of nature, tranquil, warm, and brilliant after a storm, been more attractive to her. She longed for the serenity they might gradually introduce.' *Jane Austen*, Emma, 1816

The remainder of the game is played out over two agonizing, wet days. After forty-five hours of rain, tears and intermittent tennis, Ivanisevic clinches the match. 'God wanted me to win,' he says. 'He sent the rains.'

The 'dog days' of summer begin today, according to the 1552 Book of Common Prayer – when your shirt sticks to your back, your knees are weak and lassitude envelops you. This breathless, sticky and wearying backside of summer is a phenomenon more common on the Continent, though it's just such a dog day on Merseyside (♪) today in 1957 when two pent up, teenage rock 'n' roll fanatics meet for the first time at a church fete and exchange guitar riffs. Paul McCartney is watching the Quarry Men Skiffle Group, led by John Lennon, when a thunderstorm breaks. Today in 1911, the 'dog days' unhinge one man who strips off as he walks to the local pub. And in 2006, keepers at Colchester Zoo in Essex are feeding the lions blood ice lollies to keep them cool.

34.3C Cheltenham 1976

7 JULY

11.4C min / 20.3C max

TOXTETH RIOTS It's hot and sultry in Toxteth, Liverpool (🌶) tonight in 1981, where they are, in Bob Marley's words, 'burnin' and a-lootin'. The riots have been grabbing headlines for four days and, despite the fact that the police are employing CS gas for the first time in mainland Britain, there's no sign of the trouble abating – or the weather breaking. Though no one would dispute that unemployment, racism, poor housing and the slow-burning tension between police and black communities are the root causes of the dramatic outbreaks of violence in Brixton and Toxteth this month, every riot has its spark. And hot weather can be it.

> 'Are all handsome men narcissistic? During these hot days all the young and not so young doff their shirts and work naked to the waist; and solicitous they look. They love exposing their bodies. The other day I met Mervyn the Badminton keeper in his jeep, naked to the navel. Had a chat with him. All the time he was caressing his torso, running his hands soothingly, affectionately across his breasts and navel ... Men are more interested in titillating themselves than women are.'
>
> *James Lees-Milne, writer, 1973*

Certainly, hot weather brings people on to the streets. And there is a well-established link between rising temperatures and violent crime (the murder rate in London peaks in the summer). Psychologists have also suggested that sunshine increases levels of serotonin released in the brain, one side-effect of which can be heightened aggression. Thus everything from road rage to race riots is more likely when the temperature rises. Most of the major late twentieth century civil disturbances in Britain occur either at the height of summer or in a heatwave – Notting Hill (August 1976, 19C), Toxteth (July 1981, 21C), Brixton (September 1985, 21C), Handsworth (September 1985, 23C) and Toxteth (October 1985, 22C).

Long before the local tourist board, Samuel Taylor Coleridge is in denial about the rainfall in the Lake District. In a letter written today in 1801, he lambasts William Coates 'who said to me at Bristol – "Keswick, Sir! is said to be the rainiest place in the Kingdom – it always rains there, Sir! I was there myself three days, and it rained the whole of the time." Men's memories are not much to be relied on in cases of weather; but judging from what I remember of Stowey and Devon, Keswick has not been, since I have been here, wetter than the former, and not so wet as Devonshire.' Actually the Lake District does have a high rainfall and Seathwaite (*8 May*) – all of 9 miles (14 km) from Keswick – is Britain's wettest place.

33.3C Cambridge 1893

8 JULY

11.3C min / 20.3C max

CAESAR
BLOWN AWAY
/ BRITISH
GOLF OPEN

In an easterly storm, Julius Caesar, camped on the Kent Coast near Walmer (♀) tonight in 54 BC, learns that forty of his ships have dragged their anchors and been driven ashore and wrecked. 'Neither the anchors and cables could resist, nor could the sailors and pilots sustain the violence of the storm,' he records phlegmatically. The forty-eight-year-old general may well be rueing his decision to take on the chariot-riding, chest-beating Britons once again: only a year before, his first invasion attempt is disrupted by the one thing even Rome can't control – the weather.

On that occasion, in August 55 BC, Caesar is waiting at the small encampment by the Channel for eighteen transport ships bearing his cavalry. Just as the sails come into view, Caesar records in *De Bello Gallico*, 'a storm suddenly arose that none of them could maintain their course'. The ships are wrecked or blown back to Boulogne. Facing a winter campaign without cavalry, clothing and equipment, the invasion is abandoned. Today's storm doesn't deter him, however. Rather than have a second immaculately planned expedition turn into fiasco, Caesar heads across the Thames and makes a face-saving pact with the king of the Catuvellauni tribe. The embarrassment rankles in Rome, though. When Claudius arrives, ninety years later, he lands at the head of an immense army.

In roughly the same place, almost two millennia later, a 60-knot gale tears across Royal St George's golf course on the final day of the 1938 British Open, whipping a canvas marquee into the air. One professional takes fourteen shots at the twelfth hole, giving hope to amateurs everywhere. Reginald Whitcombe makes the best of the conditions to beat Henry Cotton and win the Claret Jug.

Today in 1746, Flora MacDonald and Bonnie Prince Charlie, dressed as a lady's maid to elude the government forces after his failed Jacobite rebellion, encounter a storm as they sail a small boat across the Little Minch (☞) from the Western Isles to Skye. When they eventually make landfall, Flora is arrested. She becomes one of Scotland's heroines and the journey is immortalized in the 'Skye Boat Song'.

34.1C Camden Square 1941

9 JULY

11.6C min / 20.1C max

LONDON
LIGHTNING /
YORK
MINSTER

A day of thunder and lightning. In 1923 in London, between 1 and 5 a.m. 'men thought of France and war. The recollecting ear picked out those several sounds of musketry, machine-guns, field artillery and bursting shells . . . the birds woke without song' reports *The Times* of the 'grand terrors of the night'. It is the most dramatic lightning display ever recorded: 6924 flashes in six hours – almost one every three seconds for four hours.

More recently, in 1984, shortly after midnight, lightning ignites Britain's largest medieval cathedral, York Minster (♀). For some reason the alarm is delayed. Firemen don't arrive until 2 a.m., by which time the fire is raging through the ancient, tinder-dry oak roofing timbers of the thirteenth-century south transept. The rose window is gutted, though most of the stained glass (Britain's oldest) is salvageable. So what happens? Are smoke alarms set too low to give enough warning? Are lightning conductors (*18 June*) not properly earthed? A surprising number suggest that it is divine retribution following the controversial appointment of Dr David Jenkins, with his unconventional beliefs on the resurrection, as Bishop of Durham three days earlier (though it's noted that the Lord seems to have zapped the wrong cathedral). It's now thought that lightning rods were spaced too far apart, and the lightning jumped between them, striking the control box for the exterior floodlights. Repairs take four years, cost twice as much as estimated (£2 million) and include roof bosses designed by children in a *Blue Peter* competition.

'I certainly remember such a thing as dust: nay, I still have a clear idea of it, though I have seen none for some years, and should put some grains in a bottle for a curiosity, if it should ever fly again.' *Horace Walpole, writer, complains about the rain to the Earl of Strafford, 1770*

In 1846, a thunderstorm at Truro (⌖) floods a mine: in a 'few moments' thirty-nine drown. In 1923, the same day that London gets its lightning show, Carrbridge in Scotland loses its railway bridge for the second time (nine years earlier, in 1914, the first 'Carrbridge Cloudburst' washes away the bridge as a train is crossing, killing five). In 1959, radar coverage of the 'Wokingham Storm' in Berkshire, a hailstorm subsequently dubbed a 'supercell', leads to the formulation of new theories about how storms travel and form.

32.2C Bridgwater 1934

10 JULY

11.4C min / 20.2C max

DESERT YEAR / WETTEST DAY

Today, in 1921, it's 34C – the hottest day of one of the warmest and driest summers since the seventeenth century. Unusually, it's also windy. This spells disaster for farmers on all but the heaviest land, especially in East Anglia, where 1921 is remembered as 'The Desert Year'. The drought – almost a year in places – couldn't come at a worse time. Agriculture has been at a low ebb since 1870 (*6 September*) and though the First World War, with its need for high productivity, provides temporary respite, world prices now collapse – leading, in America, to the Great Depression.

'The slump set in during the great hot summer of 1921. I remember it well,' recalls Leonard Thompson, a farm worker, in *Akenfield* (♥), Ronald Blythe's portrait of an English village. 'We had no rain from March right through to October. The corn didn't grow no more than a foot high and most of it didn't even come to the ear. We harvested what we could and the last loads were leaving the field when

> 'One of the most beautiful days of this or any other year. A faultless morning, a blue-planet afternoon.'
> *Martin Amis, author, 1994*

we heard, "the wages are coming down this week" . . . It was the Government's fault. They ended the Corn Act less than a year after it had been made law . . . The price of wheat was quartered . . . Cattle were sold for next to nothing because the farmers couldn't afford to keep them. The farmers became broke and frightened, so they took it out on us men. We reminded them that we had fought in the war, and they reminded us that they had too! So it was hate all round. I drew 27s 6d from the farmer and after I had given my wife 24s and paid my Union 4d and my rent 3s 1d, I had a penny left! So I threw it across the field. I'd worked hard, I'd been through the war and I'd married. A penny was what a child had. I wasn't having that. I would sooner have nothing.'

In 1940, one of the sunniest-ever springs has become a glorious summer. 'Hitler's weather' holds week after week. Goering's air force takes advantage, attacking British convoys in the Channel – today is the 'official' start of the Battle of Britain (*15 September*). In 1968, heavy rain causes major disruption in the West Country. The Cheddar caves flood for the first time and all major holiday routes to the south-west are impassable – but then this is, on average, the wettest day of the year.

33.9C Hodsock Priory 1921

143.5mm Chew Stoke 1968

11 JULY

11.5C min / 20.8C max

HOW A BLIND PERSON FEELS THE WIND

'There is a certain point along my route which catches the wind,' writes Professor John Hull today in 1984, in *Touching the Rock: An Experience of Blindness*. 'As I came up the steps from the underpass and around the corner, it hit me. It was a beautiful, warm scented breeze . . . It was an unsettled wind, suggesting the break-up of rather a sultry day . . . I leaned into it and away from it and breathed it in. It was delightful.

'Can the wind mean as much to sighted people? It is invisible, so they gain nothing over the blind. Of course, the blind lose the sight of the world being blown along by the wind, the hurrying clouds and the trees swaying. On the other hand, the wind has a special beauty for the blind . . . The blind person experiences the impact of the wind upon his body and the sound of it in the trees. He knows perfectly well where it is coming from.

'Sometimes a blind person experiences a wind which is all the more exciting because it is known at long range. I hear the distant tossing of trees across the park; it comes like a wave rolling across a beach. Now it breaks upon my body in a squall, a gust, like a fist. This is very exciting.'

'A joyful day. Excessive hott.' Captain Thomas Bellingham, fighting alongside William of Orange at the Battle of the Boyne (⚔) in the middle of a heatwave, 1690

'Meant to go to London but it rained all day & I was out of order in the region of the bowels.' Ford Madox Brown, painting The Last of England, 1855

The temperature reaches 33.3C in Nottinghamshire today in 1934. It's the hottest day in the hottest spell (nine consecutive days over 27C) of a memorable summer – subsequently used by Ian McEwan as the sweltering backdrop to the life-changing events in the opening of his novel *Atonement* (*20 July*). Enjoyment of the sunshine, however, is tempered by the rising concerns of a drought in 1934 (*11 March*). The summers of both 1932 and 1933 are also hot and dry and underwater aquifers are very low. The drought grows so serious that the government passes the emergency Supply of Water in Bulk Act, which allows one statutory water company to supply another.

The three Water Engineering Associations also meet to consider a 'national water policy' for the first time. Little is achieved and 'normal service' returns the following summer.

34.4C Halstead 1921

12 JULY

11.8C min / 20.8C max

BALL LIGHTNING / WET MUSKETS / RHYL SUN CENTRE

The 'great grey ball' of lightning that enters the church during a service at Erpingham (🌑) today in 1665 passes through the chancel, leaving 'a great smoke and stink', killing one man, injuring many others and damaging the walls. Also today in 1783, two further 'balls of fire' hit a school in Essex (killing three children) and fell a chimney in Gloucestershire. In fact the summer of 1783 – following the eruption of a volcano in Iceland (*18 July*) – is characterized by these strange events.

Fireballs, ball lightning and the other curious electrical atmospherics that all tend to get lumped together are too spontaneous and unpredictable for scientific analysis. Yet the ten thousand or more recorded occurrences all have a degree of consistency. There are hundreds of hypotheses. Supercooled nonideal plasma? Electromagnetic knots? Astrophysical 'little black holes'? Anti-matter meteorites? Naturally, in the Middle Ages such incidents are ascribed to the Devil, who, reports say, often changes shape so that the ball of light resembles a black dog – or even a 'greie frier, behauing himself verie outragiouslie'.

> 'High summer continues. I shall not go to London until it breaks. This is a pleasant house in the heat. For the first time since I planted it the honeysuckle outside my bedroom window scents the room at night. My life is really too empty for a diarist. The morning post, the newspaper, the crossword, gin.' *Evelyn Waugh, author, Combe Florey, Somerset, 1955*

In 1644, through a day of alternating heavy showers and hot sunshine, the army of King Charles I and the Parliamentary forces wait at Marston Moor (⚔). At dusk, under cover of a thunderstorm, the Parliamentarians attack. The Royalist musketeers are unable to shield their elementary firearms – hard enough to use even in perfect conditions – from the driving rain and are quickly over-run by the enemy pikemen. The bloodiest battle of the Civil War is over in two hours. The Parliamentarian victory cements the growing reputation of Oliver Cromwell as a military commander.

But no worries about the weather today in 1980 – scattered showers, as it happens – when the Rhyl Sun Centre (☂) opens. Inside the £5 million plastic dome the temperature is a constant, ambient, tropical 27.77C. The indoor 'beach resort' contains a swimming pool, monorail, sun beds, water-slide and wave-machine. It's the first of its kind in the British Isles and quickly becomes Wales's foremost tourist attraction.

35.0C Clifton 1923

'BUGGER BOGNOR' CLIMOTHERAPY

On this, on average the hottest day of the year in central England, the sun shines respectfully over Weymouth (☞) as King George III arrives for his first visit. A new ritual instituted for the convalescing Monarch introduces a lighter note to the usual suffocating court rituals: 'A machine follows the Royal one into the sea filled with fiddlers who play "God save the King" as His Majesty takes the plunge,' reports the novelist Fanny Burney, travelling with the royal party as Second Keeper of the Robes to Queen Charlotte. Thus begins a royal anointing of Brighton, Weymouth and the south coast – Britain's sunniest place (*28 June*) – as *the* resort for recuperation. It's a relationship that lasts, if myth is to be believed, until 1936, when the dying King George V, in response to suggestions that he will soon be well enough to revisit Bognor Regis (the 'Regis' appended following His Majesty's visit in 1927), utters his final words: 'Bugger Bognor.'

CLIMATE ZONES

■ *Very bracing*
■ *Bracing*
■ *Relaxing*
■ *Very relaxing*

Considering its slim scientific basis, 'climotherapy' is one of the more enduring footnotes of medical history. From the eighteenth century onwards, for certain conditions doctors prescribe beneficial climates. Resort towns, quickly cottoning on to the lucrative possibilities, publish climatic guides highlighting their unique benefits (see map, left). Hastings, for example, goes so far as to distinguish five separate climates simultaneously extant in different parts of the town.

By the nineteenth century, treatments extend to sunlight therapy and dry, frosty mountain air, leading to the age of the sanatorium. One legacy of climotherapy survives until March 2007, when the daily 'bucket and spade' report – single-word weather summaries for the key resorts around our coast – is finally dropped by newspapers.

In a gale and driving rain today in 1961, the American golfer Arnold Palmer wins the British Open at Royal Birkdale, Lancashire (☞). And in 1985, after weeks of grey gloom, the sun emerges for Live Aid, watched by 1.5 billion people (85 per cent of the world's televisions) from Wembley Stadium. 'In this most dismal of English summers the sky was, for once, blue and there was not a cloud in sight,' says organizer Bob Geldof.

35.6C Camden Square 1923 ☼

14 JULY

12C min / 20.5C max

LIGHTNING STRIKES ROYAL ASCOT

After a week of stifling heat, the mercury reaches 30C today – a high for 1955 – before the skies explode across southern England. At Royal Ascot (moved from its traditional June date due to a railway strike), a bolt of lightning hits the metal railing around the enclosure on the heath in the centre of the course, fizzing along it and 'throwing out blue sparks'. The force of the strike mows down over a hundred spectators. Many are lifted off their feet and knocked unconscious; forty-nine are injured. 'It was like being stabbed in the stomach,' one man says. Two, including a pregnant woman, are killed.

It's not the first lightning death at Royal Ascot (☂). In 1930, a thunderstorm floods the racecourse and a bookmaker, sheltering under an umbrella in the Tattersall's Ring, is struck and killed. The meeting is abandoned for the first time in two hundred years. This happens to be the year the long, sweeping dress makes a return to fashion, and as news of the death spreads a journalist from the socialist *Daily Herald* picks up the story with relish: 'The real terror showed then. Men went pale ... Women pinned up their frocks in the manner of charwomen and trooped out. It was almost impossible to find their motor cars. Those who set off down the tunnel to the station found two feet of water. They went across the field ankle deep in mud. Their clothes? They were past caring for them.'

'It is vy pleasant here. The weather perfect: the garden delicious. We all bathe each morning and lie & bask on the hot rocks. How I wish you were here my dearest, & how glorious you wd look in your thinnest Venetian bathing dress!' *Winston Churchill, to his wife Clementine, from Penrhos, North Wales, attending the Prince of Wales's investiture at Caernarvon Castle, 1911*

'Having gone to see some ruins while the horses were changing at Cardiff, we found the post-boy had driven away; and on inquiring the reason on his return, he said he was afraid the horses would catch cold standing – this is delightful for the middle of July, when the people of New York are dying with heat.' *Louis Simond, American tourist, 1810*

In 1964, terrible weather strikes again. The procession to open the new Royal Enclosure has to be cancelled. Two days of racing are lost. One foreigner describes the curious scene of well-heeled, straight-backed, dressed-up, dripping racegoers as 'trees in a rain-forest'. It's the worst wash-out in Royal Ascot's history and it leads to a huge programme of drainage works. In 2000, the rain returns. Every umbrella has been sold by mid-morning – but the course drains beautifully.

32.3C Trowbridge 1983

15 JULY

11.8C min / 20.3C max

ST SWITHUN'S DAY / THE ROGER

The British legend that if it rains today, St Swithun's Day, then it will rain for the next forty days, is medieval. At the canonization of St Swithun in AD 971 the monks of Winchester Cathedral (♥) plan to transfer the body of the former bishop to a grand tomb in the choir, against his dying wish – he was a humble man and wanted a simple outdoor grave 'where the feet of passers by and rain dripping from the eaves would beat upon it'. But it rained on this day, postponing the ceremony, and kept on raining for the next forty days. Thus, the bones of the 'Watery Saint' are left where they are.

'Extremely hot . . . oranges ripening in the open at Hackney.'
Samuel Pepys, diarist, 1666

As with most legends, it has a grain of truth: when a summer weather pattern is established by mid-July, it often persists for a few weeks. Similar proverbs exist on the Continent to explain this noticeable bit of seasonal behaviour. The durability of the legend, however, says more about our desire to forecast the weather than about the power of saints.

'The day had been warm and sultry with barely enough wind to drive us,' explains Christopher Shallcross of a sailing trip on the Broads (⊘) in his Norfolk wherry today in 1998. 'We were drifting upstream on what small puffs of breeze there were, frustrated by the slow progress . . . I handed the helm over to one of my colleagues and reached down to the scuppers for a paddle. The next thing I knew I was in the water.'

Mr Shallcross has been 'Rogered'. 'The Roger', or 'Sir Rodge's Blast' is a sudden, violent whirlwind lasting up to thirty seconds that sweeps unexpectedly off the marsh. 'When the freshwaterman sees the waving of the reeds and sedges, he knows a "Roger's Blast" may hurl himself and his craft to the bottom,' records Forby in his *Vocubulary of East Anglia*, 1825, though a 'Rodjon' is documented as early as 1440. 'Even if you see one coming . . . you cannot tell whether it will strike you or not. It may blow the sail of one wherry to pieces, and another close by will be becalmed,' says *Norfolk Broads and Rivers* (1883). Arthur Ransome, in his children's novel *Coot Club*, mentions the 'loud hissing noise' it makes.

34.4C Stratfield Turgis 1881

16 JULY

11.8C min / 20.2C max

SEDGEMOOR IN THE FOG

Fog, not for the first time (*23 April, 23 September*), is the unscheduled visitor to a British battlefield today in 1685. James II's army is camped out on the Somerset Levels – a flat, low-lying landscape bisected by drainage dykes or rhynes and prone to early morning mist (✕). The Duke of Monmouth, having recently landed from Holland with a rebel army, is attempting to claim the throne. (As one of Charles II's illegitimate children, he's acting on a rumour that the King married his mother while in exile.) Relying on a local boy to guide him, Monmouth's one hope against the highly trained royal forces is surprise. He plans to approach over the marshy wastes of Sedgemoor, crossing the wide Bussex Rhyne round their encampment as they sleep.

> 'Ripon market, the thermometer in the sun at 9 am 118 [47.8C]. Heard that the heat this day in the sun ... was so great that three butchers broiled their steaks for dinner on their cleavers in the sun without fire.'
> *Reverend Benjamin Newton, 1818*

> 'Blandings Castle slept in the sunshine. Dancing little ripples of heat-mist played across its smooth lawns and stone-flagged terraces. The air was full of the lulling drone of insects. It was that gracious hour of a summer afternoon, midway between luncheon and tea, when Nature seems to unbutton its waist-coat and put its feet up.'
> *P. G. Wodehouse,*
> *Summer Lightning, 1929*

Unfortunately for Monmouth, in the darkness and fog his guide loses his bearings. First, he can't find all the 'plungeons' or plank bridges across the rhynes. Then someone's pistol goes off and the element of surprise is blown. Monmouth's untrained cavalry, unable to cross the rhyne, are fired upon and flee. After this, the royal forces make short work of Monmouth's rebels. The engagement is of little direct political consequence, but it lives on as the last pitched battle fought on English soil, as the first proper engagement for many of the most famous county regiments in the British Army, and for the command of one John Churchill – later Duke of Marlborough.

Monmouth, found hiding in a ditch in Hampshire, is later executed on Tower Hill in London, where it takes five strokes of the axe to sever his head. His followers are dealt with no more sympathetically – by the notorious Judge Jeffreys.

34.1C Chelmsford 1900

17 JULY

11.6C min / 20.3C max

FIRST FIRE OF
LONDON / THE
SWIMMING
MATHEMATICIAN

Strong southerly winds aggravate a fire in Southwark, London, today in 1212. Timber houses, packed in narrow streets on the south side of the Thames, quickly catch, and a large area of the borough, including the church of St Mary Overie (where Southwark Cathedral now stands), is razed. The houses on London Bridge (recently reconstructed in stone) also burn and, as several chroniclers state, more than a thousand die. Until 1666 (*12 September*) this is the 'Great Fire of London' and it has grave consequences for the medieval monarchy. King John is already in conflict with the Church and his barons. Now the smouldering capital, too, turns against him. Within three years he will be forced to sign the Magna Carta at Runnymede – a forerunner of all modern constitutions and a testament to his failure as king.

'We lay on the bare top of a rock, like scones upon a griddle; the sun beat upon us cruelly; the rock grew so heated a man could scarce endure the touch of it; and the little patch of earth and fern, which kept cooler, was only large enough for one at a time. We took turn about to lie on the naked rock, which was indeed like the position of that saint that was martyred on a gridiron; and it ran in my mind how strange it was, that in the same climate and at only a few days' distance, I should have suffered so cruelly, first from cold upon my island and now from heat upon this rock.' *Robert Louis Stevenson*, Kidnapped, *1886*

Stormy seas off the Channel Island of Sark (♀) today in 1998 claim the life of Sir James Lighthill, one of the great minds of the twentieth Century. The former Lucasian Professor of Mathematics at Cambridge is attempting to swim the 9-mile (14-km) circumference of the island, a feat that he himself pioneers twenty-five years earlier (describing it as 'a most pleasant way to see the scenery'). Apart from breaking new ground in various fields of pure and applied mathematics, Lighthill virtually creates the field of biofluid dynamics. This, the study of how animals move through air and water, is what he ponders while undertaking his nautical marathons. He also swims in a style of his own devising – a two-arm, two-leg backstroke, thrusting arms and legs alternately – while studying tides and currents.

32.2C Bexleyheath 1921

♀

18 JULY

11.9C min / 20.4C max

CLOUD FEVER

'The country people began to look with a superstitious awe at the red, louring aspect of the sun,' notes the naturalist Gilbert White today in 1783. 'And indeed there was reason for the most enlightened person to be apprehensive.' By day a 'sickly opalescent fog' seals in the heat. By night there are frosts. These 'horrible phænomena', as White calls them, have been going on for a month.

BRITAIN'S RECORD RAINFALL

In 1955, 279 mm (that's not a printing error: 11 inches) of rain falls on the Dorset village of Martinstown in twenty-four hours. It is – by a vast margin – Britain's record daily rainfall. Astonishingly, however, there is no damage. Because the rain falls over chalk downland, the porous rock absorbs the water like a sponge. Only when it reaches full capacity are streams released (the characteristic 'winterbornes' or steady chalk streams which flow only in winter). So no deaths, no floating cars, no drama.

Apart from inducing near-panic, the bizarre weather does something else. It draws Britain's gaze skyward. Among the many who are transfixed is a ten-year-old schoolboy called Luke Howard. 'Here was the turning point, the fulcrum of his engagement with the evolving science of weather and climate,' writes Howard's biographer, Richard Hamblyn. 'He had experienced nothing like it before, an entire summer filled with inexplicable skies, and the excitement that it generated . . . gave definition to his early ideas.' For Howard goes on to classify and decode the skies, to identify clouds and give them names – and this is the moment it all begins. (The weird weather, by the way, is caused by a 'killer cloud' of volcanic ash generated by the eruption of an Icelandic volcano, Laki.)

Howard is not the first to attempt to classify the clouds. His stroke of genius, however, is to name them in Latin – the language of science, of the Enlightenment, of Linnaean classification. Using four terms, cirrus, cumulus, stratus and nimbus, with the first three in various combinations, he defines seven key cloud forms (or 'modifications', as he calls them), many still used today. It's a supremely compelling idea. The most ethereal of all things, thin air, is finally brought to book – by rigorous scientific observation. The skies can be read. Howard's classifications inspire 'cloud fever' in, amongst others, Goethe, Shelley, Ruskin and Constable.

32.4C Hillington 1901

279.4mm Martinstown 1955

Forms assumed by clouds when gathering for a thunderstorm.

Cumulostratus forming, fine weather cirrus above.

Stratus or ground fog.

Cumulus breaking up, cirrus and cirrocumulus above.

Cumulostratus as produced by the inosculation of cumulus with cirrostratus. Cirrus above, passing to cirrocumulus.

Nimbus or rain cloud.

19 JULY

12C min / 20.6C max

HOTTEST DAY OF HOTTEST MONTH OF HOTTEST YEAR

Today in 2006 is the hottest day of the hottest month of the hottest year on record. Most parts of Britain are hotter than Casablanca and Rio. It's hotter than the hottest day of the summer of '76 (by 0.6C). Although not quite as hot as the UK's highest-ever recorded temperature (*10 August*), at 36.5C this is the hottest July day ever, recorded at Wisley, Surrey. The mean daily sunshine total of nearly 8½ hours is 50 per cent above the July average. Although these figures are remarkable, very dry summers do occur about once a decade, and it is not unusual for the hottest day of the year to be around now: on average it comes in July, and most often between the 10th and the 20th.

'Intensely hot day – left off waistcoat'.
Samuel Taylor Coleridge, poet,
Lake District, 1803

Gritters are out, spreading crushed rock on to roads in the West Midlands to stop the tarmac melting. Metal swing bridges jam in the heat, and have to be hosed by firemen to make them open and close. Railway lines buckle, requiring speed restrictions to be imposed on trains. In London, where temperatures reach 33.2C, judges and barristers remove their wigs at the Old Bailey, breaking a dress code of three centuries. Thirty-nine guests at a royal garden party at Buckingham Palace faint ('IT AIN'T HALF HOT MA'AM,' reports tomorrow's *Daily Mirror*) and temperatures in buses exceed 50C, far above the legal limit for transporting cattle. High street sales wilt, except for water butts (up 120 per cent), watering cans (up 15 per cent) and fans (Comet sells one every two seconds), air conditioning units (one every thirty seconds) and barbecues, which all reach a summer record. Basking sharks appear in large numbers off the coast and giant sunfish are officially recorded for the first time. The bookmaker William Hill shortens the odds on temperatures reaching 100F (38C) from 6/1 to 5/4. Meanwhile, parched fields crack and open up treasures of aerial archaeology, notably two 6000-year-old Neolithic enclosures in Radnorshire and the Vale of Glamorgan.

In 1460, by craftily attacking in a blinding rainstorm so that their opponents cannot fire their cannons, the Yorkists rout Henry VI's Lancastrian army and capture the King at the Battle of Northampton (⚔). But then again on 'Hot Tuesday' in 1707 men and horses die of heat stroke during the harvest – estimates suggest 38C, but this is long before official recording equipment, so 2006 holds the title.

36.5C Wisley 2006

20 JULY

12.1C min / 20.6C max

THE GO-BETWEEN

In 1900, the summer heatwave which peaks today (at 35.1C) holds vivid memories for a five-year-old boy called Leslie Hartley, because it's when his family moves to the countryside. Nine summers later, aged fifteen, Leslie goes to stay with a school-friend at a house called Bradenham Hall in Norfolk (♥). Half a century on again, in 1953, these two experiences – heatwave and country house – become a single story when Leslie Hartley, now better known as the writer L. P. Hartley, publishes his bestselling novel *The Go-Between*. More than any other British novel, this story is suffused with the heat of an English summer. And, of course, it has *that* opening line: 'The past is a foreign country; they do things differently there.'

The Go-Between glows with heat. Sweat beads trickle from every page – the heat of young passion as much as that of summer – right up to the tragic dénouement. British novelists love heatwaves, as heat does strange things to repressed Brits. It removes inhibitions as well as clothing, but in unexpected ways. It makes people do things they otherwise wouldn't. It suffocates and stifles. It shortens tempers, adds discomfort, ratchets up tension. It's a metaphor for the summer of life, for happier times, for a hotter past. But Hartley's *The Go-Between* is the all-pervading influence. 'I love England in a heat wave. It's a different country. All the rules change,' says Leon Tallis – no, not in *The Go-Between*, but in Ian McEwan's *Atonement* (another from the steamy summer heatwave genre, this time 1934).

OTHER HOT READS

• Ian McEwan's *Atonement* begins 'on the hottest day of 1934' – 11 July (33.3C), as it happens. • A mother's relationship with her daughter fries in Penelope Lively's *Heat Wave*. • The chapters of Thomas Hardy's *Tess of the D'Urbervilles* set at Talbothays Dairy rely on 'Ethiopic' heat. • *The Sandcastle*, Iris Murdoch's study of a marriage, starts in a heatwave. • The balmy heat of a midsummer house party is described in D. H. Lawrence's *Women in Love*. • Heat erodes gentility at Mr Knightley's Donwell Abbey in Jane Austen's *Emma*. • The stifling heat in Ian McEwan's *The Cement Garden* prompts brother and sister to remove their clothes and attempt to revisit childhood.

35.1C Cambridge 1900

21 JULY

12.3C min / 20.6C max

BOWLING A NEW JERUSALEM / WISLEY TORNADO

'I'm too old to be bowling uphill and into the wind,' Bob Willis complains to Mike Brearley during the Headingley Test (☝) today in 1981. When the England captain and supreme tactician switches the ageing fast bowler to the other end, one of the greatest turnarounds in Ashes history begins. Now bowling with the wind at his back, Willis tears into the Australian batting line-up like a tornado. He takes 8 for 43, the best figures of his career. Australia, who only need 130 runs to win, are all out for 111. 'England awake to a new Jerusalem,' *The Times* says next day.

'It was a lovely, sunny day and Caroline and I walked in the garden and sat on the grass and I felt we had stolen a little time from the hurly-burly of daily life.' *Tony Benn, politician, in hospital in Bristol, 1963*

Though every fast bowler prefers to bowl with the wind, it is, arguably, the aspect of weather that least affects cricket. The other seminal wind story from the annals of weather and cricket concerns W. G. Grace, that David Beckham of Victorian England. The tale tells of 'WG' opening the batting at a charity match before a large and expectant crowd. When the second ball nicks a bail off, WG picks it up and replaces it, announcing, to anyone who cares to listen: 'Strong wind today.'

At 3 p.m. today in 1965, Colin Martin, a young gardener at the Royal Horticultural Society at Wisley, Surrey is alarmed to see the sky darken dramatically. When trees start 'disappearing' and tin sheds take off into the sky, he makes a run for it out of the fruit garden. In ten short minutes, a tornado cuts a swathe through the pride and joy of English horticulture's head office, uprooting orchards and splitting trees up to 10 feet (3 metres) in girth. The adjacent weather station is untouched.

'What alarmed the lower orders more than the magnates,' Matthew Paris notes in the *Chronica Majora* in 1258, 'was the continued deluge of heavy rain, which threatened to drown the rich crops which God had given hopes of previously.' Crop failures later in the year lead to the worst famine of the thirteenth century. Recently, it has been suggested that it may have been a huge volcanic eruption in the tropics in January 1258 that caused the catastrophic weather that spring and summer (*13 June, 18 July, 13 October*).

34.0C Raunds 1911

Tewkesbury-worst floods in 60 years, 2007

22 JULY

12.1C min / 20.4C max

'PERFECT SUMMER' OF SEX The 'Perfect Summer' of 1911 records its highest temperature today at Epsom, Surrey – 36C. The heatwave begins at the start of May and doesn't let up for five months. July is the sunniest month on record (until 2006 – *19 July*). This is the summer the British discover sex, shedding whalebone corsets, petticoats, bloomers and the inhibitions of centuries as the temperature rises. French couturier Paul Poiret's new brassieres become fashionable, along with lingerie and shrinking bathing costumes – daredevil poseurs sport suits ending at the elbow rather than the wrist. But while, in the heat, indolent members of the Edwardian aristocracy play nude tennis in their country houses, in the slums the poor are dying. Social and political unrest in the stifling cities eventually detonates into mass strikes and violent demonstrations. Churchill, then Home Secretary, sends in the troops to keep order as Britain dances (the Countess of Fingall's phrase) 'on the edge of an abyss'. The shadows are lengthening – towards irreversible social change and the industrialized slaughter of the trenches.

'The wind dappled very sweetly on one's face and when I came out I seemed to put it on like a gown. I mean it rippled and fluttered like light linen, one could feel the folds and braids of it.' *Gerard Manley Hopkins, poet, Surrey, 1873*

'The heat was excessive and as I sat reading under the lime I pitied the poor haymakers toiling on the burning Common where it seemed to be raining fire.' *Francis Kilvert, Chippenham, 1873*

'Mist clearing – the top of the lighthouse an island in a creamy sea.' *Christine Evans, poet, 1997*

In 1868 the then hottest British temperature ever, of 38.1C (100.5F; eclipsed in 2003) is recorded at Tonbridge, Kent. Collected in the back garden weather station of a Dr George Fielding, it is today discounted – exemplifying the difficulties of quality control in weather data collection before the introduction of the Stevenson Screen weather station in 1869 (*12 January*).

'The hailstones ... reached the size of eggs, then, among the smooth round ones, jagged fragments started to fall ... some 4–5" [10–25 cm] long ... There was terror in Upper Plumstead,' (🌢) reports *The Times* in 1925. Amid accounts of people being lacerated by flying ice and their clothes shredded 'as if by jagged knives', there are claims that one falling ice fragment weighs over a pound (0.45 kg). In 1930, between the 20th and the 23rd four months' worth of rain falls on the North Yorks Moors, causing the rivers Derwent and Esk to flood. The Whitby lifeboat rescues families 2 miles (3-km) inland (🌢).

160.1mm Ratlinghope 1972

36.0C Epsom 1911

23 JULY

12.2C min / 20.3C max

HAYMAKING   'The old saying is, "Make hay while the sun shines". Our advice is, keep the sun off as much as possible ... Hay made this way retains the saccharine matter,' records a note in a 1921 farmer's ledger. It's haymaking time across Britain. And although silage – pickled grass – and a hybrid called 'hay-lage' have both largely usurped traditional haymaking, because they are less dependent on fine weather, hay is still an important part of ours, the greatest of all grass economies (*21 March*).

Few books recall the vanished world of horses, hay wains, and stack-making more evocatively than Adrian Bell's memoir *Corduroy*, about a year on a Suffolk farm (*). Bell details the 'monotonous zip-zip of spare knives being sharpened' and the over-anxious farmers, fearing the fine weather cannot last, cutting the hay before it is ready, and then the stacks heating:

'It is often so hot that the men can hardly stand and work on it ... If it is beyond this point ... it is fatal to open it to the air, as it immediately catches fire, and probably catches other stacks or buildings. Many an anxious week some too headstrong farmers spend watching a stack smoulder and smoulder, wondering if it is going to burst into flames. Even if it does not, it is found to be all charred inside and black, like ashes.'

THE MEANING OF CLOUDS

'I have been watching the clouds on these hills for many evenings back: they gather when I do not expect them; they dissolve when, to the best of my judgement, they ought to remain; they throw down rain to my mere inconvenience, but doing good all around; and they break up and present me with delightful and refreshing views when I expect only a dull walk. However strong and certain the appearances are to me, if I venture an internal judgement, I am always wrong in something ... So it is in life ... The point is this: in all kinds of knowledge I perceive that my views are insufficient, and my judgement imperfect. In experiments I come to conclusions which, if partly right, are sure in part to be wrong ... My views of a thing at a distance and close at hand never correspond, and the way out of trouble which I desire is never that which really opens before me.'
Michael Faraday FRS, British chemist and physicist, aged thirty-five, 1826

33.4C North Heath 1989

24 JULY

12C min / 20.3C max

It's hot and muggy at the London Olympics today in 1908 as Dorando Pietri, the diminutive Italian pastry chef-turned-marathon runner, enters White City Stadium for the final strides of his 26-mile, 385-yard run. With a Gold Medal in the bag, the cheering of the crowd rises to a roar. But suddenly, a few yards before the finishing line, opposite the Royal Box, Pietri halts. Looking dazed, he sets off back up the track in the wrong direction. Officials reorientate him, but his knees crumple and he collapses again. And again. Then again, in an excruciating ten-minute final lap. As the second-placed runner, John Hayes of the USA, enters the stadium, a kindly official gathers up Pietri and walks him gently over the line.

Unlikely as it may seem, Pietri has heat stroke – and he is not alone: only 27 of the 56 competitors finish. Not surprisingly, the American team lodge a complaint. Pietri is disqualified for receiving assistance. He is, however, presented with a special gold trophy by Queen Alexandra and becomes a popular sensation overnight. Irving Berlin writes a song in Pietri's honour and the plucky, heartrending finale sparks a worldwide marathon craze.

In 2004, after three weeks the weather finally comes good today, bringing a sigh of relief from vicars and church fete committees. Why have they been eyeing the sky more anxiously than usual? Because Zurich, the insurance giant, has stopped issuing cover for small events affected by adverse weather. For decades the 'Pluvius policy' (after Pluvius, the Roman god of rain) has provided cover against fog, frost, snow, high winds, drought, and – more commonly for garden fetes – rain causing financial loss. Premiums are calculated according to weather data and increase across the country from east to west, just as average rainfall does.

Zurich claim they are no longer offering the policies because the numbers taking them up are diminishing, following a run of long, hot summers. The cynical, however, believe it's because of record-breaking rain (2000 is the wettest year across England and Wales since 1872), meaning they have had to pay out once too often.

32.6C Barnet 1900

25 JULY

11.8C min / 20.7C max

BLÉRIOT FLIES THE CHANNEL

A gentle south-west breeze and some early morning mist that should soon clear: conditions are perfect. None the less, Louis Blériot is nervous as he climbs into his flimsy monoplane in a field near Calais at four o'clock this morning in 1909. Red flames streak the darkness as the 28hp Anzani motor shatters the calm. He takes off and, as day breaks, swings north out to sea, to England – and the supreme aviation challenge of the age.

Lord Northcliffe, proprietor of the *Daily Mail*, has offered £1000 to the first person to fly the English Channel. With height and speed records being broken almost daily, the challenge now is sustained flight. Engines are underpowered and unreliable, and flying machines seldom stay airborne more than a few minutes. Blériot knows his journey will last at least thirty minutes – over water. But of all the aviation landmarks so far, this is the big one: lucrative, international, geographically significant. Whoever claims this prize is in the history books.

Flying 250 feet (76 metres) above the water, at 40 mph (64 kph), Blériot rapidly catches and overtakes his escort, the French destroyer *Escopette*. 'The moment is supreme, yet I surprised myself by feeling no exultation,' he recalls. By mid-Channel, the wind is getting up – as is the mist. 'I turn my head to see whether I am proceeding in the right direction . . . there is nothing to be seen – neither the destroyer, nor France, nor England. I am alone. I can see nothing. For ten minutes, I am lost.' With no map or compass to follow, the strengthening wind blows him off course. After twenty minutes – about the time that the engine of his arch-rival, Hubert Latham, failed on his attempt last week – a timely shower cools the overheating Anzani. 'Then I saw the cliffs of Dover! . . . The wind had taken me out of my course. I turned and now I was in difficulties, for the wind here by the cliffs was much stronger, and my speed was reduced as I fought against it.' Clawing his way over dry land, he tries to descend but the wind catches him and whips the machine around. He cuts the engine. 'The landing gear took it rather badly, the propeller was damaged, but my word, so what? I HAD CROSSED THE CHANNEL!' And so ends, for Britain, centuries as an island fortress, protected by the greatest navy in the world. The military and political establishments are speechless.

33.9C Regent's Park 1900

26 JULY

12.1C min / 20.3C max

RAIN, RAIN, JACK MACBRYAN

Rain stops play at Old Trafford (☂) this afternoon in 1924. After only 2 hours 45 minutes of cricket, the Manchester weather sets in hard and the Test between England and South Africa is abandoned. One thing makes this match unique, however – the part played, or rather not played, by Jack MacBryan: he neither bowls, nor takes a catch. And England do not bat. As MacBryan is never picked to play for England again, the Manchester rain earns him an unrivalled place in cricket history (and pub quizzes). He is the only Test cricketer never to bat, bowl or dismiss anyone in the field.

In the history of Test cricket in England, only two games have been washed out without a single ball being bowled: Old Trafford has the unfortunate distinction of staging both, in 1880 and 1938. The chance of a Test day being lost to rain at Old Trafford is nearly one match in two, as opposed to one in six at Lord's or the Oval. Weirdly, Manchester's rainfall (*26 November*) is average for England as a whole. The clouds just seem to gather at the sound of leather on willow.

'A glorious morning ... And although the sun clouded over, it remained beautifully dry and warm so that people lolled on the grass just as we had dreamed ... At 6 p.m., as the last guests were saying they ought to go, the first raindrops fell. God's watch must have been slow!' *Barbara Castle, novelist, on her Silver Wedding anniversary, 1969*

'Still hotter. I sate with W[illiam] in the orchard all the morning and made my shoes. In the afternoon from excessive heat I was ill with the headach and tootach and went to bed – I was refreshed with washing myself after I got up, but it was too hot to walk til near dark, and then I sate upon the wall finishing my shoes.' *Dorothy Wordsworth, Dove Cottage, Lake District, 1800*

'A shower of hail as big as walnuts.' *Samuel Pepys, diarist, Essex, 1666.*

'Incessant drenching rain, beating and soaking. The Madonna lilies lie prostrate; it is disgustingly cold. Lord what a summer!' *Frances Partridge, writer and member of the Bloomsbury Group, 1954*

'This moderate climate is certainly much fitter for bodily exercise than that of America. We think nothing of five or six miles a-day on foot.' *Louis Simond, American tourist, North Wales, 1810*

33.3C Southampton 1885

27 JULY

11.7C min / 20.3C max

THE BRITISH WINE RENAISSANCE

'100F: Get Used To It' says *The Observer* today in 2006, following a week of extreme temperatures. But amid the panic about the significance of this heatwave, one group is quietly content – British wine growers. When a Sussex vineyard wins a prestigious international award for 'best traditional method sparkling wine' (outside Champagne) during this hot spell, it is apparent how the British wine industry is thriving in our new climate. Ideal conditions – no spring frosts, lots of sunshine in June and July, low humidity, sporadic summer showers and gentle breezes – in 2000, 2001 and 2003 all produce good vintages.

The rise and fall and rise again of viticulture in England and Wales correlates neatly with our climatic fluctuations. Probably it's the Romans who introduce vines. Despite the historian Tacitus pronouncing the British weather 'objectionable', pollen deposits across Britain suggest that the climate was as warm then as it is now. The arrival of William the Conqueror, and Continental abbots experienced in wine-making, starts an era of intense vine-growing, coinciding with the 'Medieval Warm Period' (*20 February*). The Domesday Book of 1085 lists forty-two vineyards, and forty years later the historian William of Malmesbury writes of the products of vineyards in the Vale of Gloucester: 'Those who drink this wine do not have to contort their lips because of the sharp and unpleasant taste, indeed it is little inferior to French wine in sweetness.'

Lower shipping costs, the Dissolution of the Monasteries in 1536, and, crucially, the worsening climate during the Little Ice Age (*3 March*) mark a gradual decline in English viticulture over the next six hundred years. But the number of vineyards has quietly expanded over the last three decades as average temperatures have increased. There are now around five hundred, many showing increasing maturity and commercial viability. If climatologists are right about global warming, there will be more. English wine is no longer a joke. Sussex and Kent share almost identical weather and soil with the Champagne region of France. In 2006 Pinot Noir and Chardonnay vines are planted near Malton (♥) – 80 miles (130 km) north of the northernmost Roman vineyard. At the current rate of climate change, the first Scottish *cuvée* can only be decades away. Château Cairngorm, anyone?

34.4C Margate 1933

28 JULY

11.8C min / 20.4C max

BIRMINGHAM TORNADO / SUN STOPS PLAY

Glass, bricks, armchairs and even roofs are circling in the skies over Birmingham today in 2005. In four minutes of mayhem, thirty are injured, a thousand trees uprooted, hundreds left homeless and £39 million worth of damage is done – in an area of the city where many cannot afford household insurance. Remarkably, no one is killed. Older residents liken the destruction to a direct hit in the Second World War. It's one of the worst tornadoes reported in the British Isles (*23 November*): with estimated wind speeds of up to 130 mph (200 kph), it travels a path 7 miles (11km) long and up to 1650 feet (500 metres) wide, though nearly all the damage is done in Birmingham.

'London has now become New York in a heat wave. Old ladies dying like flies in all directions and the Government warning us of drought and water shortage.' *Noël Coward, writer and composer, 1948*

'The rocks were loosened from the mountains ... the trees were torn up by the roots and whirled away like stubble. Two women of loose character were swept away from their own door and drowned: one was found near the place, the other was carried seven or eight miles. Hayfield churchyard was all torn up and the dead bodies were swept out of their graves. When the flood was abated they were found in several places. Some were hanging in trees: others left in meadows ... some were partly eaten by dogs or wanting one or more of their members.' *John Wesley, founder of Methodism, 'field preaches' in the rain, Derbyshire, 1748*

Today in 1995, under a perfect blue sky, to howls of derision from a capacity crowd, the players leave the field at Old Trafford (♟) during the England v. West Indies Test. The reason? Sunshine reflected off the roof of a nearby glasshouse is blinding the batsmen. It is, it must be said, the one and only time sun has stopped play in international cricket.

In 1814 today the weather, reflecting the opprobrium of the nation, does its utmost to prevent the scandalous elopement of (married) poet Percy Shelley and Mary Wollstonecraft-Godwin. However, lightning, intense heat and a violent squall in the Channel cannot stop the young lovers from reaching France safely.

35.0C Milford 1948

29 JULY

12C min / 20.6C max

THE SINKING OF THE *MARY ROSE* / WATERLOO SUNSET

It's been flat calm in the Solent all day in 1545, but towards evening an offshore breeze springs up. The King, Henry VIII, has spent the morning watching the Royal Navy engage Francis I's French invasion fleet. Now he watches, in horror, as the pride of the English fleet, the 91-gun flagship of Vice-Admiral George Carew, heels over without warning and sinks (⚓). It turns out that the inexperienced crew have forgotten to close her lower gun ports after firing broadsides all morning, and the roughening sea has poured into the lower decks. Worse, because she is rigged with netting to stop the enemy boarding, all four hundred or more men drown. The ship is the *Mary Rose*, one of the first purpose-built warships in the navy, and the very nature of her grim fate is the reason, 450 years later in 1982, that her raising makes her, once again, a household name. A Tudor time capsule, she contains clothing, jewellery and personal objects such as tools and gaming boards.

> 'Six o'clock on Saturday morning. There's a mist which blankets the valley, and the sun is starting to irradiate the mist with an expectant glow; it's all promise – of a glittering perfect English day.'
>
> *Sir Richard Eyre, Artistic Director of the National Theatre, 1995*

A month of misery, in 1956, reaches a sodden, spiteful crescendo today and tomorrow near Nairn, Scotland (💧) as 233 mm (9 inches) of rain fall in forty-eight hours. Is this Britain's worst ever summer? For unremitting cold and rain, it has few rivals. London's wettest month since records begin (in 1697) ends with gales in southern England. There are numerous fatalities and thousands of trees are uprooted. Campers have their tents torn to shreds or blown away. A cross-Channel sailing race ends in Mayday calls and *Moyana*, returning to Southampton after winning the first Tall Ships race, sinks.

Under the same miserable skies, a teenager in St Thomas's Hospital, London is brought out by hospital staff, bandages and all, to the riverside overlooking Waterloo Bridge. Following incessant rain, the swollen river has never looked higher, or so red and rusty. The teenager is Ray Davies, and his memory of the river today inspires the opening lyric of his song 'Waterloo Sunset': 'Dirty old river, must you keep rolling . . .' – number one in May 1967. The song is routinely voted the greatest ever written about London. And today in 1981 the weather smiles for the wedding of Prince Charles and Lady Diana Spencer.

181.4mm Buttermere 1938

34.4C Perdiswell 1948

30 JULY

12.1C min / 20.7C max

ASTHMA
STORMS /
PALLADIAN
SUMMERS

Thunderstorms today in 2002 coincide with a dramatic wave of asthma attacks in eastern England – more than a hundred sufferers are admitted to A&E. It's an eventuality that the Met Office correctly predicts, confirming a phenomenon first identified in July 1984: the 'asthma storm'. Research shows that certain kinds of storm trigger violent asthma attacks amongst the 5 million or more UK asthma sufferers. Big thunderstorm systems, following rain or humid weather, suck up grass pollen and fungal spores, shatter them into millions of smaller particles, then deposit them, highly concentrated, in localized areas. The result? Mayhem at the A&E.

An extraordinary run of warm, dry, sunny summers – notably July 1707 (*19 July*) – opens the eighteenth century. It swells agricultural profits and drives a country house building boom among the great landed families. But does this sunshine have a secondary effect? From around 1715, the *only* fashionable style of architecture is Classical (or 'Georgian'), borrowing from Renaissance Italian designs such as those of Andrea Palladio (above). There's just one hitch. It is a Mediterranean architecture, for a hot climate, characterized by shady colonnades, airy porticos, small windows (to keep the occupants cool) and 'articulations' revealed only by bright sunlight and sharp, contrasting shadow. In a northern climate of grey skies, rain and wind it's . . . well, absurd, really. Not that that stops us, of course. Does the memory of those glorious summers from 1700 to 1707 subconsciously predispose our acceptance of such an architecture? As the new Palladianism takes over, the poet Alexander Pope mocks the movement's cheerleader, Lord Burlington, for his 'imitating fools' and the way they inadvertently 'call the winds through long arcades to roar / Proud to catch cold at a Venetian door'.

33.3C Shinfeld 1948

31 JULY

12C min / 20.6C max

STICKY WICKETS / FLANDERS GUNS

On a rain-soaked pitch, under grey skies, the fourth Test between England and Australia at Old Trafford (☀) in 1956 is heading for a draw on the final day. At 112 for 2, the Australians are enjoying a relaxed lunch. Next door, in the England dressing room, the Yorkshireman and off-spin bowler Jim Laker is having a sandwich and a half pint when he notices the sun peek through the clouds. 'It was as though someone's prayers were being answered,' he later says.

GUNFIRE IN FLANDERS

It starts to rain across south-east England today in 1917, and doesn't stop for two days. In Flanders, where the First World War is in its fourth year, the mud and the misery of the main Battle of Passchendaele also begins. Heavy rain continues to fall across Britain – in places the monthly average is 250 per cent higher than normal – for much of August. Does the smoke generated by the bombardments on the Western Front cause all the rain? *(15 August)*

Within minutes the pitch is drying – and unplayable. Facing a spin bowler on a drying or 'sticky' wicket is the greatest test of a batsman – soft, damp patches beneath the dry surface make the ball behave unpredictably. The Aussies are duly 'Lakerized'. The Surrey spin bowler takes 7 wickets for 8 runs in 35 minutes. England win by an innings and 170 runs. Laker's remarkable figures for the match are 19 for 90 – a record that will probably never be beaten, now that cricket wickets are covered when it rains.

There are accusations that the pitch – heavily marled and devoid of grass during the wettest summer in decades – has been prepared or 'doctored' for spinners. One Australian batsman likens it to Bondi Beach because it's so sandy. Laker's first innings figures (9 for 37) suggest as much. But two days of rain have removed any advantage for England – until the sun pokes through.

The phrase 'sticky wicket' has become a byword for a tricky situation. Before wickets are first covered (even overnight), in the 1970s, a short, sharp storm can change a game completely in minutes. It is no coincidence that the world record in a first-class match of 10 wickets for 10 runs, set by the spinner Hedley Verity playing for Yorkshire against Nottinghamshire in 1932, is also achieved on a storm-affected, drying pitch.

33.9C Croydon 1943

AUGUST

1 AUGUST

11.7C min / 20.6C max

Although many of the hottest days of the year occur in August (*10 August*), it is, on average, only the second hottest month (after July). It is also, along with July, the least windy. For many eastern counties it's the wettest month, and the last third brings changeable weather and the first autumnal storms.

'*General weather probable during next two days: North – Moderate westerly wind; fine, West – Moderate south-westerly; fine, South – Fresh Westerly; fine.*' These twenty-one words, printed in *The Times* today in 1861, represent a turning point in the history of meteorology. This is the first published weather forecast. It's the work of Admiral Robert Fitzroy, 'Statist' at the nascent Meteorological Department of the Board of Trade (predecessor of the UK Meteorological Office) which, coincidentally, is founded by the government, also today, in 1854. Initially, the Department simply collates the observations sent back from ships to which it has loaned standardized instruments. In 1861, however, Fitzroy expands his mandate to weather prediction. He designs and distributes a barometer and, using the new telegraph network, starts to issue storm warnings at ports (*25 October*). Soon Fitzroy extends this to regional 'weather forecasts' (he coins the term to distance his work from the efforts of quacks), culminating in today's notice in *The Times*. His expansion of the Department's brief, however, is controversial. Not only is it expensive, it exposes the emerging science of meteorology to public scrutiny.

Fitzroy, a talented seaman, navigator and surveyor, and captain of Darwin's HMS *Beagle*, is not shy of controversy. But accuracy is not a notable feature of these early forecasts. Public scepticism quickly turns to derision. Scientists wishing to establish meteorology as credible science, feel vulnerable. The Board of Trade faces ignominy.

The Times drops the forecasts in 1864. Poor Fitzroy, weary of his critics, abandoned by his friends and deeply in debt, cuts his throat in 1865.

35.2C Boxworth 1995

And that's it for forecasting, for a while. Fitzroy, perhaps the most influential person in the history of weather prediction, is forgotten – until an area of the Shipping Forecast is named after him in 2001. Three years later, the Met Office relocates its HQ. The new address? Fitzroy Road.

2 AUGUST

11.7C min / 20.5C max

ROYAL MAIL /
BUSIEST EVER
BANK HOLIDAY

In sprays of muddy puddle-water rather than a cloud of dust, the first coach of a new mail service leaves Bristol for London at 4 p.m. today in 1784, resplendent in its pristine livery of red wheels and undercarriage, maroon body and black side panels. The time schedule demands a strict 6 mph (10 kph). The mail service will become the pride and joy of the nation and come to epitomize the coaching age, even though this arduous standard is seldom, if ever, maintained. The success of the service is largely about vanquishing the weather and its effects on our generally abysmal roads (*21 February*).

'As I drove down to dine at Greenwich with Alvanley (18 March), Foley, and the Duke of Argyll, we were overtaken near Westminster Bridge by a violent thunder-storm, and went into the House of Lords for shelter. We passed the time in the library, where the librarian showed us various curiosities; among others, the original warrant for the execution of Charles I signed by Cromwell and the other parliamentary leaders. It was found after the Restoration in the possession of an old lady in Berkshire, and formed the ground of the prosecution against the regicides. It is newly framed and glazed, and preserved in the library of the Lords as a most curious document.'
Thomas Raikes, diarist, 1832

Today's August Bank Holiday in 1926 (until 1965, at the beginning rather than the end of the month) begins 'without a cloud, and the sun shone from dawn to evening', according to *The Times*. It's the moment when the English seaside holiday (*28 June*) reaches its record-breaking peak. A tidal wave of visitors swamps Southend (☔), to take just one example, by river and rail, converting the 6-mile-long (10 km) Prom, beach and cliffs into a vast dormitory. Some find hard beds in the form of the counters of closed winkle and cockle stalls, while five hundred doss down in the stalls of a cinema. Ten thousand sleep out in the open, in the 'star hotel' – a national paper reports the line of outdoor sleepers stretching 'all the way to distant Shoeburyness', a story repeated in dozens of coastal resorts. By way of contrast, in 1906, black skies, dazzling lightning and hailstones the size of marbles herald the Guildford Tornado (☔), which kills two.

35.2C Hawarden Bridge 1990

3 AUGUST

11.7C min / 20.2C max

HAIL SMASHES HOUSE OF COMMONS / THE MUCKLE SPATE

A lively day of record heat, hail, lightning and floods. Of the numerous lists of Britain's top hailstorms, for majestic impertinence few match the storms of today in 1879, and of the day before yesterday, the 1st, in 1846. The 1879 hailstorm – notable for incessant lightning – destroys almost every glasshouse in the Thames valley. 'The war of elements in my grounds, with its destructive effects,' a market gardener complains to *The Times*, 'cannot be adequately described.' Kew Gardens suffers three thousand panes smashed in the Temperate House and seven hundred in Paxton's Great Palm House. If this sounds bad, it's nothing compared to the 1846 storm. Mid-Saturday afternoon, in temperatures of 32C, hail shatters more than seven thousand window panes in the Houses of Parliament, three hundred in Old Scotland Yard, ten thousand in Leicester Square, almost every pane of the glass arcades in Regent Street, Somerset House and the Burlington Arcade, and the picture gallery of Buckingham Palace, which, as a result, is also flooded (though, amazingly, no works are damaged).

In 1829 the River Spey in Morayshire (☔), famous for its whisky distilleries, bursts its banks after heavy rain in 'the most catastrophic severe flood'. The 'Muckle Spate' (great flood) washes away stone bridges, alters the course of the river and devastates agriculture. And in 2004, an exceptionally thundery August brings some of the most spectacular lightning displays across Britain since 1923.

'A beautiful day ... Carrie bought a parasol about five feet long. I told her it was ridiculous. She said: "Mrs James, of Sutton, has one twice as long," so the matter dropped. I bought a capital hat for hot weather at the seaside. I don't know what it is called, but it is the shape of the helmet worn in India, only made of straw.' *George and Weedon Grosmith, Diary of a Nobody, 1888*

'Our meadows are covered with winter-flood ... the rushes with which our bottomless chairs were to have been bottomed ... are gone down the river on a voyage to Ely.' *William Cowper, poet, 1782*

'To sit in the shade on a fine day, and look upon verdure, is the most perfect refreshment.' *Jane Austen, Mansfield Park, 1818, referring to 1808*

37.1C Cheltenham 1990

4 AUGUST

11.8C min / 20.1C max

ST JAMES'S
DAY BATTLE /
LONDON'S
HEAT ISLAND

Storms prevent the Dutch fleet landing at Medway (⚔) today in 1666, to fire on the British fleet in port, as they plan. Instead, they blockade the Thames. The ensuing St James's Day naval battle is a decisive English victory, with enormous Dutch losses. But the recent Great Plague (*10 November*), plus the looming disaster of the Great Fire of London (*12 September*), conspire to leave Charles II without the funds to capitalize on the victory and continue the war.

Last night, in 1990, is London's warmest on record – 24C. It's Luke Howard (*18 July*) who first identifies that cities, by night, are considerably warmer than the surrounding countryside. Buildings and roads absorb heat from the sun by day and gently re-radiate it at night, like a storage heater. Because London is our biggest city, that's where the effect is greatest, with temperatures in the centre on average 1.5C higher than in the country (contributing to an estimated 20 per cent fuel saving there). Most conspicuous during summer heat waves, the effect is sometimes used – possibly inadvertently – by novelists. The crucial opening scene of Wilkie Collins's masterpiece, *The Woman in White* (1859), takes place, coincidentally, in the heat of tonight:

'It was now a close and sultry night . . . The idea of descending any sooner than I could help into the heat and gloom of London repelled me. The prospect of going to bed in my airless chambers, and the prospect of gradual suffocation, seemed, in my present restless frame of mind and body, to be one and the same thing. I determined to stroll home in the purer air by the most roundabout way I could take; to follow the white winding paths across the lonely heath; and to approach London through its most open suburb by striking into the Finchley Road, and so getting back, in the cool of the new morning, by the western side of the Regent's Park. . . . I was strolling along the lonely high-road . . . when, in one moment, every drop of blood in my body was brought to a stop by the touch of a hand laid lightly and suddenly on my shoulder from behind me. I turned on the instant, with my fingers tightening round the handle of my stick. There, in the middle of the broad, bright high-road – there, as if it had that moment sprung out of the earth or dropped from the heaven – stood the figure of a solitary Woman, dressed from head to foot in white garments, her face bent in grave inquiry on mine, her hand pointing to the dark cloud over London, as I faced her.'

35.2C Kew Botanic Gardens 1990

5 AUGUST

11.9C min / 20.7C max

PUDDLETOWN AND PIDDLE TOWN

A dramatic fall of rain – 48mm (1.9 inches) in 75 minutes – justifies Puddletown's name today in 1931. The village, in the Piddle valley (♥) near Dorchester, is the inspiration for Thomas Hardy's fictional village of Weatherbury in *Far from the Madding Crowd*.

As temperatures nudge over 30C, Network Rail imposes speed restrictions across the Midlands and southern England today in 2003, fearing tracks may buckle. 'BRITAIN'S RAILWAYS BLAME THE WRONG KIND OF SUNSHINE,' say several newspaper headlines, inevitably. Steel does contract in cold weather and, yes, it expands in heat. But passengers are left wondering why trains are restricted to 60 mph (96 kph) between London and Birmingham, while in exactly the same temperatures they thunder across France at 215 mph (345 kph).

'It is a lovely morning ... There is not a cloud in the sky except a few toy white ones; and yet it is thundering away like mad, peal after peal, over eastward. And now they are just dropping a gauze over the horizon, and the tree under which I am sitting has shivered and sighed, as if it were catching a cold.' *George Bernard Shaw, dramatist, Monmouthshire, 1897*

'I left Stonehaven soon after eight o'clock. The wind was coming in feeble dog-breathes off the land, and the sea outside the harbour was riddled with curlicues of morning mist. I set a course of 040 to clear Girdle Ness and waited for the sun to show out of a sky that was evenly luminous from horizon to horizon ... By nine o'clock, with only seven more miles to go, I realised that the boat was swaddled in thick fog. It had happened invisibly, the damp air slowly turning white as if it was ageing round me. It was impossible to tell how deep the fog was. Sometimes it seemed to stand like a bright cliff, a mile away across the still water, sometimes the bow of the boat appeared to be gouging a hole for itself in the swirling wall of fog. Peering, or trying to peer, ahead, I saw the boat's head slowly swivel round against the lumps and ridges of the fogbank. We were turning in a wide circle. I pulled the wheel round to make the boat point straight again, but then found the compass was reading 105 – on a heading to somewhere in the Frisian Islands. I brought it back to 040, and again saw the boat's bow begin to spin against the fog, while the compass card remained as if glued in its bowl. In fog, you have to trust your instruments.' *Jonathan Raban, Coasting, 1986*

32.2C Cambridge 1897

169.7C Waen Sychlwch 1973

6 AUGUST

12C min / 20.3C max

LARGEST STORMCLOUD / TUNBRIDGE WELLS 'RICE PUDDING'

Today, in 1981, is vividly remembered by Londoners as the day the sky turns black. At midday, in sweltering heat, the sky suddenly starts darkening. And, like some science-fiction nightmare, it keeps on darkening. Street lamps automatically come on. Motorists have to switch on their headlights. Eventually, it's like the dead of night. But this is no solar eclipse; as people are wondering if the apocalypse is nigh, the cloudburst begins. 'WHEN DAY BECAME NIGHT,' headlines London's *Evening Standard* next day. The explanation is a giant cumulo-nimbus thundercloud (*1 July*) that develops when a trough of low pressure moving north from France encounters hot moist air over southern England. Experts subsequently reckon the cloud to be 8 miles (13 km) high – if true, it's one of the largest storm clouds ever. Remarkably, something similar has been happening further north: in Manchester, 95.9 mm (3.8 inches) of rain – six weeks' worth – falls in a few hours last night.

In 1956, Britain's most miserable summer (*29 July*) reserves its *coup de grâce* for today's August Bank Holiday: 1.2 metres – yes, *4 feet* – of hail falls on Tunbridge Wells (☁). The great icy mass is described by one spectator as 'rice pudding on an unimaginable scale'. At other times in history, too, today's weather has surprises up its sleeve. A 'bright red globular ball of fire', 60 cm (24 inches) in diameter, is observed for twenty minutes above the Glendowan Mountains in County Donegal (☁) in 1868. And a rain of dead, but apparently fresh, sprats is delivered upon Great Yarmouth in 2000 – fishes in their dishes without the boats coming in (*18 August*).

'I ascended close under Scafell, and came to a little village of sheep-folds ... Here I found an imperfect shelter from a thunder-shower - accompanied by such echoes! O God! What thoughts were mine! O how I wished ... that I might wander about for a month together, in the stormiest month of the year, among these places, so lonely and savage and full of sounds!' *Samuel Taylor Coleridge, poet, on his nine-day laudanum-inspired 'circumcision' of the Lakeland Hills, 1802*

'It was so hot at Chequers that I took very little exercise after the first day, but I was able to mark out sites for the magnolia laburnums and other trees that I want to plant in the autumn. Out of 18 forest trees that I planted last year in Crow's Close 14 have died! But what could one expect in such a phenomenal drought?' *Neville Chamberlain, Prime Minister, 1938*

33.9C Wisley 1933

7 AUGUST

11.8C min / 20.2C max

MENDELSSOHN'S STORM / HUMPHRY CLINKER

A Hebridean storm inspires Felix Mendelssohn's most famous overture today in 1829. Reaching dry land, after a boat trip to see Fingal's Sea Cave (known in Gaelic as Uamh Binne – 'the cave where the sea makes music') on the uninhabited Isle of Staffa (♥), the twenty-year-old composer writes down the opening theme of 'The Hebrides, overture for orchestra in B minor ('Fingal's Cave'), Op. 26'. Mendelssohn tells his father, in a letter, of his attempt to capture the spirit of the ocean and the weather in musical form: 'In order to make clear what a strange mood has come over me in the Hebrides, the following has occurred to me:

'We took boat again on our return to Leith (♥), with fair wind and agreeable weather, but we had not advanced half-way when the sky was suddenly overcast, and the wind changing, blew directly in our teeth . . . In a word, the gale increased to a storm of wind and rain, attended with such a fog . . . And at the same time, most of the passengers were seized with a nausea that produced violent retchings . . . Mrs Winifred Jenkins made a general clearance with the assistance of Mr Humphry Clinker, who joined her both in prayer and ejaculation. As he took it for granted that we should not be long in this world, he offered some spiritual consolation to Mrs Tabitha, who rejected it with disgust, bidding him keep his sermons for those who had the leisure to hear such nonsense. My uncle sat, recollected in himself, without speaking; my man Archy had recourse to a brandy-bottle, with which he made so free, that I imagined he had sworn to die of drinking any thing rather than sea-water: but the brandy had no more effect upon him in the way of intoxication, than if it had been sea-water in good earnest. As for myself, I was too much engrossed by the sickness at my stomach, to think of any thing else. Mean while the sea swelled mountains high, the boat pitched with such violence, as if it had been going to pieces; the cordage rattled, the wind roared; the lightning flashed, the thunder bellowed, and the rain descended in a deluge.' *Tobias Smollett*, The Expedition of Humphry Clinker, *1771*

34.0C Bromley 1975

8 AUGUST

11.7C min / 20.4C max

ABBEY ROAD /
DRACULA

On a sunny Friday morning in 1969, photographer Iain Macmillan puts up a step-ladder in the middle of a North London road near the EMI recording studios. From about 10 feet (3 metres) up he briskly snaps the four Beatles walking over a zebra crossing. With no text, band listing or title, his photograph is destined to become one of the most iconic album covers in rock history. Myth and folklore immediately begin to accrue around the image. A Detroit DJ starts the 'Paul is dead' rumour, which rapidly builds into a frenzy. Why are Paul's eyes shut? Why is he holding his cigarette in his right hand when he's left-handed? The number on the VW Beetle in the background reads 'LMW 281F', or '28 IF' – *if* Paul were still alive, he'd be 28. (Actually, he'd be 27. But then many cultures count you as one when you're born, don't they . . . ?) Most damning of all, Paul is *not wearing shoes* – he, *alone of the four*, has bare feet. Clearly he is dead. The truth? The truth is that it's a bloody hot day in August, so Paul takes his shoes and socks off.

Today is also the day the novelist Bram Stoker (1847–1912) summons one of the most dramatic storms in fiction to deliver his anti-hero, Count Dracula, to the Yorkshire port of Whitby (♥) where his prey, Mina Murray, awaits. The storm follows a spectacular sunset, after which Stoker, like a madman at the controls of a weather machine, presses into service every conceivable theatrical effect, however contradictory. Despite wind that 'roared like thunder', there is 'masses of sea-fog . . . so dank and cold that it needed but little effort of imagination to think that the spirits of those lost at sea were touching their living brethren with the clammy hands of death'. At length the 'strange schooner' *Demeter*, deserted save for, lashed to the helm, 'a corpse, with drooping head, which swung horribly to and fro at each motion of the ship', makes port. The instant it touches the shore, 'an immense dog sprang up on deck from below . . . and running forward, jumped from the bow on the sand' – the inestimable Count departing in the form of a wolf.

34.2C Heathrow Airport 1975

9 AUGUST

11.8C min / 20.3C max

The hailstorm which strikes Norfolk around 4 p.m. today, in 1843, is one of the worst ever recorded. Almost certainly a convective supercell, it has a fury akin to something out of Tornado Alley in the US – indeed, this *is* a tornado. Ranking H7 ('Very destructive . . . risk of serious injuries') on the TORRO Hail Scale, an intensity achieved only ten times in British recorded history, it devastates a 158-mile (255-km) 'hail swathe' from the North Sea near Horsey (☝) to Stow-on-the-Wold in Gloucestershire (☝). In places, 75-mm (3-inch) stones pile 1.5 metres (5 feet) deep.

'Heat perfectly awful!' *Queen Mary, during the 'Perfect Summer' of 1911 (22 July)*

'"A boiling, sultry day," says Michael Fish on the radio – but here thick fog, murk and drizzle.' *Christine Evans, poet, Lleyn Peninsula, Wales, 1997*

'The lightning and hail were terrific, the former like sheets of fire filled the air and ran along the ground, the latter as large as pigeon's eggs,' writes the Rector of Wimpole, Cambridgeshire. 'All the windows on the north side of the Mansion [Wimpole Hall] were broken . . . The corn over which it passed was entirely threshed out, boughs and limbs torn off trees, pigeons and crows killed, many sheep struck by lightning, and what the hail and lightning did not utterly destroy, the rain which fell in torrents finished . . . Such was the violence of the rain that a stream . . . washed men off their feet, and carried away 30 or 40 feet of the Park wall.' Cambridge suffers wholesale destruction of glass, chimney pots and slates. Another contemporary letter records that '9 days after the Hail Storms at Tew, [Mr Wilkinson] took up Hail stones 6½ inches diam. and sent 2 Cartloads of Ice to the Ice House.'

The agricultural damage is so colossal that, in November, the General Hail Insurance Company is founded. It's not the first company to offer hail insurance. That was the Farmers & General Fire & Life Insurance Company (first issuing policies in 1842), which suffers massive losses following this storm. But, in another way, the storm catalyzes the market, emphasizing hail risk. As the directors of the Farmers & General note, 'It became manifest that a very extensive business could be done.' The General Hail absorbs other hailstone insurance societies, in 1898 amalgamating with the Norwich and London, then, in 1908, with the Norwich Union Fire. As the Norwich Union, it is now one of the largest insurance groups in Europe – all from today's cannonade of hail.

36.7C Raunds 1911

10 AUGUST

11.9C min / 20.4C max

HOTTEST EVER DAY

This afternoon in 2003, the usual muted murmur of the National Meteorological Centre at Bracknell is suddenly broken by 'a suppressed cheer'. News from the London Heathrow weather station has arrived. A UK temperature record has been set: the thermometer has hit three figures – 100.2F (37.9C) – for the first time since reliable records begin in 1875. 'It's an honour to be working on such a day,' one forecaster says. The joy is not shared by the bookies: William Hill alone has to pay out £250,000 when the magical 100F is hit.

> 'A great swell was rolling in from the Atlantic, and in the Race there was such a tremendous sea that neither the boy nor myself could look outside the boat after the first few minutes without turning giddy . . . I felt, besides, the wretched weakening sensation in the spine which most people feel when tossed in a high swing against their will.'
> *R. T. McMullen, sailing off Lizard Point, 1857*

The record is from one of the 250 official Met Office observation stations, where temperatures are recorded to the nearest 0.1C every hour. The Heathrow figure only holds the top spot for a few hours, however. Later the same day, the mercury creeps even higher at Gravesend, Kent where 38.1C (100.6F) becomes the national heat record. London is hotter than Madrid, Cairo, Rome, New York or Nairobi. 'A psychological barrier has been broken,' one newspaper commentator says. 'Global warming is no longer mere theory.'

Elsewhere, this Sunday, there's 'no spare sand' on the beaches along the south coast. Roads melt. Crops catch fire. Shops sell out of paddling pools, air conditioning units and fans. Peals of thunder boom across the Midlands, where twenty people are struck by lightning. In Carlton-in-Cleveland, near Middlesbrough, 30mm (1.2 inches) of rain falls in five minutes, setting a new 'short duration rainfall record'.

At Berwick-upon-Tweed, in 1975, a cricket umpire is struck by lightning, welding solid an iron joint in his artificial leg. In 1718, two lovers are killed by a lightning strike at Stanton Harcourt near Oxford. Following 'so loud a crash as if the heavens had split asunder', they are discovered, so the tale goes, black and stiffened, in tender embrace. Alexander Pope, living nearby, writes them an epitaph which includes the lines: 'Hearts so sincere th' Almighty saw well pleased, / Sent his own lightning, and the victims seized.'

38.1C Gravesend 2003

11.7C min / 20.3C max

COWES WEEK AND THE FASTNET RACE

The sun is shining. It's a beautiful Saturday in 1979, and Cowes Week closes – as always – with the Fastnet Challenge Cup. This is the most prestigious event of the week, a timed, 608-mile (970-km) race from Cowes, west down the English Channel, across the Irish Sea, round the Fastnet Rock (known as 'Ireland's teardrop') off south-west Cork, and back to Plymouth via the Scilly Isles. Taking three to five days, this biennial event (odd-numbered years only) has been run since 1925. For yachtsmen, it's one of the supreme ocean-going challenges, largely because of the long section of open sea crossing the Western Approaches where there's no shelter from transatlantic weather – precisely why, in three days' time, 'Fastnet' will be the most notorious word in offshore racing history.

THE YACHT *AMERICA*

It's fitting that Britain hosts the most weather-dependent of all sporting events – we did invent yacht-racing (*10 October*). Cowes Week, the longest-running regular sailing regatta in the world, is first held in 1826. The week runs from the first Saturday after the last Tuesday in July. It is timed for 'good sailing breezes' (Force 4, or 15–25 mph/24–40 kph) and minimal chance of calm, as well as to fit into the summer social calendar, and the Solent teems with thousands of competitors. In 1851, on a day of little wind, an elegant American-owned schooner effortlessly wins the prestigious 100 Sovereign Cup Race. The yacht is called *America*, and ever afterwards the cup – the oldest trophy in international sport – is known as the America's Cup.

Although it all changes after today, the Fastnet epitomizes a treasured British amateur marine tradition – that any individual can buy a yacht or motor boat and put to sea without legally requiring any kind of certificate of competence. To enter the Fastnet, all you have to do is apply and pay the entry fee. You don't even need a radio.

So, in conditions of almost flat calm, at 1.30 p.m., 303 boats and 2700 crew depart. The 6 a.m. shipping forecast has promised good weather – not that the competitors regard it as 'good'. For thirty-six hours they curse light winds, fog, even patches of total calm. Meanwhile, unknown to them, off the American East Coast a low – 'Low Y' the forecasters call it – is deepening. Already it is generating 40-foot (12-metre) waves as it begins to cross the Atlantic.

By dawn on Monday, most yachts have finally cleared the South Coast. After a frustrating time playing cards and drumming their fingers in fog banks, they're praying for weather. Most of all, they're praying for wind . . . (continued *13 August*).

34.1C Hillington 1884

Fastnet Course

12 AUGUST

11.8C min / 20.4C max

FOOTBALL'S
WINDIEST
GOAL /
GLORIOUS
TWELFTH

The most famous wind-assisted goal in football is scored today in 1967 during the FA Charity Shield match between Tottenham Hotspur and Manchester United. After eight minutes Tottenham's goalie, Pat 'Guardian of the Gates' Jennings, famed for his hands rather than his feet, posts a hefty left-foot punt into orbit from his own penalty area. 'Wafted by a favouring gale,' *The Times* reports, the ball carries far over the halfway line. The crowd wait for Alex Stepney, the opposition goalkeeper, to complete the formality of collecting it. Confounded by the way the wind carries the ball, and the arc of its descent, Stepney is caught far from his goal line. The ball bounces once, high on the bone-hard turf, over his head – and straight into the back of the net. Stepney holds his head in despair, before giving Jennings, according to *The Times*, 'the sort of look a fast bowler gives another when he bowls a bumper at him. "Against union rules, old man, surely?"'

'The queen and I were going to take the air this afternoon, but not together, and were both hindered by a sudden rain. Her coaches and chaises all went back, and the guards too: and I scoured into the market place for shelter.' *Jonathan Swift, writer and satirist. Windsor, 1711*

'We had a heavy fall of rain the whole night. Miss Carney poisoned herself.' *John Fitzgerald, Cork school-teacher, 1793*

Meanwhile, a high wind is what the aristocrats and plutocrats who gather to shoot grouse on the moorlands of northern England and Scotland on this, the Glorious Twelfth, first day of the grouse-shooting season, usually want. It makes birds more 'sporting'. But in 2005, the sun is shining and there's a different problem. Where are the grouse? They are highly susceptible to the weather, and the unusual conditions in 2004 have decimated the population. A good spring – leading to a bumper breeding season – is followed by gales and heavy rain during the shooting season, and few birds are shot. A wet and mild winter then provides ideal conditions for the build-up of the fatally destructive gut parasite strongyle worm. The grouse are almost wiped out.

Back in 1948, the grouse population in the Borders gains a reprieve from the weather: torrential rain keeps all but the hardiest off the Lammermuir Hills, and the ensuing flash floods destroy seven mainline railway bridges over the River Eye.

33.9C Epsom 1911

13 AUGUST

12C min / 20.3C max

FASTNET DISASTER

On this normally innocent August day – when, on average, you can pick the season's first fully ripe blackberries – nineteen die at sea in the 1979 Fastnet Race. It's the worst disaster in ocean-racing history. All 303 competing yachts, after days of frustrating calms, are now in the unsheltered Western Approaches – just as 'Low Y' (*11 August*) arrives from across the Atlantic. Suddenly, unpredictably, everything changes. The barometer plunges to the second-lowest reading around the British Isles for 150 years. Gale-force winds pick up at precisely the point where two wave trains from different directions meet. The result is the storm of the century. Wind speeds reach Storm Force 11, in mountainous seas, as the race sinks into chaos and tragedy. Yachts outside the eye of the storm – those with radios – listen to Mayday after Mayday as the freak event rages for twenty hours. Only after 6.30 a.m. on the 14th, when winds drop to Force 9, can the largest-ever coordinated peacetime air-sea rescue commence, involving the Royal Navy, a Dutch destroyer, four RAF Nimrods, eight helicopters and seven lifeboats, plus assorted tugs, trawlers and tankers.

CHANGES AFTER FASTNET

Yacht stability – yachts should not be able to turn upside down, and should be self-righting. Crews must stay with stricken yachts rather than take to life-rafts. Of twenty-four abandoned ships, only five sink, yet numerous life-rafts come to grief. Also, life-rafts are harder to spot by rescue services. Mandatory VHF long-range radios. Mandatory 48-hour emergency position-transmitting beacons or life-raft 'scramble bags'. Mandatory entry experience qualifications

How bad is it? 'It was like standing at the bottom of the White Cliffs of Dover and looking up to the top – that was the height of the waves,' recalls Nick Ward, left for dead on the yacht *Grimalkin*. Smaller yachts are not just knocked sideways but 'pitch-poled', when the boat cartwheels stern over bow. At one point the 3000-ton Dutch destroyer *Overijssel*, hit by a wave as it turns to collect a life-raft, lies so far on its side that its mast touches the water. In the ensuing mayhem, the life-raft disappears and a man goes overboard. When the ship rights itself, by chance the life-raft is scooped on board. As for the lost man, cries are heard from above, and there he is – two-thirds of the way up the rear mast, tangled in the radar antennae. After the storm, *Overijssel* is written off. Of the original 303 boats, 85 finish; 136 are rescued.

33.9C Seaford 1911

Fastnet Course

14 AUGUST

12C min / 20.4C max

RECORD RAINBOW

A rainbow lasting three hours is seen off the North Wales coast (🌂) today in 1979. Rainbows are notoriously ephemeral, seldom lasting more than minutes. As this is just after the Fastnet storm (*13 August*) passes through the Irish Sea, some read significance into the record duration of this chameleon of the air.

RICHARD OF YORK GAVE BATTLE IN VAIN

Though the rainbow spans a continuous spectrum, the sequence most people cite is: red (outside), orange, yellow, green, blue, indigo and violet. The human ability to distinguish between bluish colours is poor, and it's thought that indigo was only included with the six primary colours in Newton's time because of the preferable religious connotations of the number seven. The traditional way to recall the seven colours in sequence is the mnemonic 'Richard Of York Gave Battle In Vain'.

A rainbow is an optical illusion, which forms when sunlight passes through a 'sheet' of millions of raindrops and is refracted or bent, splitting white light into its component colours in each individual drop, and reflecting it back. Rainbows appear directly opposite the sun from the observer when the angle between the rays of sunlight, the raindrops and the human eye is between 40 and 42, so they are most commonly seen late in the afternoon when the sun is getting lower.

Thinkers from Aristotle to Descartes sought to understand the mysterious rainbow, but it is the young, colour-blind Isaac Newton who first explains how they are formed, in 1665. Not that he is thanked for it. He falls out bitterly with the physicist Robert Hooke over his explanation, and 155 years later in his poem 'Lamia', John Keats accuses Newton of 'unweaving' the poetry of nature:

> '*Philosophy will clip an Angel's wings,*
> *Conquer all mysteries by rule and line,*
> * Empty the haunted air, and gnomed mine —*
> * Unweave a rainbow, as it erewhile made*
> * The tender-person'd Lamia melt into a shade.*'

Despite Keats's fears the rainbow remains, in painting and literature, a symbol of hope and optimism. As D. H. Lawrence writes at the end of his novel *The Rainbow*: 'She saw in the rainbow the earth's new architecture, the old, brittle corruption of houses and factories swept away, the world built up in a living fabric of Truth, fitting to the over-arching heaven.'

35.0C Cambridge 1876

170.8mm Hampstead 1975

15 AUGUST

11.6C min / 20.2C max

LYNMOUTH CONSPIRACY THEORY, AND 'OPERATION CUMULUS'

A series of violent thunderstorms, jettisoning several months' worth of rain – 228mm/9 inches – in twelve hours on the steep-sided, already saturated northern slopes of Exmoor causes death and mayhem this evening in 1952. More than 200,000 tons of boulders, trees, mud and accumulated debris swills off the moor, engulfing the pretty coastal resort of Lynmouth (♥). It's the most disastrous flash flood in decades, killing 34 and leaving hundreds homeless. The East and West Lyn rivers break their dams, smashing 28 bridges and skittling 38 cars out to sea. Nearly 100 houses have to be demolished. Several bodies are never found.

'The lovely weather has Alas broken & this morning we woke to gray skies & torrents of rain, which has now degenerated into a fine drizzle almost like a Scotch mist. However the British do not allow anything so trivial as the Climate to interfere with their Summer Sports & I understand we are to play through the thick of it.' Clementine Churchill to her husband Winston, from Cromer on the Norfolk coast where she's playing in a tennis tournament, 1923

Although this is not the record rainfall for the south-west – a region prone to occasional very heavy rainfall (*18 July*) – the flood leads to conspiracy theories. Nearly fifty years on, a BBC radio programme attempts to show that the Lynmouth disaster was caused by a post-war 'rain-making' experiment that went wrong.

It's true that, in the late 1940s and early 1950s, the Met Office and the RAF do carry out trials in 'cloud-seeding,' in the hope of 'making' rain for the benefit of agriculture. Operation Cumulus experiments with injecting large quantities of silver iodide or dry ice into individual cumulus clouds. The theory is that this will stimulate the formation of ice crystals, causing rain to fall. However, the tests are, at best, equivocal, and in a country that hardly seems short of rain the process never looks economically viable. So is a Ministry of Defence experiment responsible for the Lynmouth flash flood? It's highly unlikely. The trials are always directed at a single cumulus cloud, whereas Lynmouth is hit by rain from a depression which tracks off the Bay of Biscay and drenches the whole of south-west England and South Wales. In short, it's the wrong kind of rain.

33.6C Stratfield Turgis 1876

228.6mm Longstone Barrow 1952

16 AUGUST

11.3C min / 20.2C max

BOSCASTLE: ANOTHER FLASH FLOOD

Today's flood at Boscastle in Cornwall (), in 2004, is the classic flash flood. And because it's recent, happens in daylight and is widely filmed – dramatic images of cars sweeping like Pooh-sticks through the picturesque funnel valley – it's the example we remember. For all the drama, however, the long-term consequences are not serious. Yes, for eight hours this tiny village has the equivalent of the Thames washing through it, and, yes, in that short time, over a hundred are rescued in one of the biggest airlifts since the Second World War. But, compared to the tragedy of the flash flood in nearby Lynmouth (*15 August*) fifty-two years earlier, the consequences are trivial. At Boscastle, no one is killed. What's going on, with two such similar disasters so close? In both instances the storms follow days of heavy rain which have already waterlogged the ground. The result is identical: an immense amount of water with nowhere to go, finding the shortest route to the sea.

Today in 1722, in the north of England, snow on the Pennines between Rochdale and Blackstone Edge (the route of the M62) surprises Daniel Defoe on his *Tour Through Great Britain* (1724) especially when it is accompanied by a clap of thunder ('the first that ever I heard in a storm of snow'). The store of good ale, however, 'seems abundantly to make up for the inclemencies of the season'.

'There has certainly been some serious mistake about this summer. It was intended for the tropics; and some hot country is cursed with our cold rainy summer, losing all its cloves and nutmegs, scarcely able to ripen a pine-apple out of doors, or to squeeze a hogshead of sugar from the cane.' *Sydney Smith, clergyman and wit, Taunton, Devon, 1842*

'The sun still burns & the whole earth glows ... took the bus to Heytesbury to have tea with Siegfried [Sassoon]. The Park looked extraordinarily beautiful – with large clumps of trees ... And the grey square house – cool as a vault as we walked into it out of the sun. At first I thought it was empty – tho' all the lovely rooms were open. Then just as we were going down to the cricket field – S. appeared in white shirt & flannels, looking I thought rather younger & less harassed. We sat outside gazing up at the "hangar" of woods beyond – heavy with summer beauty.' *Violet Bonham Carter, political hostess, 1947*

200.4mm Otterham 2004
33.9C Bournemouth 1947

17 AUGUST

11.4C min / 20.2C max

THE SPANISH ARMADA: 'GOD BREATHED AND THEY WERE SCATTERED'

'Never anything pleased me better than seeing the enemy flying with a southerly wind,' says Sir Francis Drake in 1588, when a violent storm, rather than British seapower, defeats the Spanish Armada. Briefly, the story so far: initial skirmishes off the south coast; the fleets drift into the Channel; Lord Howard, the English commander, launches 'fireships' – unmanned, flaming vessels – towards the Spanish anchorage off Calais, on a perfect westerly breeze. Next day, there's an inconclusive encounter at Gravelines; the wind abruptly backs to the south-east and strengthens; the Spanish fleet puts to sea; for the next few days, the English Navy chase them up the Channel. And that's it. Few ships are lost. It's all thoroughly inconclusive.

DRAKE, BOWLS AND THE WIND

The myth about Drake being alerted to the approaching Armada while playing bowls on Plymouth Hoe, and calmly deciding to finish his game, in fact has more to do with weather and tide than with Drake's *sang-froid*. The great circumnavigator has to wait for an ebb tide and a change in wind direction before his ships can clear the harbour.

Today, the Duke of Medina Sidonia (an unlikely choice for commander – he has never fought at sea and suffers chronically from seasickness) holds a council of war in the North Sea. The southerly wind, attended by 'squalls, rain and fog with heavy seas,' as one Spanish captain records, has dragged the fleet far north. The Duke resolves to cut his losses. Rather than beat back down into the wind to the coast of Flanders, to link up with the Spanish army and revive the proposed invasion, he decides to embark on 2000-mile (3200-km) journey home via the north-west coasts of Scotland and Ireland. It's a fatal decision.

Coming round north-west Scotland, the fleet encounters one of the worst storms ever recorded in the area. It blows for weeks. 'It would overwhelm and destroy the whole world,' records one witness. The Armada is annihilated: some 50 of the 130 ships sink (*9 November*).

Had the weather held and the fleet reached home intact, the Armada would have been hailed as a Spanish triumph. Instead, as the inscription on the medallions minted by Elizabeth I reads: 'God breathed and they were scattered.' In a war fought over religion, England's Protestants feel that God wanted them to win. Drake becomes a national hero, though he has little to do with it.

33.9C Southampton1947

18 AUGUST

11.7C min / 20C max

RAINING
SPRATS AND
FROGS / CHITTY
CHITTY BANG
BANG

'I was amazed to see hundreds of tiny frogs falling with the rain on to the concrete path surrounding the pool,' a swimming pool attendant in Trowbridge (♀) is reported saying today in 1939. The phenomenon whereby showers of frogs, sprats, winkles, starfish or sundry other sea and land creatures rain from the skies (never cats and dogs, actually) has baffled weather-watchers ever since Athanasius wrote *De Pluvia Piscium* (*On the Rain of Fishes*) 1500 years ago. Pepys notes 'that frogs and many insects do often fall from the sky, ready formed', and most weather books have a dose of similar examples. In fact, the explanation is perfectly simple. Small creatures are sucked up by tornadoes (over land) or water-spouts (over sea), then suspended for a time in the updraughts of a thundercloud before being deposited during a heavy shower in some localized spot. Not so interesting now, eh?

HOW TO DEAL WITH BANK HOLIDAY TRAFFIC

'The next day was a Saturday ... and the sun positively streamed down. It was a roaster of a day, and at breakfast Commander Pott made an announcement. "Today," he said, "is going to be a roaster, a scorcher. There's only one thing to do, and that's for us to take a delicious picnic and climb into CHITTY-CHITTY-BANG-BANG and dash off down the Dover road to the sea."'

And so, as everyone knows, the Pott family all pile into the car ('hood down of course') and head for the main road to Dover. 'But, but, but! And once again but!! 22,654 other motor-cars full of families (that was the number announced by the Automobile Association the next day) had also decided to drive down the Dover (☞) road to the sea on that beautiful Saturday morning, and there was an endless stream of cars going the same way'. As they crawl along, getting more and more hot and impatient, 'even CHITTY-CHITTY-BANG-BANG began steaming angrily out of the top of her radiator'. Eventually, of course, jammed solid outside Canterbury, Commander Pott notices an 'angry red knob' reading 'PULL IDIOT'. He pulls down the 'little silver lever' and everyone knows what happens next. CHITTY CHITTY BANG BANG 'tilted up her shining green and silver nose and took off! Yes! She took off like an aeroplane ... soaring over the long line of cars in which the poor people were roasting in the sunshine and sniffing up the disgusting petrol fumes of the cars in front'. *Ian Fleming*, Chitty Chitty Bang Bang, *1964*

238.8mm Cannington 1924

35.0C Jersey Harbour 1932

19 AUGUST

11.6C min / 19.9C max

MORAY
FISHING
DISASTER

'At daybreak this morning the scene that presented itself along the shore between the Buchanness Lighthouse and the entrance to the south harbour, was of the most appalling description. The whole coast was for a mile and a half strewed with wrecks and the dead bodies of fishermen,' reports *The Times* in 1848. The Moray Firth fishing disaster (✠) is the worst maritime catastrophe ever to strike the notoriously storm-torn east coast of Scotland. As is so often the case, the afternoon before is disarmingly fine. Around a thousand herring boats (each manned by five fishermen) put out, as usual, from the villages between Wick and Peterhead, fishing close to shore. Around midnight, the weather deteriorates rapidly. As the wind strengthens and seas mount, many fishermen haul their nets and make for home.

RAIN, WALES AND *INSPECTOR MORSE*

Rain, today in 1972, is ruining the Dexter family holiday in Trefor on the Lleyn Peninsula in North Wales (♀). The father, Colin, has had enough. As *The Archers* finishes, he locks himself in the kitchen with a couple of detective novels. As he reads, all he can think is: *I could do better than this*. So, with the rain rattling against the windows, he starts to try . . .

The supreme tragedy of the ensuing gale is that many encounter disaster within reach of land, but the tide is too low for them to enter Wick harbour: 124 boats are lost and 100 drown, leaving 47 widows. How does it happen that Scottish fishermen are constantly exposed to such calamities, while English fishermen generally weather the storm? asks *The Times*. The answer is that Scottish boats are open, while English boats are decked. In bad weather, Scottish boats provide no shelter and are lethally prone to swamping. Also, the boats' shallow draft (which means they can be launched from beaches and small harbours) makes them easy to capsize. So concludes the report of the Admirality inquiry into the tragedy, chaired by Captain John Washington.

Its recommendations boil down to improving harbours and introducing decks. The latter is resisted – partly because of a fear that decks increase the risk of being swept overboard. Gradually, however, 'forecastles' appear in the bows, and by the end of the century fully decked vessels – the *Baldie* and the *Zulu* – have been introduced: not that this helps much when tragedy strikes again in thirty-three years' time (*14 October*).

36.1C Halstead 1932

20 AUGUST

11.5C min / 19.9C max

HARVEST PROBLEMS So persistent has the rain been throughout the summer of 1860 that divine intervention is deemed necessary to save the harvest, and today the Church of England distributes a 'Prayer for Fair Weather'. It works. The autumn has below-average rainfall and at least part of the crop is salvaged. Disaster is averted.

'I am at present grovelling among the thistles & bees on the brink of a sandpit. The rain was trying to dislodge me an hour ago, & now the sun is having a turn. Nothing could have been less successful with me than the country air ... The heat grows insufferable: I must up & away. Yet last night the country was covered in icebergs.' *George Bernard Shaw, dramatist, Surrey, 1891*

It's a paradox of agriculture in the British Isles that the principal month for harvesting cereals, particularly wheat, is also, for many eastern counties (the Borders, Lincolnshire, East Anglia), the wettest. Elsewhere, it's the second or third-wettest. Too much rain and wheat simply rots, leading in medieval times to famine (*12 May, 25 September*). More recently, very wet Augusts in 1912, 1917, 1956 and 1966 have destroyed harvests. The fatal rainfall figure is 4 inches (101.6mm) – any more than this and gathering it in becomes both difficult and expensive. Even with modern combine harvesters and grain-drying equipment, weather can still ruin the wheat crop: in 2004, following the wettest August for almost fifty years, the price of bread and biscuits is forced up.

August has, over the last fifty years, become drier and, despite the old maxim that 'Drought never killed the British farmer', lack of rain can also affect yields. Without sufficient moisture, wheat prematurely dies off. An ideal summer for an arable farmer goes something like this: May – damp, with plenty of sun; June and July – occasional days of light rain, with long hours of sunshine; August – blazing hot.

Today in 1795, sitting in a garden in Somerset, Samuel Taylor Coleridge listens to the mysterious tones created by an Aeolian harp, a stringed instrument played by the wind alone. This 'transposing of the spirit of the wind' inspires a poem:

> 'And that simplest Lute,
> Plac'd length–ways in the clasping casement, hark!
> How by the desultory breeze caress'd,
> Like some coy maid half–yielding to her lover,
> It pours such sweet upbraiding.'

32.3C Cheltenham 1995

21 AUGUST

SWALLOWS AND AMAZONS

'The storm broke with a crash of thunder that woke the whole camp (♪). With it came a flickering light as bright as day. There was a wild shriek of a parrot as if it were one of a flock screaming through the palm trees in a tropical hurricane. Then darkness and quiet. Then heavy drops of rain pattering down on the tents . . .'

The storm which shatters the charmed, holiday weather in Arthur Ransome's *Swallows and Amazons* breaks today in 1929. For the last two weeks, 'the skies had smiled on the Swallows and the Amazons. There had been a few hours' drizzling rain, a few hours of fog . . . But day after day had been dry and clear and, even when there had been clouds, there had also been sunshine and wind to drive their shadows, chasing each other, over the bright heather and bracken of the hills.'

SNAP, CRACKLE AND POP – THE SOUND OF THUNDER

The sound of thunder is commonly audible 5–8 miles (8–13 km) away (*12 June*). The *type* of noise we hear, however, depends both on distance from and 'structure' of the lightning strike: Rumbles – a rumbling means the lightning is distant; Bangs – a short, violent bang or crack means it is nearby; Crackles – a lightning strike with many branches makes a continuous crackling if it is close by, and a long, low-intensity rumble if far away.

Not any more. The enchanted period of sailing, camping, exploring, lemonade and piracy ends with a bang.

'One flash followed another and then there were three tremendous crashes and a lot of little ones as if the sky were breaking into solid bits and rattling down a steep iron roof.

"There's a broadside for you," called Nancy Blackett from her tent. "Pieces of eight," said the parrot . . .

There was a glare of lightning and a crash of thunder all in one, and after that for a long time the thunder and lightning came so close one after the other that no one knew which flash belonged to which clap of thunder. The camp was full of light and the rolling, crashing thunder overhead made things seem hurried . . .

It was dark again and suddenly quiet . . . as if the storm were holding its breath. Then there was a deep, rushing noise, far away, louder and louder every moment.

"What's that?" said Titty.

"Wind," said Susan.

"I say," said Titty, "this *is* a storm."'

33.0C Cheltenham 1995

22 AUGUST

11.5C min / 19.9C max

WELSH RAIN INSPIRES POETRY In relentless rain, today in 1924 A. A. Milne is on holiday near Porthmadoc (♥) with his three-year-old son, Christopher Robin. 'Screaming with agoraphobia,' Milne escapes to the summer-house with blank exercise-book and pencil. In this 'heavenly solitude', which he determines not to leave until it stops raining, he gazes 'ecstatically at a wall of mist which might have been hiding Snowdon or the Serpentine for all I saw or cared. I was alone . . .' In eleven wet days in the summer-house, Milne writes his first eleven sets of children's verses. Later in that year, the verses are published as *When We Were Very Young*. It becomes one of the best-loved books ever. (Uncannily, Milne is not the only writer to whom Welsh rain proves inspirational – *19 August*).

'Lord, what can one do such Weather as this, continual Rains. My Genius is so dampt by it that I can do nothing to please me.' *Thomas Gainsborough, artist, Bath, 1768*

'First it was a fine evening, with the swallows high and cirro-cumulus clouds, so that I thought it would be a fine day to-morrow. But the clouds got ragged, the gnats were blown away (I smoked cigarettes till my tongue was sore), the swallows went to bed, and even dozens of bats got less enthusiastic. The rain came.' *T. H. White, author, 1934*

'Dense mists swilled from hill to hill and I was by now "mokado", a Romany word meaning "soaked to the skin" that George Borrow once casually threw into conversation to establish his credentials as he sought shelter from a downpour in a Cornish gypsy encampment above Rosewarne. My feet, by this time, had been for a distance swim in my boots. I sensed the imminent onset of trench foot under the composting socks, and incipient blisters boiling up like party balloons. Somehow, I found my way back to the only shelter for miles, my tent, and lay in it nibbling chocolate like an adolescent, contemplating the eight-mile walk back to Ardlussa and the post-bus, and thinking wistfully of the Ardpatrick bath.' *Roger Deakin, Waterlog, 2000, on the Isle of Jura*

SOME WELSH WORDS FOR RAIN

Bwrwglaw raining
Pigo spotting
Gwlithlaw drizzle
Brasfwrw big spaced drops
Sgrympian short sharp shower
Tollti pouring
Byrlymu pouring very quickly
Taflu throwing
Hegar law fierce rain
Lluwchlaw sheets of rain
Pistyllio fountain rain
Curlaw beating rain
Tywallt bucketing rain
Tresio maximum-intensity rain
Mae hi'n bwrw hen wragedd a ffyn
it's raining old women and sticks.

33.9C Canterbury 1918

23 AUGUST

11C min / 19.4C max

THE THREE
HOSTAGES
In pouring rain, on an outside live broadcast today in 2006, ITV Central News weather girl Joanne Malin offers an unusually candid summary of prevailing conditions: 'It's pissing down,' she informs viewers.

Meanwhile, wind plays a crucial part in a *Boys' Own* adventure classic, John Buchan's *The Three Hostages*, today in 1920. The protagonist, Richard Hannay, finds himself being stalked on a Scottish moor (♥) by his arch-enemy, Dominic Medina. 'I remembered a trick which Angus had taught me – how a stalker might have his wind carried against the face of an opposite mountain and then, so to speak, reflected from it and brought back to his own side, so that deer below him would get it and move away from it up *towards* him. If I let my scent be carried to the Pinnacle Ridge and diverted back, it would move the deer on the platform up the corrie towards me. It would be a faint wind, so they would move slowly away from it . . .'

The Cloudspotter's Guide, *1934*
'Clouds are an interesting hobby . . . because they have a direct bearing upon the weather, like the wind. Clouds which live separately in the sky mean dry weather, and clouds which extend, run into each other, or cover the sky, mean wet. There are many other simple rules . . . If there is a good dew in the summer there won't be rain, and vice versa . . . If your own senses become sharpened it generally means wet. Thus country people should redouble their efforts at harvest if they can hear distant railways, smell extensively, or see clearly far. Primed with all this wisdom I asked Tom Bourne how he predicted disaster during harvest, and he replied: "When the wind comes from Gorble Farm, it's going to rain."' *T. H. White, author*

Standing on the skyline, he duly lets the strengthening wind ruffle his hair, then, after five minutes, lies down to watch. 'Presently I saw them become restless, first the hinds and then the small stags . . . then by a sudden and simultaneous impulse the whole party began to drift up the corrie . . . Medina must see this and would assume that wherever I was I was not ahead of the deer . . . I moved a little to the right so as to keep my wind from the deer, and waited with a chill beginning to creep over my spirit . . . I knew now that the beauty of earth depends on the eye of the beholder, for suddenly the clean airy world around me had grown leaden and stifling.'

32.2C Chivenor 1955

24 AUGUST

11.5C min / 19.5C max

LIGHT METERS AT LORD'S / DARTMOOR FOG / PILOTAGE

On a 'really lovely cricketing day' at Lord's, in 1978, light meters are introduced to international cricket. Bad light suspending play has always been (and still is) a contentious issue, as it usually suits only one team to go off. The theory is that a light meter gives umpires an electronic reading that they can use as a benchmark to offer both teams throughout a five-day Test, eliminating disputes between players. It does little to placate spectators, of course, for whom seeing an umpire check his light meter has become an irritating ritual of the game. Not today in 1978, though. The sun blazes down, as it does for the wole Test match between England and New Zealand, and the new device has no chance to prove its merits.

> 'We are in great alarm here for the harvest. It is all down, and growing as it stands. It is Whig weather, and favorable to John Russell's speeches on the Corn Laws'. *Sydney Smith, clergyman and wit, Taunton, Devon, 1841*

Today in 1964, in mist and low cloud with visibility down to 75 feet (less than 25 metres), prison officers and police comb Dartmoor (🌧) (*22 October*). Their quarry is Walter 'Angel Face' Probyn. Having already escaped from seventeen approved schools and prisons, the 'Boy Houdini' slips his guards in filthy weather while on an outside working party. He is rearrested in East London two months later.

'Heavy gale north-north-east,' a Cornish seaman notes in his diary today in 1842. 'At 1 a.m. a large barque discovered in the bay . . . succeeded in boarding her after great risk . . . brought the vessel to anchor to the eastward of the Carrack. Proved to be the *Bosphorous* for Jamaica, with a general cargo. Pilotage claimed £400, settled for £150.' During the nineteenth century, the profits for pilotage – the fee for conducting a ship along a dangerous coast or in and out of a harbour – can be enormous, particularly in bad weather. For the people of Cornwall and the Scilly Isles, the worse the weather, the harder the bargain they can drive. In the 1870s, the advent of iron steamships means this supplementary income largely dries up, though several Cornish ports still operate a policy of compulsory pilotage.

Today in 1940 London records its earliest frost. In 1976, Denis Howell is appointed Minister for Drought (*3 July*). He calls for the country to halve water consumption, and appoints an emergency advisory committee. Five days later, Exeter is flooded.

31.7C Bude 1955

25 AUGUST

WHEN IS A
HURRICANE
NOT A
HURRICANE?

The remains of a tropical cyclone, Hurricane Charley, reach the British Isles today in 1986 – just in time for the Bank Holiday weekend. Forming off the American east coast ten days earlier, it heads across the Atlantic. Over cold water, however, hurricanes lose their energy – and their official hurricane status. No matter how wild the weather that arrives here, it is never, by definition, a hurricane. However, it's still powerful enough to cause the worst flooding in Dublin for a century. In Birmingham, the first inner city 'Superprix' road race – an event sixteen years in the making – is cancelled when conditions are so bad that drivers can't see the car in front, let alone the flag marshals.

'Black cliffs and caves and storm and wind, but I weather it out and take my ten miles a day walks in my weather-proofs'. Alfred, Lord Tennyson, composing Idylls of the King, *at Tintagel in Cornwall, 1860*

'Blue sky and sunshine do so much to warm my heart ... But here they don't exist. The summer is gone, and without having sent a single summer day. Yesterday was a good day i.e., I only got soaked three times ... Bad days are beyond imagination.' Felix Mendelssohn, composer, Llangollen, North Wales, 1829

Only when maximum sustained surface winds reach 74mph (118 kph) does a tropical cyclone – a generic term for a 'low-pressure system over tropical or sub-tropical waters with thunderstorm activity' – become a 'hurricane' (in the North Atlantic) or a 'typhoon' (in the North-west Pacific). Hurricanes generally require sea surface temperatures of 27C or above. Since 1922, at least ten storms in the British Isles (mostly in late August and September) have originated as hurricanes. By the time these storms arrive, they are known as 'extra-tropical storms'. Hurricanes are given names for easier reference and the five most dramatic have been:

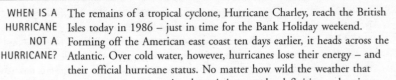

- Gordon (September 2006) – 100,000 without electrical power in Northern Ireland
  - Jeanette (2002 – *27 October*) – the Lerwick ferry is at sea for eighty hours
  - Lili (October 1996) – 80mph (130 kph) winds across Scotland, £150 million of damage
  - Flossie (September 1976) – gusts of 104mph (166 kph) on Fair Isle
    - Debbie (1961 – *17 September*) – entire forests are felled in Antrim, Northern Ireland

32.5C North Heath 1976

26 AUGUST

11.2C min / 19.4C max

GREAT
NORFOLK
FLOOD /
BANK HOLIDAY
WEATHER

Monsoon-like downpours over East Anglia today in 1912 cause the Great Norfolk Flood – the worst flooding by rain (as opposed to the sea) on record. With the land so saturated after a month of rain that it can't absorb another drop, three times the average monthly rainfall arrives in a single day. Forty bridges wash away. Norwich (♥) is completely cut off by road, rail and river. Rescue parties in rowing boats find that innocent street features such as bollards and iron railings have turned into deadly underwater obstructions, capsizing boats and trapping unwary limbs; 3600 houses are destroyed and much of the Fens remains underwater throughout the following winter. It's a grand finale to the worst summer month ever – August 1912 scoops all three records: for wet, cold and absence of sunshine. London has only one day over 21C; Birmingham and Manchester have nothing higher than 19C (it doesn't even make 16C in Aberdeen).

'Here I sit in rather melancholy state saving myself for Lear, which is no joke to plod through these rather boiling days, even though we have wonderful audiences which, after the awful press and the long, unprecedented heat wave is something of a miracle ... The National Gallery is so packed with shorts, lederhosen, corduroy and other distracting gentlemanly attributes that one is tempted to spend long hours there – almost the only place in London with air-conditioning too.'
John Gielgud, actor, 1955

And what of August Bank Holiday weather generally? Is it really the bringer of foul weather we've come to believe? More than two months past the longest day, day length is decreasing markedly and, with it, the potential for sunshine and high temperatures. Thunderstorms, such as the one that wakes Samuel Pepys in the early hours of this morning in 1664 ('put into a great sweat with it, could not sleep till all was over'), are fairly routine. The 1986 Bank Holiday is the wettest on record (*25 August*). On the positive side, sea temperatures are at their highest of the year, afternoon 'highs' average around 21C and, all in all, today is as good a time as any for a break from work and a day on the beach.

32.8C Cromer 1964
185.7mm Brundall 1912

27 AUGUST

RIPE UNTO
HARVEST

'Day after day, as harvest time approached, the children at the end house would wake to the dewy, pearly pink of a fine summer dawn and the *swizzh, swizzh* of the early morning breeze rustling through the ripe corn beyond their doorstep,' Flora Thompson writes in her chronicle of a rural childhood at the end of the nineteenth century, *Lark Rise to Candleford* (1945). 'Then, very early one morning, the men would come out of their houses, pulling on coats and lighting pipes as they hurried and calling to each other with skyward glances: "Think weather's a-gooin' to hold?" . . . There were night scents of wheat-straw and flowers and moist earth on the air and the sky was fleeced with pink clouds.

'For a few days or a week or a fortnight, the fields stood "ripe unto harvest". It was the one perfect period in the hamlet year.'

'A gloomy, dull day. It is astonishing what an impression gloomy weather has on the mind.' *Strother, draper's assistant, Hull, Yorkshire, 1784*

'There is something depressingly mucky about English sea resorts. Of course, the weather is hardly ever sheer fair, so most people are in woollen suits and coats and tinted plastic raincoats.' *Sylvia Plath, American poet, Whitby, Yorkshire, 1960*

A freak rainstorm during the lunch interval floods the pitch on the last day of the fifth Ashes Test at the Oval in 1968. Only when the crowd help the ground staff clear the outfield does any hope of further play become a possibility. Finally, with the pitch ready, there are only 75 minutes of play left. Australia are 86 for 5. For 35 minutes, there is no bounce or spin to help the bowlers: the game is clearly heading for a draw. Then, 'when hope had almost gone', *The Times* reports, the sun comes out (*31 July*) and a ball from Ray Illingworth fizzes past the shoulder of the bat of the Australian opening batsman, Inverarity, setting up a dramatic climax.

With Derek Underwood bowling, and all ten fielders in catching positions near the bat, England sniff the faintest chance. Wickets begin to tumble. With five minutes to play, the suspense breaks as the final wicket falls: Inverarity is out leg before, offering no stroke to Underwood. Australia lose by 226 runs. England, already defied twice this summer by rain, level the series. The victory begins an unbeaten Ashes run that only ends at Lord's in 1972.

33.9C Sprowston 1942

28 AUGUST

10.8C min / 19.5C max

BRITISH
MALARIA /
HARDY'S
LANDSCAPES

Today in 1658, as malarial fever spreads across England, Oliver Cromwell, born in the Fens (♀), is on his back with the 'bastard tertian ague', enduring 'burninge fits violent'. In sixteen days, he is dead. The post-mortem indicates septicaemia, though malaria is contributory. After a wet and muggy August, there's another outbreak in 1848. The highest incidence is reported in Cambridgeshire, Essex and Kent – coastal or marshy areas where the *Anopheles* mosquito thrives.

Although malaria has never been a serious killer here, it's always been present. The Venerable Bede mentions it in the seventh century. The literature of the Middle Ages contains many references to 'ague' or 'intermittent fever', as malaria is known. Certainly Shakespeare is aware of the link between the disease and wetlands. King Lear curses Goneril: 'Infect her beauty / You fen-suck'd bogs, drawn by the pow'rful sun/ To fall and blast her pride!' It is thought that Charles II, James I, Elizabeth Stuart and Cardinal Wolsey all suffer from malaria.

'The faint summer fogs in layers, woolly, level, and apparently no thicker than counterpanes, lay spread about the meadows in detached fragments of small extent ... The cows sat, the snoring puffs from their nostrils, making an intenser little fog of their own amid the prevailing one ... and the meadows lay like a white sea, out of which the trees rose like dangerous rocks.' *Thomas Hardy*, Tess of the D'Urbervilles, *1891*

In the seventeenth century it is known as the 'harvest ague' because it strikes in hot summers. In 1638 it 'raged so fiercely that there appeared scarce hands enough to take in the corn'. One traveller in Essex in the early nineteenth century reports meeting with men 'that have had from six to fourteen wives' because of the custom of going to the uplands to find them. When they return, the women 'came into the fogs and amps in the marshes changed complexion, got the ague.' And died, of course.

The last two distinct epidemics – 1848 and 1859 – when the death toll is in the hundreds, are both in unusually hot, wet summers. There has been a long-term, steady decline in incidence since the 1860s, probably linked to non-climatic factors such as the increase in livestock populations, drainage of the Fenlands, better medical care and wider use of quinine. Some, including the UK Chief Medical Officer in a 2002 report, argue that global warming will lead to a resurgence of malaria in Britain.

33.9C Rickmansworth 1930

29 AUGUST

10.8C min / 19.2C max

MARY QUEEN OF SCOTS' HAAR / CRICKET'S ASHES

A timely haar, the early summer sea fog characteristic of Britain's north-east coast, saves Mary Queen of Scots from capture in 1561. Returning from thirteen years' exile in France, she creeps unnoticed through the English ships awaiting her. 'Borne up the Firth of Forth on a fresh east wind, the fog settled for miles along the shoreline, heavy and impenetrable ... All night long the ships floundered outside the harbour,' reports a contemporary chronicler. Although several of Elizabeth I's biographers assert that she never issued adequate orders for Mary's capture, the fact is that Mary sails unobserved through the haar to land safely at Leith (♀). The political turmoil between Scotland and England will eventually cost her her life.

In 1882, sodden and damaged by two days of rain (with more rain during the match), England play Australia in the Test match on an uncovered pitch at The Oval. Although England have the edge the first day, after more rain on the second they go to pieces completely – and for the first time an England side loses a Test in England. The *Sporting Times* runs a mock obituary: 'In Affectionate Remembrance of ENGLISH CRICKET ... The body will be cremated and the ashes taken to Australia.' Thus international Test cricket's greatest contest gets its name – the Ashes.

TOLL FOR THE BRAVE

A sudden gust of wind today in 1782 tips off balance one of the largest ships of her day. The 100-gun *Royal George* is leaning over for repairs in the Solent when the gust topples her completely. Unthinkable disaster follows, reminiscent of the *Mary Rose* (*29 July*), as not only all her crew of 850 but hundreds of visitors, family and friends – nearly 1400 in all – are drowned. The disaster inspires William Cowper's poem 'Toll for the Brave'.

*'Toll for the brave
The brave that are no more,
All sunk beneath the wave,
Fast by their native shore.*

*'Eight hundred of the brave,
Whose courage was well tried,
Had made the vessel heel,
And laid her on her side.*

*'A land breeze shook the shroud,
And she was overset,
Down went the Royal George,
With all her crew complete.'*

33.3C Kensington Palace 1930

30 AUGUST

10.8C min / 19.1C max

SCUFFLES AT LORD'S / HARDY'S THUNDERSTORM

Rain inspires perhaps the all-time least likely incident of crowd violence today in 1980. The scene is the Centenary Test between England and Australia at Lord's. Supposedly it's a celebration of cricket, but two rain-soaked days have dampened proceedings. By this afternoon, Saturday, the rain has stopped and the sun is shining brightly, but there's still no cricket – areas of the ground remain damp. The frustrated crowd begins to interpret the dithering of the umpires as a malign desire to keep the players off the field.

Mid-afternoon Messrs Dickie Bird and David Constant, described by the commentator, John Arlott, as 'young, conscientious and capable officials', make their fifth pitch inspection. Returning to the pavilion, the captains and umpires are 'assaulted', or at least 'jostled' (Constant is grabbed by the tie) outside the Long Room by no lesser figures than . . . members of the Marylebone Cricket Club. The 'language of the terraces comes to Lord's,' *The Times* exclaims. And to prove that being roughed up by Sancerre-charged MCC members compares to being attacked by lager-loaded Millwall fans, when the umpires eventually come out to restart play they are escorted by four policemen. Despite two declarations by the Australians, the game inevitably ends in a draw. Dickie Bird pronounces it the worst day of his cricketing life.

'Heaven opened then, indeed. The flash was almost too novel for its inexpressibly dangerous nature to be at once realized, and they could only comprehend the magnificence of its beauty. It sprang from east, west, north, south, and was a perfect dance of death. The forms of skeletons appeared in the air, shaped with blue fire for bones – dancing, leaping, striding, racing around and mingling altogether in unparalleled confusion. With these were intertwined undulating snakes of green, and behind these was a broad mass of lesser light. Simultaneously came from every part of the tumbling sky what may be called a shout; since, though no shout ever came near it . . . meantime one of the grisly forms had alighted upon the point of Gabriel's rod, to run invisibly down it, down the chain, and into the earth. Gabriel was almost blinded, and he could feel Bathsheba's warm arm tremble in his hand – a sensation novel and thrilling enough; but love, life, everything human, seemed small and trifling in such close juxtaposition with an infuriated universe.'
Thomas Hardy, Far From the Madding Crowd, *1874*

31.7C Jersey 1906

31 AUGUST

10.5C min / 18.9C max

MUSHROOM 'WHITE-OUT' 'The skies themselves seemed to make way for her arrival, the clouds parting like an honour guard,' records Jonathan Freedland of the BAe 126 jet landing at RAF Northolt with the body of Diana, Princess of Wales, the day after her death in a car crash in Paris in 1997. 'Then . . . the sky darkened, and the wind whipped harder. It felt like the last day of summer, and the beginning of a long winter.'

'This is the last day of August and like almost all of them of extraordinary beauty. Each day is fine enough and hot enough for sitting out; but also full of wandering clouds; and that fading and rising of the light which so enraptures me in the downs; which I am always comparing to the light beneath an alabaster bowl ... The clouds, if I could describe them I would; one yesterday had hair flowing on it, like the very fine hair of an old man.' *Virginia Woolf, writer, 1928*

'This morning I woke to autumn – a warm autumn morning rather than the cool summer morning of yesterday. It's like the difference between a fiddle and a 'cello both playing middle C: the same note, but another quality. There was a light veil of mist on the meadow, dew heavy on the rough grass ... and nearby in the shorter grass the first autumn crocuses.' *Wilfrid Blunt, writer, 1963*

If it rained yesterday, after a dry spell, then this, the last official day of summer, is the ideal moment for mushroom-hunting in woodlands, pastures and meadows. Fungi may be found throughout the year, but it's the damp, temperate autumn months between now and the first frost (*25 October*) when the common field mushroom, and scarcer ceps and chanterelles, are most abundant. Forming a mushroom takes masses of moisture (they are 93 per cent water), so hot, dry summers are hopeless. But timely rain can be like flicking a switch – suddenly they are everywhere. Years beginning with a slow, fine summer, continuing wet and mild into autumn, are best. A recent vintage year was 2000, and 2005 wasn't bad. The perfect combination of temperature, humidity and soil can occasionally produce the sought-after 'white-out', when so many field mushrooms appear simultaneously that the grass appears covered by snow. 'I have heard of this happening to lucky villages and of every portable container up to the size of the local bus being commandeered to take the crop in,' writes Richard Mabey in *Food for Free.*

146.8mm Ditchingham 1994

34.9C Maidenhead 1906

SEPTEMBER

1 SEPTEMBER

10.7C min / 18.8C max

It's touching 34C today in 1906 when Jimmy Conlin makes his debut for Manchester City – sporting a knotted hanky on his head. Conlin will go on to play for England and become only the second player in the world to be transferred for over £1000. His debut is remembered, however, not so much for the way he plays as for the way he doesn't play: he passes out in the first half. And he's not alone: Manchester City lose five men to heat exhaustion, finishing the game (against Woolwich Arsenal, who win 4–1) with just six. It's probably the hottest day on which a football League programme has ever been completed. Down the road in Liverpool, the first-ever game to be played in front of Anfield's new open stand, the Kop (named after Spion Kop, the hill in South Africa where so many Liverpudlians die in the Boer War) is nearly postponed because of the heat.

'The fog was so thick and white along some of the low land, that I should have taken it for water, if little hills and trees had not risen up through it here and there.' *William Cobbett, radical journalist and MP, Kent, 1823*

This freak Saturday is part of one of the most remarkable heatwaves ever to bake the British Isles. Low pressure sucks burning hot, southerly winds across the land and temperatures rocket: 35C in London on the 1st; 35.6C at Bawtry in Yorkshire on the 2nd is the hottest temperature of 1906 and a still unbeaten record for September; on the 3rd, 34.2C is recorded at Westley in Suffolk. As the whole summer has been hot and the countryside is tinder-dry, fires break out on heaths and in haystacks. Infant mortality peaks at five times the average rate, mainly on account of worsening sanitary conditions in the cities.

Though this is, for meteorologists, the first day of autumn, September often feels like summer. At least thirty Septembers in the twentieth-century have been as warm as June. Normally, though, the horse chestnut trees start to change colour now as the leaves thin and make space. The days begin to shorten noticeably, with morning fog and afternoon haze: shady places stay damp all day. There is still a risk of big thunderstorms in the first half of the month, and the prospect of gales increases after the 20th.

35.0C Collyweston 1906

2 SEPTEMBER

16.8C min / 19C max

WEATHER AND THE AGE OF ENLIGHTENMENT

'Rainy morn,' Robert Hooke notes in his diary today in 1667. It may be just one small entry for the acclaimed physicist, but it's a giant leap for meteorology. Read in conjunction with Hooke's diary entries from mid-July, this innocuous note is tremendously important: it disproves the universally held fallacy that if it rains on St Swithun's Day, 15 July, it will rain for the next forty days. Hooke notes on 15 July 'a pretty deal of rain', then forty days of drought, followed by rain again today.

'Over the radio I heard the Germans excitedly calling to each other ... I looked down. It was a completely cloudless sky and way below lay the English countryside, stretching lazily into the distance, a quite extraordinary picture of green and purple in the setting sun. We were at twenty-eight thousand feet ... Sheep yelled "Tally-ho" and dropped down in front of Uncle George in a slow dive in the direction of the approaching planes.'
Richard Hillary, Spitfire pilot during the 1940 Battle of Britain, The Last Enemy, 1942

This is the Age of Enlightenment and Hooke, 'the father of modern science', is pioneering a new methodology for objectively observing the weather. In 1663, his paper to the Royal Society – *A Method for Making a History of the Weather* – sets out his template for keeping a weather journal with eight immutable criteria. It encourages several other members of the Society, including the philosopher John Locke, to do so.

Hooke is convinced that by recording and studying daily accurate, comparable observations it will be possible to deduce the laws of weather, and thereby forecast it. He is way ahead of his time. Modern meteorology defined by scientific rigour and institutionalized discipline will not emerge for two hundred years. But the balance is tipped. Weather is no longer read as divine punishment, predicted by lore and superstition, but as a science requiring steady accumulation of empirical data. The prospect of predicting the unpredictable is now real. This is also, incidentally, the time when the weather first becomes a favoured topic of conversation for British people.

In 1875 today, in thick fog off Wicklow Head (✛) on the Southern Irish coast a 'lamentable disaster' befalls the British Navy when two armour-plated steamers, HMS *Iron Duke* and HMS *Vanguard*, collide. With a 'chasm in her side', the *Vanguard* sinks in an hour, though all hands are saved. Nearly a century later, in a gale touching Force 9, politician Edward Heath's yacht *Morning Cloud III* sinks off Shoreham-by-Sea in Sussex (✛) when it's hit by a 26-feet (8-metre) wave, washing two crew overboard.

35.6C Bawtry-Hesley Hall 1906
171.0mm Cowlyd 1983

3 SEPTEMBER

10.8C min / 18.9C max

DISCOVERY OF PENICILLIN

It's unseasonably cold and has been for days, well below September's 13.6C average, when the Scottish bacteriologist Alexander Fleming returns to London from his two-week summer holiday today in 1928. He's habitually untidy, and the benches in his lab at St Mary's Hospital are in the same appalling clutter that he left them in. Piles of culture dishes smeared with the *Staphylococcus* bacteria he's been experimenting with lie around, and the room's chilly since he left the window open. Tidying up ready for a visitor, he notices that one of the dishes has been contaminated by greenish-yellow mould – probably from a spore blown in through the open window. In the cold it's thrived. (Moulds like the cold, which is why food left in fridges goes mouldy.) But the sharply observant Fleming notices something odd. Around the mould, between it and the bacteria, there is a clear halo where bacteria have not grown. What follows is one of the most brilliant 'Eureka!' moments in science, which will change medicine for ever. Fleming realizes that the mould must be releasing a substance that inhibits the growth of the bacteria.

Fleming names the active ingredient of his mould 'penicillin', after the *Penicillium notatum* mould from which it comes. He lacks the expertise to purify it, however, and it's only when, a decade later, Howard Florey and Ernst Chain at Oxford isolate the active ingredient, that usable penicillin is produced. In 1941, a doctor at an Oxford hospital with a patient dying from bacterial infection tests some of Chain and Florey's new, purified penicillin. The results are hard to believe. Immediately and spectacularly the patient begins to recover. Sadly, the penicillin runs out and the patient relapses and dies, but its potential is revealed. Pharmaceutical companies hasten to perfect stable doses of the new 'wonder drug'. By 1944, in time for D-Day, doctors have the most powerful weapon in their armoury that they have ever had, something that can conquer many of mankind's oldest and deadliest scourges, from syphilis and TB to gangrene. It's the first of an almost magically effective new kind of drug, the 'antibiotic': the most effective life-savers ever invented. Fleming, Chain and Florey share the Nobel Prize in 1945. All because of a messy, brilliant scientist – and an unseasonably chilly London September.

34.2C Westley 1906

4 SEPTEMBER

10.4C min / 18.9C max

On this bright, sunny morning in 1966, the remains of Hurricane Faith (*25 August*) arrive off the Cornish coast in the form of a giant wave. 'A small crowd gathered to watch, open-mouthed,' report Rod Holmes and Doug Wilson in their surfing memoir *You Should Have Been Here Yesterday*. 'This looked more like Waimea Bay [Hawaii] than Cornwall.'

'It was hard to estimate the exact size,' writes John Conway, late Editor of British surfing magazine *Wavelength*. 'But when Jack Lydgate paddled up one wave, he left three hand-hole paddle marks in the face – and he wasn't even at the top'. One spectator describes the wave as 'the size of an average house.'

This phenomenon is the Cribbar, a wave up to 35 feet (nearly 7 metres) high. Put in perspective, that's as high as two and a half double-decker buses, and considerably higher than the 2004 Asian tsunami. Curiously, one of Britain's most spectacular weather features is almost unknown, except to surfers. Partly this is because the monster is elusive. It appears only two or three times a year, for a couple of hours at a time. It takes its name from the Cribbar Reef which stretches out for half a mile from Towan Head at Britain's surf central, Newquay (🌧).

The Cribbar arrives only under highly specific weather conditions – very low pressure (as caused by dying tropical hurricanes) and a low spring tide, with a gentle offshore breeze to hold up the walls of water. These are all met this morning in 1966, when the Cribbar is conquered for the very first time by three Australians: Jack Lydgate, Johnny McIlroy and Pete Russell. Riding giant waves is still a daring novelty, first attempted in Hawaii only a year or two earlier.

Europie Castle Reef

The Cribbar has been ridden only rarely since, largely because – as the title of Holmes and Wilson's memoir suggests – it is so unpredictable. Vintage recent Cribbars include those of December 2004 and the 'Big Wednesday' of September 2006. In September 2001, during a month of spectacular surfing weather off Devon and Cornwall, a dozen top surfers gather for Britain's first-ever Big Wave contest, all hoping to become the Cribbar King. But the wind flattens the sea, and there is one critical absentee: the Cribbar itself.

Bundoran Reef

The Cove

Cliffs of Moher

32.3C Rickmansworth 1940

Cribbar
Porthleven

4 SEPTEMBER

'I became an official elite Cribbar surfer ... at 8.30 am. I was using a 10ft. board, but when I caught that big wave it felt like a toothpick under my feet. It was like jumping out of a fourth floor flat and then being chased at 40mph down the road.' Simon Jayham, surfer and ex-international swimmer, 2006

Waimea Bay, Hawaii? No, Newquay, Cornwall.

GIANT WAVES IN THE BRITISH ISLES
Porthleven, south Cornwall (☞)
Castle Reef, Thurso, Scotland (☞)
Europie, Isle of Lewis, Scotland (☞)
The Cove, Staithes, Yorkshire (☞)
Bundoran Reef, Co. Sligo (☞)
Cliffs of Moher, Co. Mayo (☞)

5 SEPTEMBER

10.8C min / 19C max

THE HORSHAM
HAILSTONE

This (see below) is the actual size of one of the hailstones which fall on Horsham, Sussex (♀) today in 1958. At 6.5 oz (190g), it weighs more than a cricket ball and is the heaviest hailstone ever recorded in Britain. Orchards are destroyed and lawns left pitted along the south coast by this storm. Giant hailstones (*24 January*) begin as snow or ice crystals in thunderclouds (*15 May*), and increase in size rapidly by colliding in violent air currents with supercooled water droplets. Cut open, a large hailstone resembles an onion. Eventually they become too heavy for the air currents to keep them airborne, and fall from the cloud to the ground.

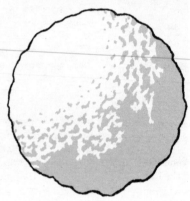

'Some girls up the road spend a very wet Sunday morning playing leap-frog in their pyjamas around the tennis lawn. It makes me envious. To think I never thought of doing that! and now it is too late [he is dying] ... I once hugged myself for undressing in a cave by the sea and bathing in the pouring rain, but that seems tame in comparison.'
W. N. P. Barbellion, aka Bruce Frederick Cummings, Journal of a Disappointed Man, *1917*

'The wind shifted a point to the east, the mist furled up, the rain ceased, and a world was revealed from which all colour had been washed, a world as bleak and raw as at its first creation. The grey screes sweated grey water, the sodden herbage was bleached like winter, the crags towering above them might have been of coal. A small fine rain still fell, but the visibility was now good enough to show them the ground behind them in the style of a muddy etching.' *John Buchan,* John MacNab, *1925*

32.8C Maldon 1949

6 SEPTEMBER

10.6C min / 18.8C max

END OF THE HOLIDAY FOR AGRICULTURE

After a fleeting dry period rain resumes today in 1879, during the wettest and coldest summer of the century. 'Remarkable,' the *Journal of the Royal Agricultural Society* records, 'both for its frequency and for the amount' of rain, as it dawns on the rural community that the harvest is a disaster. The seemingly endless, effortless tide of prosperity enjoyed by Victorian agriculture for half a century is over.

The consequences of this meteorological *annus horribilis* – rain on 179 days, below-average temperatures in every month – cannot be overstated. The agricultural depression, which begins in 1879, changes the countryside for ever and shatters Victorian self-belief. As the historian G. M. Young writes, 'Great Wars have been less destructive of wealth than the calamity which stretched from 1879, the wettest, to 1894, the driest, year in memory.'

'The sun was setting as I crossed Waterloo Bridge, a red bubble behind the Houses of Parliament, but in Waterloo station the sunlight had still been intense, though of that thick, almost palpable radiance that low sunbeams have from autumn suns seen through glass.'
Arthur Graeme West, soldier and anti-war poet, 1916

Of course, the weather alone is not responsible for the Great Depression. Rising costs of production, lamb arriving in refrigerated ships from New Zealand and cheap wheat imports from the USA all contribute to diminishing returns for British farming. Yet it is the trigger. 'The lessons such seasons teach us are really few. We cannot contend with Nature in such mood,' one Forfarshire farmer notes. A meteorologist in Wales concludes, 'The loss was that of money, crops, rent, and all that goes to make up the sum of material prosperity, and even the little reputation for climate this country previously enjoyed.'

The artist J. M. W. Turner is aboard a steamer on the Thames at Rotherhithe (♀) today in 1838, sketching the *Temeraire*, second ship-of-the-line at Trafalgar, as she's towed away to the breakers' yard. His painting, *The Fighting Temeraire*, comes to represent the passing of the age of sail, as the critic John Ruskin later notes: 'Never more shall sunset lay golden robes on her, nor starlight tremble on the waves that part at her gliding.' However, as one witness of the event later observes, the sunset that Turner decides to paint – to symbolise the end of the era, and, perhaps unconsciously, the decline of his own life – does not take place. It's actually cloudy and dull.

32.1C Brookwood 1911

7 SEPTEMBER

10.2C min / 19C max

GRACE DARLING / VAPOUR TRAILS

A violent storm off the Northumberland coast at first light this morning in 1838 leads to the most famous lifeboat rescue of them all. When the passenger steamer *Forfarshire* wrecks on the Harcar Rocks (⊕), the keeper of the Inner Farne lighthouse and his beautiful twenty-two-year-old daughter Grace launch into the mountainous sea to pull seven survivors from certain death – dumbfounding the local lifeboatmen who arrive later. Unquestionably a tale of conspicuous courage, it rapidly grows into far more than this, ticking, as it does, every box of mid-Victorian romantic sensibility – wild setting, raging sea, a vulnerable maiden, even her name. In a fit of Grace-mania, 'the girl with the windswept hair' swiftly finds herself exalted to Florence Nightingale status as the *ne plus ultra* of Victorian heroines: a proto-feminist phenomenon, epitomizing every idealized virtue of the age. She's mobbed by portrait-painters and those wanting scraps of her clothing or locks of her hair. She's the subject of two novels and at least three biographies. Her likeness soon adorns tea caddies, china, postcards, chocolate boxes and girls' annuals. From working-class households to Queen Victoria, the genuflecting is universal – and, apparently, inexhaustible. Wordsworth leaps on the bandwagon with some toe-curling verse. In 1888, she even receives that twenty-first-century badge of authentic celebrity status when her sister publishes the debunking *Grace Darling: Her True Story*. Needless to say, it doesn't sell half as well as the sensationalized versions. Most fitting of all, just four years after the rescue she dies of consumption in her father's arms – untarnished by marriage, unwithered by age.

A NEW KIND OF CLOUD

'Suddenly we were gaping upwards. The brilliant sky was criss-crossed from horizon to horizon by innumerable vapour trails. The sight was a completely novel one. We watched, fascinated, and all work stopped.' So notes Desmond Flower today at the start of the London Blitz in 1940. The condensation trail or 'contrail', left by hot engine gases at high altitude, is now so familiar we hardly notice it. But to British civilians at the start of the Second World War they are a bizarre novelty. 'The little silver stars sparkling at the heads of the vapour trails turned east. This display looked so insubstantial and harmless; even beautiful,' continues Mr Flower. 'Then, with a dull roar which made the ground across London shake, the first sticks of bombs hit the docks ...'

33.4C Norwood 1898

THE GREAT
HUNGER

It is mild, grey and wet today in Dublin in 1845. Nothing newsworthy in this, as it's been mild, grey and wet all summer. What makes today's *Dublin Evening Post* is indeed momentous news, however. The potato blight has arrived. Ireland is about to be crippled for a century and, though no one realizes it yet, the weather is responsible.

The potato comes to the Emerald Isle around 1585, and Irish farmers discover that this strange tuber thrives in their mild climate and peaty soil. 'Universally palatable from the palace to the pig-sty', it becomes the staple diet. By the mid-nineteenth century, the country is dangerously close to monoculture. The prevalent variety is the watery 'Lumper' – highly productive and easily grown, but susceptible to blight. Disaster is poised: it just needs the right weather to trigger it.

The fungus *Phytophthora infestans* probably comes to Europe in 1844 and reaches Ireland on a breeze in August 1845. Reports of decaying potato leaves with black spots and white mould reach Dublin today. To germinate, the fungus spores need moisture. When mature, spores are detached by splashing rain or gentle winds and carried to surrounding plants. In this way, a single plant can infect thousands in hours – but conditions must be exactly right. Cool, wet nights and cloudy days with high humidity, a breeze and a temperature of over 10C, are ideal.

There is no effective antidote to blight, and during the damp autumn of 1845 the scale of the crop failure emerges. Though Ireland is accustomed to agrarian crisis, even famine, this is new territory. By October, millions of ripe tubers have turned black and a nauseous stench pervades the country. In the wettest regions, 40 per cent of the crop is ruined. The famine begins in spring 1846, when every last scrap of edible potato has been consumed. In August, amid torrential rain and wild thunderstorms, the blight drifts with the prevailing winds across the country again and the potato crop is annihilated. Deaths from starvation, disease and, during the unusually severe winter of 1846–7, exposure become the norm.

34.6C Raunds 1911

An Ghorta Mor, 'The Great Hunger', continues until 1849. At least a million die. Another million emigrate on the 'coffinships' to the New World. The famine enters the Irish psyche deeply. Though the weather in Ireland is still a constant source of humour, no one jokes about the potato.

9 SEPTEMBER

10.1C min / 18.4C max

NEWTON'S
WIND-
JUMPING

A mighty gale rocks Britain today in 1658, toppling church steeples, damaging harbours and beaching a whale in the Thames at Dagenham in Essex. Such a great wind, according to a superstition of the time, predicts the death of a great person – and Oliver Cromwell lies on his deathbed (*28 August*). 'Behold', say Cromwell's supporters as the wind howls, 'the rush of archangels to marshal into paradise my lord's illustrious soul!'

'Mark', say his enemies, 'how the demons of the air battle for the mastery of his spirit, and assemble to grasp it when it glides away!' Sure enough, Cromwell dies four days later.

While the metaphysical poet Andrew Marvell composes verses about the storm ('of huge trees, whose growth with his did rise, / The deep foundations opened to the skies'), a sixteen-year-old Lincolnshire schoolboy amuses himself trying to calculate the wind's force. First he jumps with the wind, then directly against it. After measuring the length of each leap, and comparing it with the length he can jump in perfectly still weather, he rates the force of the storm in feet. It is, he claims later, one of his first-ever scientific experiments. He is Isaac Newton.

In an average year, the first ripe sloes can be picked today, though to make sloe gin country lore dictates that you must wait until the first frost (*25 October*). Of course, there's always the freezer.

'Rain has a way of bringing out the contours of . . . previously invisible things . . . I hear the rain pattering on the roof . . . dripping down the walls to my left and right, splashing from the drainpipe at ground level on my left, while further over to the left there is a lighter patch as the rain falls almost inaudibly upon a large leafy shrub. On the right, it is drumming, with a deeper, steadier sound upon the lawn . . . Further out . . . I can hear the rain falling on the road, and the swish of the cars . . . I think that this experience of opening the door on a rainy garden must be similar to that which a sighted person feels when opening the curtains and seeing the world outside . . . If only rain could fall inside a room, it would help me to understand where things are . . . This is an experience of great beauty. I feel as if the world, which is veiled until I touch it, has suddenly disclosed itself to me . . . I am no longer isolated.' *Professor John Hull*, Touching the Rock: An Experience of Blindness, *1990*

32.2C Geldeston 1898

OWAIN
GLYNDWR

In a wild rainstorm tonight in North Wales in 1402, Henry IV's tent collapses and nearly crushes the King to death. It's the last straw: during Henry's three-week campaign to put down the Welsh revolt led by Owain Glyndwr, 'never did a gentle air breathe on them, but throughout the whole, rain mixed with snow and hail afflicted them with cold beyond endurance,' his chronicle records. How Henry must hate Wales. This is his third expedition in successive years that has been ruined by the weather. Rumours that Glyndwr – characterized by Shakespeare as 'not in the roll of common men' – is using powerful magic to control the elements against the King are difficult to suppress. As the sodden English march back towards their border, the tide of revolt spreads rapidly through the green, green hills. Glyndwr, proclaimed 'Prince of Wales' in 1400 but really just a fugitive guerrilla at the head of a band of 'bare-footed rascals of small reputation', begins to acquire the authority of a national leader.

ADVANCED FOUL-WEATHER FETISHISTS

'All day long towering curtains of rain swept down Buttermere (**♥**) ... and shattering gusts of wind whipped the waters of the lake, rattling the canvas of the marquee. A good day for snoozing over the fire with the Sunday papers but 150 people, in their twenties to their sixties ... spent the day racing round the mountain tops, fighting their own private battles in the mist, rain, wind and piercing cold. It was the annual mountain trial, the country's most formidable test for fell runners, over a secret course of more than 20 miles ... This year's course went over, or near, the widely-scattered summits of Crag Hill. Dale Head, Seathwaite Fell and Red Pike ... In a race, in appalling weather, with perspiration running into your eyes, your body shivering with cold and the map streaming with water, which way would you go through the boiling cloud from, say, Crag Hill to Dale Head? There were problems like this all the way round, with thousands of feet to be climbed and much scree and rough ground to be descended. The winner, a 22-year-old Borrowdale farm lad – his third victory in a row – did so in an incredible 4¼ hours, nearly 25 minutes ahead of the second man. For this superb effort he holds a trophy for a year and picked up a piece of outdoor gear ... In 40 years this annual event has never been cancelled, no matter what the weather.' *A. Harry Griffin, climber and journalist, Lake District, 1994*

30.0C Loughborough1891

11 SEPTEMBER

9.9C min / 18.2C max

9/11 2001

Bad light and rain stop play in the 2005 Ashes series at the Oval today; 54 crucial overs are lost. Time is running out for Australia, who must win to retain the Ashes. As the England captain, Michael Vaughan, says, 'A lot of time was lost to rain, but it didn't seem to matter to those who had paid to watch cricket. The umbrellas went up and they were singin' in the rain – literally.' The sight of 23,000 spectators, some of whom have paid £1000 for their tickets, standing and cheering as the players leave the field, is the most bizarre sporting sight of the summer. As one fan says, 'We were merely giving thanks at the altar of the weather gods.'

No one is singing in the rain, however, today in 1800. In fact, it's rained every day since 19 August. Following an intense July drought – no rain falls in London all month – and severe hailstorms in August, the wheat harvest is ruined. Between January 1799 and March 1801, the price of bread trebles and Bread Riots – the 'evil necessarily arising from unfavourable seasons', as George III puts it – follow. William Pitt's government bans the distillation of spirits and the powdering of wigs with flour, and offers a bounty to importers of grain. Meanwhile, the military are on the streets. At the back of the government's mind, of course, is the French Revolution twelve years earlier. Crop failures there led, ultimately, to the storming of the Bastille.

The international consequences of the failed harvest become apparent in 1800. Tsar Paul of Russia imposes an embargo on British ships in the Baltic, from where much of Britain's grain is imported. The Royal Navy is despatched to Denmark, to break the blockade. After numerous shenanigans comes the moment, the next spring, when Vice-Admiral Horatio Nelson holds his telescope to his blind eye, ignores the signal from his commander to withdraw and destroys the Danish navy at the Battle of Copenhagen, reopening the Baltic. The price of bread falls and Nelson becomes a hero.

At Raunds in Northamptonshire in 1919, today is the hottest day of the year. The mercury hits 32.2C (90F). Nine days later, there is snow in Wales, Scotland and northern England. Autumn, it seems, has shrunk.

32.2C Raunds 1919

12 SEPTEMBER

9.9C min / 17.9C max

THE GREAT FIRE OF LONDON

'Pish, a woman might piss it out,' declares Sir Thomas Bloodworth, Lord Mayor of London, in the early hours of this Sunday in 1666 after surveying from his window the fire that's started in Pudding Lane. With this, he returns to bed. By dawn, three hundred houses have burned down, the flames are halfway across old London Bridge and, in Samuel Pepys's words, the 'infinite great fire' rages out of control.

'Gloom: damp: fear: worry: perplexity: depression.' *Harold Nicolson, diplomat and writer, 1931*

'An exciting morning here, with real thunder and lightning, stage management up there must be working overtime – Heaven does these things so well; I thought there had been a bit of skimping lately, but this morning rescinds all former verdicts. The skies boomed, the lightning flashed with a year's vengeance, and the rain streamed down like everlasting tears.' *Kenneth Williams, actor, 1951*

What has this to do with the weather? Well, everything. For weeks, a summer drought has reduced the flow of London's springs to a trickle, wells are low and the capital has been blow-dried by arid north-easterly breezes. By September, with its densely packed warrens of thatched, timber houses and warehouses containing pitch, tar, oil, spirits, hemp, flax, tallow, straw and coal, London is a powder keg.

The fatal spark, as every schoolchild knows, comes from Charles II's royal bakery. From there, flying embers ignite hay and straw in the yard of the adjacent Star Inn. Thence to St Margaret's church, and thereafter, fanned by a fierce east wind, the fire spreads – well, like wildfire along the narrow, crooked streets, leaping from one projecting upper storey to another. Pigeons, 'loath to leave their houses . . . hovered about the windows and balconies till they . . . burned their wings, and fell down,' notes Samuel Pepys. With the wind 'mighty high and driving it into the city', Pepys warns the King that demolishing houses to create firebreaks is the only answer. 'But the fire overtakes us faster than we can do it,' wails the hapless Bloodworth ('like a fainting woman', records Pepys). But Bloodworth has a point. The wind is now so strong that the fire can jump gaps of as many as twenty houses.

By evening, 'the streets full of nothing but people and horses and carts loaden with goods, ready to run over one another', the fire is 'still increasing and the wind great . . . one entire arch of fire . . . above a mile long. It made me weep.' And so London burns on into the night . . . (continued, *14 September*).

30.6C Tunbridge Wells 1911

THE FIRST SPY STORY / CROMWELL AT DUNBAR

A North Sea gale today in 1902 is the centrepiece to Erskine Childers' classic novel *The Riddle of the Sands*. Into the lethal, mist-shrouded sandbanks of the Frisian Islands, where the waters are shallowest and the waves a 'wall of surf,' the sinister Captain Dollman deliberately lures the cruising yacht *Dulcibella*.

Appearing around the same time as the first modern intelligence agencies, Childers' only novel invents a new literary genre: spy fiction, or 'spy-fi'.

> 'I dream of cottages in the country, but of course the damp always deters me; all those wet leaves drearing about like Henry James.'
> *Kenneth Williams, actor, London, 1969*

Joseph Conrad, John Buchan, Graham Greene, John Le Carré, Ian Fleming, Len Deighton and Frederick Forsyth all follow in his wake. Closely based on a North Sea cruise undertaken by Childers in 1897 – large parts of his journal appear verbatim – the book draws the attention of the British Admiralty to the danger posed by the foggy and forgotten Baltic, which could easily be used for invading Britain. Winston Churchill later credits the book as the major reason why the Admiralty decides to establish naval bases at Invergordon, the Firth of Forth and Scapa Flow.

Today in 1650, under cover of high winds, hail and darkness, Oliver Cromwell manoeuvres his New Model Army into position before the Battle of Dunbar (⚔). At first light, his cavalry sweeps into the enemy camp and slaughters the startled Scots Covenanters as they wake, snatching his greatest military victory. Up to now, the New Model Army's Scottish campaign has been a fiasco. Hampered by weather at every turn, cut off from England by road, exhausted, demoralized, dying fast from disease and short of supplies (a series of storms has prevented reprovisioning by sea), the English are utterly depleted. Of the orginal 16,000 troops, only 11,000 remain fit for duty. With their backs to the sea, they contemplate a Dunkirk-style evacuation before devising a plan to strike under cover of the stormy night.

157.7mm Broadford 2005

28.3C Camden Square 1934

By 7 a.m., as the sun burns off the last curlicues of mist, General Leslie's Covenanter army has been scattered: 3000 killed and 10,000 captured. Cromwell, 'drunken with the Spirit and filled with holy laughter', marches unopposed to capture Edinburgh. Parliament issues the 'Dunbar Medal' – the first of its kind – to all ranks.

BURNED: 'LONDON WAS, BUT IS NO MORE'

'The noise and crakling and thunder of impetuous flames, the shrieking of Women and children . . . a resemblance of Sodome, or the last day . . . London was, but it is no more' – John Evelyn describes the Fire of London, this Tuesday in 1666. With no respite from the east wind, London's been burning since Sunday (*12 September*). The flames can be seen from Oxford. Melted lead from the roof of St Paul's pours down Ludgate Hill. Burning embers rain on Kensington.

OUT OF THE ASHES
• Insurance (*9 August*) • Fire 'brigades' – organized by insurers who realize it's cheaper to extinguish than to rebuild. • Brick and stone buildings – for decades to come, the newly rebuilt part of the City will contrast bizarrely with the remainder of London's timber buildings • Wren's St Paul's Cathedral (and 51 City churches). • Eradication of the plague – which only the previous year kills 17,440 (*10 November*) • Wider, clearer streets – though the medieval plan remains • Philadelphia, USA – yes, one of the new rejected grid plans for the new London (by Richard Newcourt) is used to lay out the US city.

Tomorrow it reaches – finally and literally – a brick wall at Temple. Elsewhere gunpowder checks its march. It won't go out, however – fully out – until the first rain for weeks falls next Sunday. Even then, the ground remains too hot to walk on for days and charred embers reignite in coal cellars until March. Everywhere stinks of ash and smoke, as Samuel Pepys surveys the surreal scene: 'All the town burned, and a miserable sight of St Paul's church, with all the roofs fallen . . . nor could one possibly have known where he was, but by the ruins of some church or hall, that had some remarkable tower or pinnacle standing.' Unburnt houses are looted. Whitehall is deserted. Over thirteen thousand houses have gone, together with eighty-seven churches, fifty Livery Halls, the Royal Exchange, Newgate jail, the Guildhall and four bridges. A hundred thousand have been made homeless.

Amazingly, however, there are almost no casualties. Unlike the first 'Great Fire' in 1212 (*17 July*), or the Blitz, the wind has given clear notice of the fire's course, allowing people to escape. A new City of London will rise from the ashes to become the hub of Britain's empire.

27.2C East Bergholt 1947

15 SEPTEMBER

9.5C min / 17.2C max

BATTLE OF BRITAIN

Exceptionally early snow across north-east Scotland today in 1782 becomes known as 'the Black Aughty Twa' – Scots for 'eighty-two'. It sets the oat crop back weeks. 'The corns which had milky juices in the ear were totally ruined; those which had only watery juices wanted season,' a witness writes. It's Christmas before all the corn is cut and most of it has to be given, unthreshed, to cattle. The Duke of Gordon saves his tenants by rent rebates and imported meal.

'The dripping weather has lasted this day nine weeks, all thro' haying & harvest.' *Gilbert White, naturalist and writer, Selborne, Hampshire, 1785*

A century and a half later 'Hitler's weather' – the perfect summer of 1940 – finally breaks today. Despite this, 400 German aircraft pour from the clouds as the RAF launches a massive counter-attack. More than fifty are shot down. It's the turning point in the Battle of Britain, one of the few campaigns in history remembered for gloriously fine weather (*10 July*). Goering's threat to destroy the RAF and crush the morale of civilians has failed, and Hitler abandons his invasion plans (*22 September*).

'As for the sunshine, it goes on and on, unbelievably – Macmillan weather, which makes it impossible for the public to take politics seriously. So we are left with an awkward predicament. In order to destroy a complacent mood of euphoria, we need thunder, rain and hail. But will thunder, rain and hail deter the Labour voters? There's a nice choice for you.' *Richard Crossman, Labour politician, on the first day of the election campaign, 1959. The fine weather continues, affirming the feeling that people have, in Harold Macmillan's words, 'never had it so good'. On 9 October, 'Supermac's' Conservatives are duly re-elected (9 April).*

Today in 1643, persistent rain floods the siege trenches and tunnels under the walls of Gloucester (⚔) during the Royalist siege of the city in the English Civil War. 'The most miserable, tempestuous, rainy weather, that few or none could take little or no rest,' Captain John Gwynne writes. 'The creasing winds the next morning soon dried up our thorough wet clothes we lay pickled in all night.' The siege has to be abandoned. And on a clement day in 1784, only 'a little depressed by the wind', Vincent Lunardi, secretary to the Neapolitan ambassador, carries out the first manned balloon flight in England. He is accompanied by a dog, a cat and a pigeon.

28.9C Southend on sea 1947

16 SEPTEMBER

'CLOUDY, OUTBREAKS OF RAIN'

If the 'Ten Greatest Weather Forecasting Cock-ups of All Time' has Michael Fish's reassurance that there will be no hurricane in 1987 (*16 October*) at No. 1, then the forecast for today in 1968 ranks a close second. 'Generally cloudy, outbreaks of rain,' the Met Office proclaims for this Sunday. What follows, over the next five days, is complete paralysis of the south-east due to flooding. From Norfolk to the Hampshire–Sussex border is, a rescue worker says, 'one giant lake'. Full-scale mobilization of the army, complete with thirty-two assault craft, is required to evacuate the Molesey area in Surrey (♀). Several drown. All rail services are cut. Gatwick airport is closed. 'Everywhere there's a dip in the road there's water, sometimes feet deep,' says the AA. It's the worst flooding since the East Coast sea-surge of 1953 (*31 January*) and the last occasion before the Boscastle Flood of 2003 (*16 August*), when more than 8 inches (200 mm) of rain falls in a day. 'Cloudy, outbreaks of rain' doesn't really cover it.

Unfortunately, the Met Office has just taken delivery of a fancy new computer – the Atlas – costing taxpayers £475,000 (about £12 million in 2007 prices). The BBC quickly disassociates itself from its forecast providers – pointing out that, only an hour after the forecast, they were reporting floods in Devon – as questions are raised in the House of Commons. The Director-General of the Met Office, Dr B. J. Mason, issues a statement to the effect that the Met Office is really pretty good at forecasting these days. 'There's no computer sufficiently powerful anywhere to make an accurate forecast of the rainfall on a day-to-day basis,' he says, disappointing those who naively believed that this was the principal function of the Met Office. 'We shall not be able to do this daily unless we acquire a computer twenty times faster.' A computer eighty thousand times faster duly arrives in 1972, but for the era of genuinely accurate forecasting we have to wait until the era of the supercomputer.

In 1620, on a 'prosperous' wind, a 'fine, small gale' from the east-north-east, the *Mayflower* sails from Plymouth (☝) with 102 Pilgrim Fathers. After 'many difficulties in boisterous storms' the tiny vessel (just 90 feet x 26 feet/27 x 10 metres) reaches Cape Cod, New England, in two months, with only a single casualty.

31.1C Norwich 1947

100mph gust
Scilly Isles 1935

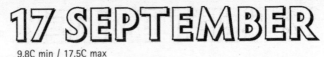

9.8C min / 17.5C max

HURRICANE DEBBIE

Hurricane Debbie – or at least the remains of it – joins the annual Northern Ireland Forestry Division tree-felling competition today in 1961. Sparing no thought for the feelings of the Emerald Isle's sturdiest lumberjacks, hundreds of acres of forests are thrown down in winds gusting over 100 mph (160 kph). Sections of the coast are eroded; the network of telephone lines is shredded like cobwebs. In Londonderry (🌧), scarcely a house is undamaged. Thirteen are killed. Though Debbie does not feature in the grand pantheon of American hurricanes (it veers north-east before hitting the US coast), it's one of the most destructive storms ever to hit Ireland (*25 August*). Crossing the North Channel, the wind speeds moderate a little before it strikes Scotland, but the demolition continues from Ayr, where a baby is killed by a falling chimney stack, all the way to Caithness where a 59-ton fishing vessel is thrown on to rocks at Dwarwick Head (❄). But in 1944 it's a 'brilliant' day, and a giant air armada crosses the Channel heading for Holland. Operation Market Garden is the (ultimately, unsuccessful) attempt to capture the road and rail bridges over the Lower Rhine at Arnhem – the bridge too far. The day is so clear that the glider pilots, many of whom never return, can see people far below on their way to church, waving encouragement.

'The creeping fogs in the pastures are very picturesque and amusing and represent arms of the sea, rivers, and lakes.' *Gilbert White, naturalist and writer, Selborne, Hampshire, 1777*

'The weather last night & this morning has been lovely. The rose garden outside my window is full of colour and one cannot believe in this marvellous silence and stillness that we are at war.' *Neville Chamberlain, Prime Minister, Chequers, 1939*

'What shall I write about? Shall I write about the bright morning with the sharp bird notes and the delicious spongy cooings of the pigeons on the roof of this house? Shall I write about the noises of the aeroplanes, the last flower of the wisteria that I can see mauve and pitiable out of the window? Shall I write about the war ending?' *Denton Welch, novelist, 1944*

31.9C Cambridge 1898

18 SEPTEMBER

9.7C min / 17.3C max

IRELAND'S SOFT RAIN / STORM IN THE CHANNEL

A staggering 9½ inches (243.5mm) of rain falls at Cloore Lake in County Kerry (🖊) today in 1993: a daily rainfall record for Ireland. The dominant influence on the climate of Ireland is the Atlantic. Because of the Atlantic, it suffers none of the temperature extremes of countries at similar latitudes. As the Venerable Bede notes in the eighth century: 'Ireland, for wholesomeness and serenity of climate, far surpasses Britain.' There is, of course, a hitch. The isle is emerald for a reason: it rains a lot. As the agricultural reformer Arthur Young puts it in 1776: 'I have known a gentleman in Ireland deny their climate being moister than England's; but if they have eyes let them open, and see the verdure that cloathes their rocks.' Tropical deluges like today's at Cloore Lake are not common, however. More often the rain falls gently: 'A fine soft day in the spring,' as the narrator in John Ford's film *The Quiet Man* puts it. It is said that the Gaelic lexicon for rain – covering everything from the misty drizzle that won't spot a cigarette paper to driving pitchforks – is as comprehensive as the Inuit vocabulary for snow.

'I am convinced there is sun and blue sky enough in England to satisfy any reasonable person.' *Louis Simond, American traveller, 1811*

'Rose at seven. Soft morning inclined to rain. Went the rounds after breakfast ... Scolded Thomas for growing fat ... Spent half an hour in the shrubbery. Mild grey day.' *Lady Eleanor Butler, Llangollen, 1785*

'What dreadful Hot weather we have! – It keeps one in a continual state of Inelegance.' *Jane Austen, author, Kent, 1796*

A storm that's penned the invasion fleet of William, Duke of Normandy on to the French coast for weeks finally breaks today in 1066. William, conscious that the wait is sapping the ardour of his army, wastes no time in launching his four hundred or more galleys and hundreds of transport ships. But just as he does so, the storm returns. During a perilous sail 100 miles (160 km) east along the Normandy coast, several sink. Putting into harbour at St-Valéry-sur-Somme, according to the Norman chronicler William of Poitiers, the Bastard Duke strives 'with holy prayers ... to quell the contrary wind'. He does not realize it yet, but it's precisely these 'contrary winds' that will help William's successful invasion of England (*3 October*).

28.9C Aber 1926

243.5mm Cloore Lake 1993

19 SEPTEMBER

9.4C min / 16.8C max

KEATS'S ODE 'TO AUTUMN' / FLODDEN

Today in 1819 (appropriately the day, on average, on which conkers first appear) sets the standard by which all future oh-so-perfect autumn days will be judged. This is because the poet John Keats selects it for literary immortality. Trying to work in his rooms in Winchester, Keats's muse has to compete with the landlady's daughter practising the violin. Finally, maddened by the din, the poet decides to escape and walk through the water meadows behind Winchester College (☂). 'How beautiful the season is now – How fine the air. A temperate sharpness . . . – Aye better than the chilly green of the spring,' he later writes to his friend John Reynolds. 'This struck me so much in my Sunday's walk that I composed upon it.' Back in his rooms, he dashes off nothing less than his 'Ode to Autumn' – along with Wordsworth's 'Daffodils' (*15 April*), the most famous poem of the Romantic movement.

'A most glorious, cloudless sky; the weather brilliant but cool . . . I went out and painted a beautiful river in the afternoon light with crimson and golden hills . .'. *Winston Churchill, Dunrobin Castle, Scotland, 1921*

'An incredible autumn morning, still and hazy. At seven or thereabouts, before going to the station, I walked the dogs . . . All the fields are yellow with corn stubble, but in the valley the trees are dark, dark green; in that last cycle before they start to shed their leaves.' *Alan Clark, politician, 1988*

Keats's odes are often acclaimed as his greatest works. It's a poignant time of life for him (his brother has recently died, he himself is suffering from TB, and his love for Fanny Brawne seems doomed because he can't afford to marry her), so 'Ode to Autumn' is infused with a sense of mortality, melancholy and time passing that many feel is emblematic of the season. Professor Harold Bloom calls it 'as close to perfect as any shorter poem in the English Language'.

Three centuries earlier, conditions are so wet at the Battle of Flodden in Northumberland (⚔) in 1513 that advancing Scottish pikemen remove their shoes for better grip on the slippery hillsides. It doesn't save them. With the Scots' momentum and footing lost, the English billmen surge through the gaps in the front ranks and are soon lopping the heads off their pikes. The battle is one of the bloodiest defeats inflicted on the invading Scots (under James IV) by the English. Today can be frosty or frying. In 1952, it's -3C in East Anglia and the Midlands; in 1926, a late heat-wave pushes temperatures in Camden, North London to 32.2C – the latest date that the magic 90F happens in the twentieth century.

31.7C Hunstanton 1926

John Keats (1795–1821) *To Autumn*

Season of mists and mellow fruitfulness!
Close bosom-friend of the maturing sun;
Conspiring with him how to load and bless
With fruit the vines that round the thatch-eaves run;
To bend with apples the mossed cottage-trees,
And fill all fruit with ripeness to the core;
To swell the gourd, and plump the hazel shells
With a sweet kernel; to set budding more,
And still more, later flowers for the bees,
Until they think warm days will never cease,
For Summer has o'erbrimmed their clammy cells.

Who hath not seen thee oft amid thy store?
Sometimes whoever seeks abroad may find
Thee sitting careless on a granary floor,
Thy hair soft-lifted by the winnowing wind;
Or on a half-reaped furrow sound asleep,
Drowsed with the fume of poppies, while thy hook
Spares the next swath and all its twined flowers;
And sometimes like a gleaner thou dost keep
Steady thy laden head across a brook;
Or by a cider-press, with patient look,
Thou watchest the last oozings, hours by hours.

Where are the songs of Spring? Ay, where are they?
Think not of them, thou hast thy music too, —
While barred clouds bloom the soft-dying day
And touch the stubble-plains with rosy hue;
Then in a wailful choir the small gnats mourn
Among the river sallows, borne aloft
Or sinking as the light wind lives or dies;
And full-grown lambs loud bleat from hilly bourn;
Hedge-crickets sing, and now with treble soft
The redbreast whistles from a garden-croft;
And gathering swallows twitter in the skies.

20 SEPTEMBER

9.3C min / 16.8C max

SENSE AND SENSIBILITY The earliest widespread snow of the year falls today in 1919 – especially remarkable considering that, nine days earlier, at 32C (90F), it's the hottest day of the year. In 1846 gales ravage Ireland, already devastated by the potato famine (*8 September*). So while this can be the finest, most settled time of year, the equinox, with all its weather implications, looms.

'We have had noble clouds & effects of light & dark & colour – as is always the case in such seasons as the present.'
John Constable, artist, 'sky-ing' (23 October), Hampstead, 1821

'An absolutely perfect windless autumn day, as we went up the Dee Valley. Of course the Grampians are nothing like as beautiful as the west coast. Balmoral was chosen by Prince Albert merely because the weather was drier than in the west.' *Richard Crossman, Labour Leader of the Commons, driving to Balmoral, 1966*

Jane Austen (*6 July, 12 November*) uses today's weather to advance her plot in *Sense and Sensibility* of 1811. (Austen fans have laboriously plotted timelines for every detail of her books.) On a country walk, 'attracted by the partial sunshine of a showery sky', the flighty, emotional Marianne Dashwood ('Sensibility' – as opposed to 'Sense', her down-to-earth, practical elder sister Elinor) is duly caught in a shower. 'They pursued their way against the wind, resisting it with laughing delight for about twenty minutes longer, when suddenly the clouds united over their heads and a driving rain set full in their face.' Obliged, unwillingly, to turn back, 'one consolation remained for them, to which the exigence of the moment gave more than usual propriety, – it was that of running with all possible speed down the steep side of the hill which led immediately to their garden gate. They set off. Marianne had at first the advantage, but a false step brought her suddenly to the ground . . .'

Her ankle twisted, leaving her unable to stand, the 'uncommonly handsome' Mr Willoughby is – very fortunately – passing but yards away. 'Perceiving that her modesty declined what her situation rendered necessary,' he sweeps her up, and down to the house. There, greeted with amazement by Mrs Dashwood, 'the influence of youth, beauty, and elegance, gave an interest to the action which came home to her feelings'. Mr Willoughby, we hardly need tell you, is a vile rogue and libertine – with whom, naturally, Marianne remains entranced for the rest of the story. Reader – we regret to tell you, she doesn't marry him.

29.4C Shoeburyness 1926
190.7mm West Stourmouth 1973

21 SEPTEMBER

8.7C min / 16.5C max

Today (or, more commonly, tomorrow) is the autumn equinox, the day when day and night are momentarily the same length. In practice, this is not the case. Because sunlight bends as it passes through the Earth's atmosphere, this gives an extra few minutes of daylight in the morning (when we can see the sun rising before it actually does) and in the evening (when we can still see it setting after it's actually set).

'Beyond the orchards the lone aspen was rustling loud and mournfully a lament for the departure of summer.'
Reverend Francis Kilvert, diarist, 1870

'Walking along South Audley Street this afternoon, the streets half-filled with soft autumnal sunshine, I got a glimpse of the reality of my recent life; the meaning of it permeated my mind in the same way as the sunshine that was making the quiet purlieus of Mayfair so pleasant and so discreetly enchanted. Yes, I soberly accepted the fact that I have spent the rainless months since April in ruminating the final flavours of my prolonged youthfulness.'
Siegfried Sassoon, anti-war poet aged 35, 1921

The skies are clear for the first-ever parachute descent, by André Jacques Garnerin, in a field at Marylebone in 1802. But after he's cut the cord from the hot air balloon ('with a hand firm from a conscience void of reproach'), wind begins to oscillate Garnerin's parachute violently. With equal violence, on landing the French pioneer is sick over his admirers.

Finally, no equinox would be complete without an equinoctial gale (*22 September*) story. Today in 1881 the writer Joseph Conrad, aged 24 and a Second Mate aboard a ship sailing from London for Newcastle to collect coal, runs into a gale between Great Yarmouth and Dogger Bank (⊕):

'It was wind, lightning, sleet, snow, and a terrific sea. We were flying light . . . she shifted her ballast into the lee bow . . . There was nothing for it but go below with shovels and try to right her . . . in that vast hold, gloomy like a cavern, the tallow dips stuck and flickering on the beams, the gale howling above, the ship tossing about like mad on her side; there we all were, Jermyn, the captain, every one, hardly able to keep our feet, engaged on that gravedigger's work, and trying to toss shovelfuls of wet sand up to windward. At every tumble of the ship you could see vaguely in the dim light men falling down with a great flourish of shovels. One of the ship's boys (we had two), impressed by the weirdness of the scene, wept as if his heart would break.'

27.2C Jersey 1989

22 SEPTEMBER

8.8C min / 16.5C max

THE MYTH OF EQUINOCTIAL GALES

The equinoxes, spring and autumn, have become indelibly associated with gales since the Norman Conquest (*18 September*), Julius Caesar (*8 July*) or even earlier. Yet this is despite data unequivocally indicating that gales are no more likely now than at other times (although gales *are* commoner in winter). Why does the idea persist? Certainly, as is readily noticeable from the surrounding pages, there is no shortage of gales at sea at this turn of the year. But is this just self-fulfilling prophecy, because mariners are superstitious and like to have identifiable 'bad' periods? Whatever the reason, centuries of sea lore have compacted into an equinoctial gale myth. One of the reasons that Hitler never seriously considers Operation Sealion, his plan for invading Britain, is because German High Command is warned how bad equinoctial weather in the Channel and North Sea can be in late September. The tradition is so ingrained that any storm around this time of year is automatically called an 'equinoctial gale'.

Two other extremes: in 1810, at Fernhill Heath, Worcester (**♀**), a titanic T8 tornado (*see Appendices*), one of the most violent British tornadoes ever, has a reported track-width of up to a mile (1.6 km); if so, making it the *widest* recorded here. In 1935, meanwhile, an H6 hailstorm with a track-length of 208 miles (335 km) is the *longest* recorded here. Finally, it's the unofficial end of summer: the last swallows, on average, fly south today.

'Autumn begins for me with the first day on which the stags roar. Because the wind is nearly always in the west, and because the fences keep the bulk of the stags to the higher ground above Camusfeàrna (♀) ... I hear them first on the steep slopes of Skye across the Sound, a wild, haunting, primordial sound that belongs so utterly to the north that I find it difficult to realize that stags must roar, too, in European woodlands. It is the first of the cold weather that leads in the rut, and the milder the season the later the stags break out, but it is usually during the last ten days of September. Often the first of the approaching fall comes with a night frost and clear, sharp, blue days, with the bracken turning red, the rowan berries already scarlet, and the ground hardening underfoot; so garish are the berries and the turning leaves in sunshine that in Glengarry a post-office-red pillar-box standing alone by the roadside merges, for a few weeks, anonymously into its background.'
Gavin Maxwell, Ring of Bright Water, *1960*

26.1C Collyweston 1901

23 SEPTEMBER

8.9C min / 16.5C max

In 'a cloud o'mist them weel concealed', General David Leslie manoeuvres his Parliamentarian army out of Melrose and over the Ettrick river (⚔) to within musket range of the enemy today in 1645. The autumnal mist is so thick that none of his men are detected by the Marquess of Montrose's scouts, and by the time the alarm is raised the Battle of Philiphaugh has begun. The advantage of surprise is decisive. In just two hours, the Royalist army is demolished and Montrose's brilliant Scottish campaign is abruptly halted. And so Charles I's hopes of salvaging something by military means in the Civil War are snuffed out.

'Black snails lie out, and copulate. Vast swagging clouds.' *Gilbert White, naturalist and writer, Selborne, Hampshire, 1783*

In 1810, a 'grand . . . wonderful . . . sublime' thunderstorm passes over Chevin Hill (⚡), where J. M. W. Turner is staying with his friend Walter Fawkes at Farnley Hall, in Yorkshire. Furiously, the artist makes notes of its form and colour on a scrap of paper. As Fawkes's son, Hawkesworth, later recalls, 'He was entranced. There was the storm rolling and sweeping and shafting out its lightning over the Yorkshire hills. Presently the storm passed and he finished. "There Hawkey," said he, "In two years you will see this again, and call it Hannibal Crossing the Alps."' Though some art historians have suggested that the anecdote is a little too neat, the painting, one of Turner's most celebrated works, is duly exhibited in 1812 and now hangs in the Tate Gallery in London.

The harvest moon – the nearest full moon to the Autumn Equinox – rises around sunset for a few nights and, given clear skies, bathes the British Isles in soft, silvery light all night long. Before the twentieth century farmers worked by this light to gather in the last of the harvest – hence the name. If the harvest moon seems impossibly large, it's just an optical illusion. Because it sits so low in the sky, it acquires telephoto scale from the terrestrial trappings through which it shines. The days draw in quickly at this time of year: August to late September is the period of accelerated daylight loss as a few minutes are shaved off each day.

27.3C Norwood 1895

24 SEPTEMBER

9C min / 16.3C max

GALE IN THE CHANNEL

What's it like to face a gale in a small sailing boat? This is what Claud Worth does when, sailing his 28-foot (8.5-metre) *Tern* from Falmouth (♀) to the Thames Estuary today in 1896 with the help only of a single hand hired in a pub ('obese and alcoholic-looking'), he is caught in a notorious Force 9 gale which claims hundreds of lives. His graphic account, *Yacht Cruising*, has become a sailing classic.

'In the heat of summer I forget how cold winter was, and in December I forget about the languorous heat of summer. I don't really forget. I mean I could, if pressed, recall these things, but the fact is that there are still these moments of surprise – that awful "Oh it's like THIS" feeling – which experience should have prepared you for ...'
Kenneth Williams, actor, 1973

By the small hours of this morning he has almost lost control: 'Waiting for a chance to heave her to, a great sea came over the taffrail, completely burying the vessel. I was nearly taken overboard, and the cockpit was filled to the coamings. She broached-to, and I expected that the next sea would finish her.'

His account is almost as fascinating for its incomprehensibility to non-sailors as for its undoubted drama. Through an interminable night and day, he tirelessly 'led warp through hawsepipe ... let go throat halyard ... paid out drogue warp as she gathered sternway, slacked topping-lifts ...' But one gets the gist. 'I don't want to pile on superlatives, but I have been in a pampero off the Plate [the Atlantic coast of Uruguay] and a heavy gale of wind in the South Pacific, and I have never known the wind so strong or the seas so steep.'

At the other end of the country, a century earlier, the diarist John Aspinwall crosses by ferry from Kinghorn to Leith (♀) in 1795: 'The wind blew very hard, and being the only boat that was going for 10 hours was very full & mostly women who came from the Highlands to get work in the harvest. We ... tack'd in order to clear the wharfe but the wind was too much ahead & we ran bowsprit against the wharfe ... The rope ... gave way, and came thundering aft and in its career met a poor covered cart which was on deck. This it demolish'd ... The poor highland women were almost frightend to death.'

30.0C Loughborough 1895 ☼

25 SEPTEMBER

8.7C min / 16.4C max

WHEN BAD WEATHER MEANT REBELLION

A thunderstorm rocks the monastery at Canterbury (♀) today in 1271, and Walter de Hemingburgh later describes 'such a flood of rain, with thunder, lightning and tempest'. Another chronicler records 'such swelling of waters that the crypt of the church and the cloisters . . . filled.' The flood destroys the harvest, and the following year there is 'a very great famine . . . throughout the whole kingdom.'

'When people walk in a deep white fog by night with a lanthorn, if they will turn their backs to the light they will see their shades impressed on the fog in rude, gigantic proportions.'
Gilbert White, naturalist and writer, Selborne, Hampshire, 1780

'A Remarkable appearance was seen in Rutland, which I suspect was of the same nature as Spouts at Sea. Though there was no wind it moved apace from S by W to N by E, being all along divided into two parts, and making a great noise, like a distant wind, or a great flock of sheep galloping along on hard ground.'
Thomas Barker, Lyndon Hall, Rutland, 1749

The period 1250–72, if medieval chroniclers are to be believed, is dominated by very wet autumns and summer droughts. In thirteenth-century England, life hinges precariously on the success or failure of the annual harvest. Harmony in nature means harmony in the state; bad weather often means rebellion. The incessant crises that characterize the second half of Henry III's reign (1216–72) are all precipitated by weather. After Llewelyn ap Gruffydd declares himself ruler in North Wales in 1255, harvest failures, floods and famine in the next few years lead to a revolt of the barons and the Provisions of Oxford – an unprecedented constitutional landmark whereby the king agrees to limit royal authority. In 1264, outright civil war tears the country apart.

Strong winds two centuries earlier in 1069 help William the Conqueror sack York (⚐) by fanning the flames of the burning Minster (*9 July*). The city, one of the richest in Britain, is razed by the inferno, allowing William to turn to the 'Harrying of the North'.

179.3mm Croy 1915

30.6C Stratfield Turgis 1895

26 SEPTEMBER

8.6C min / 16.4C max

BLUE MOON /
HARVEST
FESTIVAL

'No natural phenomenon has ever caused such intense interest, speculation, and even alarm,' Mr G. Bain Ross of Melrose (♀) informs *Weather* magazine of today's occurrence in Scotland in 1950. 'Following the spectacle of the blue sun, the moon when it rose was observed also to be coloured blue.' A pilot from RAF Leuchars in Fife also reports that the sun appears blue to a height of 6 miles (nearly 10 km), after which the aircraft entered thick brown haze. Other pilots report an accompanying smell of burnt paper and their aircraft being coated with an oily resin.

'I have never seen anything so wonderful as the sun climbing over our view in golden mist. I see now where Turner found such sights as Norham Castle, etc.' *Edward Elgar, composer, Brinkwells, Sussex, 1918*

'It was so cold on the stage that one wanted to close the French windows, and the autumn barrage of coughers in front made our witty remarks sound rather explosive.' *John Gielgud, actor, appearing in Somerset Maugham's* The Circle, *Glasgow, 1944*

Although the saying 'Once in a blue moon' usually refers to the astrological quirk whereby, occasionally, an 'extra' full moon appears in the lunar cycle, there are occasions – and today's example is one – when the moon actually appears blue. While the saying conveys the notion of an event so infrequent that it hardly ever happens, in fact blue moons occur . . . well, a lot more often than that. The explanation? As with Gilbert White's 'horrible phænomena' of 1783 (*18 July*), or 'The Year Without a Summer' of 1816 (*13 October*) or Ruskin's 'Storm Cloud of the Nineteenth Century' of 1841 (*4 February*) and innumerable other tricky-to-explain atmospherics, the culprit is atmospheric dust. In this instance, the dust is smoke particles from vast forest fires raging in the Canadian Rockies, carried to Europe by the jet stream (see *Definitions*). Visibly blue moons invariably follow such intercontinental pyrotechnics (they are observed for weeks, for example, following the Krakatoa eruption of 1883). Impatient observers, the meteorologist Philip Eden suggests, can enjoy the same effect by viewing the moon through bonfire smoke.

-6.7C Dalwhinnie 1942
(record low)

28.4C Brighton 1895

COLERIDGE'S
ROMANTIC
RHAPSODY /
FILMING
*MICHAEL
COLLINS*

'The river is full, and Lodore [a waterfall] is full, and silver-fillets come out of clouds and glitter in every ravine of all the mountains; and the hail lies like snow, upon their tops, and the impetuous gusts from Borrowdale snatch the water up high, and continually at the bottom of the lake it is not distinguishable from snow slanting before the wind – and under this seeming snow-drift the sunshine gleams, and over the nether half of the lake it is bright and dazzles, a cauldron of melted silver boiling! It is in very truth a sunny, misty, cloudy, dazzling, howling, omniform day, and I have been looking at as pretty a sight as a father's eyes could well see – Hartley and little Derwent running in the green where the gusts blow most madly, both with their hair floating and tossing, a miniature of the agitated trees, below which they were playing, inebriate both with pleasure – Hartley whirling round for joy, Derwent eddying, half-willingly, half by the force of the gust, – driven backward, struggling forward, and shouting his little hymn of joy.'
Samuel Taylor Coleridge, poet, in the Lake District (♀*) with his sons, aged 6 and 2, 1802*

'I have managed to squeeze a helicopter from the production for a series of shots that will show the group of ambushers running over the heather towards the heights ... The winds are howling by now. The pilot, an Irishman, with experience in Vietnam, decides to give it a go, though the vortex the winds are creating above the valley makes it extremely dangerous. I go up with them and take them through the shot, bouncing around the skies like a puppet whose strings are about to snap. The shot is spectacular, and the problem as always is trying to get the aircraft low enough, within the margins of safety. Most of these unit helicopter pilots have brief lives and, being up there with him, I can understand why. The extras down below only have so much running in them ... Leaping over mounds of heather, at full speed, with a helicopter trying to breathe down your neck in a force 9 gale is not the best way to earn your living.' *Neil Jordan, directing* Michael Collins *in Ireland (*♀*), 1995*

'The weather is cold today, and the winter is coming. I hate it so much. One always feels like dying as the winter comes on in England. It is so cold and lugubrious.' *D. H. Lawrence, author, London, 1915*

30.6C Stratfield Turgis 1895

WHY HOBBIT WEATHER IS BRITISH WEATHER

A 'thick, cold and white' fog envelops Frodo Baggins and his companions when they fall asleep at the Hill Fort today in 1418 (Middle Earth time). J. R. R. Tolkien was fascinated with the weather. We know this from the countless beautifully observed references to it in his letters (*1 March, 16 March, 26 December, 30 December*). Devoted *Lord of the Rings* aficionado Mr Pike of Hungerford reveals just how exactly the weather in the opening sequences of the novel fits a late autumn pattern that is eerily familiar to any resident of the British Isles. Frodo departs, on 23 September, on a 'clear, cool and starry' night with 'smoke-like wisps of mist'. The first light showers and the arrival of a westerly front then bring a day of 'straight, grey rain', while the hobbits shelter at Tom Bombadil's house a couple of days later. And there is fog today. Even the first mention of a frost, when the troop reach Weathertop Hill on 6 October, is about right.

The debate about which landscapes inspired Tolkien when he was writing *The Lord of the Rings* – the Swiss Alps, the moorlands of Wales, the battlefields of the First World War, the Malvern Hills – will rage for as long as people read the stories. The inspiration for the seasons and the daily weather that Tolkien uses for the tales of Middle Earth is less equivocal. As Tolkien writes, during a wrangle with the film company proposing to make an animated version of the trilogy in 1957, '*The Lord of the Rings* may be a "fairy story" but it takes place in the Northern Hemisphere of this earth; miles are miles, days are days, and weather is weather!' Alas, such singularly British weather is missing from the movies: Peter Jackson filmed his trilogy in New Zealand.

'A storm of wind raged with such destructive violence that, without mentioning other incalculable and irreparable damage, more than twenty ships were sunk at Portsmouth.'
Matthew Paris, historian,
Chronica Majora, *1238*

'Strange, with what freedom and quantity I pissed this night, which I know not what to impute to but to my oysters – unless the coldness of the night should cause it, for it was a sad rainy and tempestuous night.'
Samuel Pepys, diarist, 1666

'Rainy weather, Nothing material occurred. The Russians and Germans are at war with the Turks & Sweden.'
John Tennent, grocer, County Derry, 1788

29.4C Loughborough 1895

29 SEPTEMBER

8.5C min / 15.9C max

GERMAN
GEORGE'S
FOGGY START
/ THAMES
BARRIER

The dense fog finally lifts from the Thames today in 1714, and at 6 p.m.
George, Elector of Hanover and King of Great Britain and Ireland to be,
is finally able to land at Greenwich. Unfortunately, the lords and
courtiers assembled to greet their new sovereign, his son (the future
George II), their retinue and assorted mistresses have all grown tired of
waiting. So George steps on to a deserted quay. It's a most inauspicious
start to his reign. Though the majority support the succession of a Protestant foreigner (as opposed to a native Catholic), they are not impressed
by the short, irascible German who speaks scarcely a word of English.
George, we must presume, is equally unimpressed with the British
weather.

A gale in Scotland creates a storm surge in the North Sea today in 1969.
Much of Hull (☞) is already under a metre of water and forecast sea levels
for the Lincolnshire coast are higher than for the 1953 flood (*31
January*). By chance, just as the flood warning for London reaches
'Danger' level, Eric Johnson, Chief Engineer for the Ministry of Agriculture, Fisheries and Food, is sitting in a policy committee meeting on the
design and construction of a Thames Flood Barrier. Londoners have felt
vulnerable to flooding since 1928 (*7 January*), and a flood barrier is first
proposed following the 1953 disaster. Fortunately, wind velocities in the
North Sea are not strong enough to cause serious problems on this occasion, and sea levels recede with the tide. But the sense of disaster averted
is a timely spur. Within four months, the Greater London Council
(which has had responsibility for dealing with flooding in London since
1965) submits a report to the government and work on the giant engineering project begins in 1970. The Thames Barrier is completed in 1982,
at a cost of £440 million.

The now familiar concrete piers and steel gates spanning the river
at Woolwich Reach (☞) are first used in February 1983, though an
emergency does not arise until 2003, when the barrier has to close a
record fourteen times in eight days. As London is sinking at a rate
of 8 inches (20 cm) a century and sea levels are rising, it is estimated that the present barrier will be redundant by 2030.

27.8C York 1895

30 SEPTEMBER

8.3C min / 15.7C max

SNOW-BEDS   For the rent of one snowball, to be presented, on request, to the reigning king at midsummer, the Foulis estate, overlooking the Cromarty Firth (☞), is granted in the eleventh century to the chief of the Munro clan. But is such a picturesque rent guaranteed? Does permanent snow lie *anywhere* in the British Isles? The answer is, or was: sometimes. Today is the test. The end of September is the challenge that the vestiges of the previous winter's snow-drifts must survive before the weather turns colder again. Until recently,

'Immensely cold, a kind of solid cold outside the windows ... I have been, in the good, old-fashioned way, feeling my skin curl.'
Katherine Mansfield, author, 1918

above 3000 feet (915 metres) deep drifts lingered until July and even August. In two places snow often lasted through the year. Both at 3750–3800 feet (1140–1160 metres), these were the north-east flank near the summit of Ben Nevis and the sheltered, deeply shaded gullies on Braeriach (☞). In the early 1930s, the Nevis snow-bed was considered so consolidated that there were moves to classify it as an incipient glacier. Inconveniently, it melted in 1933. Then again in 1935, 1938, 1945, 1953, 1958, 1959 ... by which time the thesis was abandoned (though, curiously, the snow then lasted from 1960 to 1996). The key to snow surviving is drought. Nothing removes snow like rain.

In England, on Helvellyn, there is evidence that in the nineteenth century snowdrifts were much more persistent. The Manchester-based scientist John Dalton used to climb Helvellyn every July between 1805 and 1823, making humidity experiments at the top for which he used melting snow as a cooling agent. He always found it 'in the usual place, about a quarter of a mile north of the summit'. In late June 1817, Dorothy Wordsworth notes drifts above Red Tarn (it has been suggested that her brother's line about the recess 'that keeps till June October's snow' derives from this excursion). In August 1818 Keats crosses snow-beds as he walks up the *west* side of Ben Nevis.

The late Professor H. H. Lamb estimates (in 1964, so considerable allowance is needed for global warming) that, at Britain's latitude, our mountains would have to be at least 5300 feet (1615 metres) high in the Ben Nevis region, 5900 feet (1800 metres) high in the Lake District and 6300 feet (1920 metres) high in Wales and the Pennines for there to be a permanent snow-line. Despite this talk of snow, heatwaves also occur this late in the year: in 1895 the month ends with temperatures of almost 30C (86F).

172.5mm Seathwaite 1890

27.8C Maidenhead 1908

OCTOBER

1 OCTOBER

SUN-BATHING AT STAMFORD BRIDGE / VULCAN CRASH

The highest-ever October temperature (29.4C) is recorded in Cambridgeshire today in 1985, though the month is traditionally more about falling leaves and first frosts than applying Factor 15. Fine spells do occur mid-month, in the guise of Indian summers and St Luke's little summers (*18 October*). A perfect October might begin with thunderstorms, include a warm, stable period in the middle and end with a frost. In truth, it's more likely to rain – hence the farmers' advice:

'Dry your barley in October, Or you'll always be sober.'

The weather is, according to *Heimskringla*, the chronicle of the kings of Norway, 'uncommonly fine' and the men 'very merry' today in 1066. King Harold Hardrada's Viking army disperses along the River Derwent, near York (⚔), to make the most of the sunshine. And why not? Safe in the knowledge that the English army is on the south coast awaiting an invasion from Normandy (*18 September, 3 October*), the King of Norway lets his soldiers top up their tans. It's so hot that they have even left their 'brynies' – the hard leather jackets they use as armour – aboard their longships, 15 miles (24 km) away on the Humber. Suddenly, they catch sight of something flashing and glinting in the distance at Stamford Bridge: it's sunshine reflecting off arrow points and swords. Harold, King of England, has just force-marched his army 200 miles (320 km) in four days. The battle is a slaughter. Harold Hardrada is killed. Of the three hundred ships which have brought the Norwegians over, only twenty-four are required to bear the survivors home. The Vikings will never invade the British Isles again.

Nine centuries later, Henry Wild, a local villager, witnesses the RAF's first Avro Vulcan bomber, arriving from Australia on approach to Heathrow today in 1956: 'A very severe thunderstorm was approaching with a lot of lightning and strong wind. There was a tremendous noise and there appeared a triangular aircraft from the low clouds. It was so low it was obvious it could not reach the runway. There were two cracks then a loud bang as it crashed then silence. The cracks were the ejector seats. It fell in a field of brussels sprouts about half a mile southwest of the village.' The pilot and one passenger survive but four others, including the co-pilot, die instantly.

-6.4C Eskdalemuir 1928 ❄

29.4C March 1985 ☀

2 OCTOBER

8C min / 15.3C max

FLIGHT OF THE EARLS

In an 'excessive storm and dangerous bad weather', a ship bearing the 'flower of Gaelic nobility' – the earls of Tyrone and Tyrconnell, their families and associates – is lost off the south coast of England today in 1607. The Irish leaders, fleeing the Protestant authorities in fear of their lives, are bound for La Coruna in Spain and the bosom of the Catholic court. But the ship, battered by storms every day since leaving Lough Swilly in Donegal (♥), is now so disoriented that not even the captain knows 'what particular coast was nearest to them'. This is the 'Flight of the Earls' – one of the most enigmatic and defining events in Irish history, marking the end of the medieval Gaelic order. The ship finally makes landfall at Quilleboeuf on the Normandy coast, with only one barrel of drinking water to spare. A demand for the extradition of the earls escalates into an international crisis, involving the governments of France, Spain and England. They eventually make their way to Rome. Neither will see Ireland again. Had they made it to Spain, and returned at the head of a Spanish invasion force, the next four hundred years in Ireland might have been very different. As it is, with the earls gone, so has the focus of Irish resistance to the English government. The colonization of Ulster now proceeds with little opposition.

'A very rainy morning. We walked after dinner to observe the torrents.'
Dorothy Wordsworth, Lake District, 1800

'The weather is bad. (The weather is always a foe, rain or shine. It doesn't matter what it is, it causes problems unless it remains absolutely constant.)' *Simon Callow, actor and writer, 1987*

'Sometimes the Lake District seems to move into autumn almost overnight, so that you wake up one morning to discover that new, warm colours of yellow, orange and red have been painted while you slept.' *A. Harry Griffin, climber and journalist, 2001*

Today in 1912 is the wettest day of the wettest October of the twentieth century. In 1980 an ice meteor crashes through the roof of a house in Plymouth. And in a great gale in 1697, the slow process of burying the community of Udal on North Uist (♥) is completed: the village, continuously inhabited for four thousand years, is overwhelmed by wind-blown sand and abandoned.

185.2mm Ben Nevis 1890

-4.9C Stapleton 1888

28.1C Whitby 1908

1066 AND ALL THAT ... WEATHER

After weeks of storms and heavy seas in the Channel (*18 September*), a fair southerly wind carries the massive invasion fleet of William, Duke of Normandy, to England today in 1066. He lands at Pevensey (♥) completely unopposed. Crucially, the long delay and the roaring northerly gales have convinced Harold, King of England, that the invasion he's long expected will be put off until spring. Thus he's disposed of his flotilla of ships, demobilized the *fyrd* – the militia-like backbone of the Anglo-Saxon army – and departed north to fight the Vikings (*1 October*). Sinking to the ground as he steps ashore, William seizes the earth of the kingdom he believes is rightfully his. Had the *fyrd* still been in position along the coast, it would have been a different story.

The invasion force – including soldiers, horses, engineers and carpenters plus all their provisions, tents and three small forts – disembarks at leisure before moving through the Sussex countryside, pillaging freely. They find an elevated position along the coast, build themselves a permanent encampment above the village of Hastings and await the arrival of Harold.

'The wind was still against us. Dr Johnson said, "A wind, or not a wind? that is the question"; for he can amuse himself at times with a little play of words ... We set sail very briskly about one o'clock. I was much pleased with the motion for many hours. Dr Johnson grew sick, and retired under cover ... Finding myself not affected by the motion of the vessel, I exulted in being a stout seaman, while Dr Johnson was quite in a state of annihilation. But I was soon humbled; for after imagining that I could go with ease to America or the East Indies, I became very sick, but kept above board, though it rained hard.'
James Boswell, departing the Isle of Skye, The Journal of a Tour to the Hebrides, *1773*

28.3C Whitby 1908

-4.8C Berkhamsted 1928 ❄

'A good way of naming a day, is to call it by its wind ... A north-westerly day in spring. The sun bright and the clear spaces of the sky a teeming blue; but ice in the quick air, and the clouds, which move rapidly, going from white at the blue-sky edge to a gravid indigo. The easterly day is an individual, a relentless Borgia searching the townsman's bones; but to those who can stand up to him, in a kind of country equality, a stimulus. He is a pirate; but if you can wear the knife between your teeth also, a boon companion. So with the other winds, the blusterous, the snow-carrying, the corn-rippling, the becalmed. They make the day.'
T. H. White, author, Wiltshire, 1934

☼ 4 OCTOBER

7.9C min / 15.2C max

ROMAN SOCKS AND SANDALS

'Tomorrow I will provide some goods . . . by means of which we may endure the storms even if they are troublesome.' So reads a fragment of a letter from one Flavius Cerialis to a friend, dated today, in around AD 100, and discovered at Vindolanda fort on Hadrian's Wall (☀). Flavius, prefect of the third cohort of Batavians, is most likely offering *superariae* or overcoats (price '13 denarii'). Also known as the *birrus Britannicus*, this hooded woollen rain cloak – the British national dress of the time and original 'hoodie' – is an internationally renowned garment. Two centuries later, it's the only British item to make it on to an edict of the Emperor Diocletian, listing the finest goods and services traded across the Empire.

'Waked last night by cats, and really frightened for a minute or two by the wild yowls; not sure even when I got up if cats, or the wind.'
John Ruskin, critic, 1873

'A violent rain storm on the pond . . . a helter skelter rain and the elms tossing it up and down; the pond overflowing on one side; lily leaves tugging . . . Now light from the sun; green and red; shiny; the pond a sage green; the grass brilliant green; red berries on the hedge; the cows very white; purple over Asheham.'
Virginia Woolf, writer, Sussex, 1934

W. H. AUDEN, ROMAN WALL BLUES
'Over the heather the west wind blows,
I've lice in my tunic and cold in my nose,
The rain comes pouring out of the sky,
I'm a wall soldier, I don't know why!'

For a young Roman used to Mediterranean sunshine, the prospect of winter on Hadrian's Wall must have seemed appalling, even with a woollen coat. So bad, in fact, that there is a suspicion that the Northumbrian weather led to the perpetration of fashion's supreme crime. Another papyrus letter fragment found at Vindolanda reads: 'I have sent you pairs of socks from Sattua, and two pairs of sandals.' Socks with sandals . . . and nearly two millennia before the Germans made it their own.

The climate during the Roman occupation of Britain isn't, in fact, that harsh, as the historian Tacitus notes in AD 94: 'The sky is overcast with continual rain and cloud, but the cold is not severe.' The numerous vineyards (*27 July*), some as far north as Yorkshire, confirm this. There is a theory, however, that the climate deteriorates around AD 400–415, causing famine and rebellion, particularly along Hadrian's Wall, which is finally abandoned by Roman garrisons around this time.

-6.7C Eskdalemuir 1912 ❄
28.3C Whitby 1908 ☼

5 OCTOBER

FOG, LONDON AND GOTHIC HORROR

'The first fog of the season . . . lurid brown, like the light of some strange conflagration . . . like a district of some city in a nightmare.' Thus Robert Louis Stevenson describes London's Soho in *The Strange Case of Dr Jekyll and Mr Hyde* (1886).

It's hard to emphasize just how foggy London is before the Clean Air Acts of 1952 (*6 December*). For seven hundred years, the atmosphere just grows smokier and denser, due to industrial pollution and coal fires, until fogs are *the* defining feature of the city. (Even in the 1960s, tourists can still buy souvenir tins of 'London Fog'.) Buildings turn black with deposited soot. Fog shrouds and blurs, with visibility ranging from murky, at best, to completely impenetrable (*16 December*). The air, never, ever as clear as we know it today, gives the city an unreal, ethereal, other-worldly aspect that is seized upon by writers like Dickens (*23 December*) and, later, painters like Monet (*23 February*). Drawing on the fascination and horror of events like the Ratcliffe Highway murders of 1811 – where a serial killer escapes into the fog (*19 December*) – a new genre of 'Urban Gothic' evolves. Stevenson now takes this a step further, making it darker and more psychological. His 'gruesome tone', writes novelist Henry James the following year, 'is like the late afternoon light of a foggy Sunday, when even inanimate objects have a kind of wicked look'.

> 'Weather bad. Attacked by inflammation of the testicles, and groaned all day.'
> George Gissing, writer, 1895

Stevenson's story, published as London approaches the foggiest period of its foggy history, in 1890 (*16 December*), is hugely successful – forty thousand copies sell in six months and it's read by everyone including the Prime Minister and Queen Victoria. Within eighteen months, Jack the Ripper adds to the grim mythology (though, ironically, not *one* of the 'Canonical Five' Whitechapel murders by the Ripper occurs on a foggy day). Stevenson has 'an immediate influence on writers like Oscar Wilde [and] Arthur Conan Doyle,' observes the cultural historian Dr Robert Mighall, 'and is perhaps largely responsible for creating the late-Victorian London of our cinematic imaginations; a foggy, gaslit labyrinth where Mr Hyde easily metamorphoses into Jack the Ripper, and Sherlock Holmes hails a hansom in pursuit of them both.'

-7.2C Balmoral 1966

28.9C Kensington Palace 1921

The Nemesis of Neglect – this cartoon from this week's *Punch*, in 1888, followed the first two Jack the Ripper murders. Such images helped nurse the grim association between fog, London's East End and serial murder.

HILLARY'S 'SHIPWRECK INSTITUTION'

A gale howls round the Isle of Man this morning in 1822. As the wind shifts from west to east the captain of a Royal Navy cutter, *Vigilant*, sheltering in Douglas Bay, rashly decides to make a break for the open sea. As he noses out, however, a sudden gust blows the ship on to Conister Rocks (☼). Watching this drama from his house overlooking the bay is local worthy Sir William Hillary. Hurrying down to the pier, he organizes two pleasure craft to set off for the *Vigilant*. Throughout the night they ply to and fro, taking off passengers and crew until, by dawn, 97 are saved. The night's work plants a seed in Sir William's mind: lifeboats may dot the coast of the British Isles, but there is no coordinated, nationwide institution to help ships in distress.

> 'No sun, and much rain. In bed all day, groaning.'
> *George Gissing, writer, 1895*

Ten weeks later, the ship sent to escort the patched-up *Vigilant* back to England founders on the Skionnes, off southern Man, and three Manx rescuers drown. Sir William secures Admiralty pensions for their families, but he's acutely aware that, were the vessels privately owned, no such thing would happen. And so his idea for a national lifesaving organization takes root. It would have three key responsibilities: to design and construct suitable lifeboats; to man them with trained crews; and to compensate any volunteers in those crews who come to harm – or their families if they die. With that, the following February he launches his catchy *Appeal to the British Nation on the Humanity and Policy of forming a National Institution for the Preservation of Lives and Property from Shipwreck*.

The Admiralty is politely indifferent. Sir William, however, did not earn his title (he raised England's largest private army to help George III repel Napoleon) accepting indifference. He rallies wealthy merchants, royalty and politicians, including no lesser reformer than William Wilberforce – who declares it 'the duty of the opulent to provide'. In 1824, the 'National Institution for the Preservation of Life from Shipwreck' duly comes into being. King George IV is patron. The Prime Minister is President. Over the next 183 years the Royal National Lifeboat Institution, or RNLI (as it becomes known), will save more than 137,000 lives at sea, mainly from disasters brought about by bad weather. It is a charity dependent on voluntary contributions, providing a twenty-four hour rescue service up to 100 miles from Britain and Ireland's

-6.2C Carnwath 1972 ❄

28.9C Kensington Palace 1921 ☀

6 OCTOBER

shores. 'Its spirit,' Churchill later says, 'drives on with a mercy that does not quail in the presence of death. It drives on as a proof, a symbol, a testimony that man is created in the image of God and that valour and virtue have not perished in the British race.'

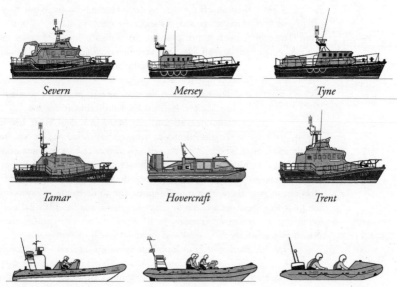

Severn   *Mersey*   *Tyne*

Tamar   *Hovercraft*   *Trent*

E class   *B class*   *D class*

THE RNLI TODAY
• Thirty-minute arrival time from launch (for 90 per cent of casualties within 10 nautical miles of shore). • 233 stations • Five classes of all-weather lifeboats, some capable of up to 25 knots in extreme weather, plus two classes of inshore lifeboat • 4800 crew, mainly volunteers • 66 per cent funded by legacies

LIEUTENANT SAXBY'S FLOOD

Waterside warehouses are empty. Ships are secured in dockyards. Sea walls are reinforced. Thousands of people with 'lively antediluvian apprehensions' of an 'eruption of the waters', as *The Times* reports, line the embankments, piers, quays, bridges, wharves, headlands and clifftops of the British Isles today in 1869, anxiously awaiting a great storm and flood. Everyone is poised on account of a much-publicized prophecy by Lieutenant Saxby, naval engineer and 'lunarist'.

'A real Indian summer afternoon – not a breath of wind, a sun as warm as a mid-summer day, the wasps busy over orchard windfalls.' BB, pseudonym of Denys Watkins-Pitch-ford, author and illustrator, 1978

Saxby holds that the moon's position relative to the equator, along with various other astronomical occurrences, can foretell 'extreme atmospheric disturbances'. In November 1868 he warns, in the London press, of coastal flooding with a storm of exceptional severity, between 5 and 9 October 1869, whereby all are 'in danger of engulfment by wind or wave'. Today's 'Saxby Flood' never happens, needless to say; it's just another high spring tide. To the disappointment of all who travel to the coast, there is no giant wave, no death, no destruction. The seas are tranquil and the weather is bright, with a gentle breeze.

'The road led through meadows and fields, along hedges of hawthorn and traveler's joy . . . It was so beautiful when the sun set behind the grey clouds and the shadows were long . . . The clouds stayed red long after the sun had set and the dusk had settled over the fields.' Vincent Van Gogh, artist, Isleworth, London, 1876

'Dull Weather. My ailment improved.' George Gissing, writer, 1895

Scientific meteorology and weather forecasts are still in their infancy (*1 August*), so it's not surprising that people heed Saxby's prophecies. Scientific rigour, accurate measuring instruments and empirical procedures do not convincingly replace astrology until the twentieth century. Even today, national newspapers carry the prophecies of quacks.

-6.7C Braemar 1919

Today in 1829 snow lies in London – the earliest in the year that it has settled on the capital – taking everyone by surprise. *The Times* reports how this 'sort of anticipation of Christmas' contributes 'in some degree to make the inactivity of the money-market more complete'.

183.9mm Horncastle 1960

26.7C Southend on sea 1921

8 OCTOBER

7.8C min / 14.7C max

GRAHAM
SWIFT'S
WATERLAND

Today, in 1903, 100mm (nearly 4 inches) of rain causes severe flooding in the north-east during the wettest October since records begin. Few writers capture the quiet, steady, relentless rise of a river in flood better than Graham Swift in *Waterland*. Although his fictional flooding of the Fens (🖉) is set around now, in October 1874, curiously this is the one year, in a phenomenally wet nine-year phase, that is not especially wet. The two occasions most reminiscent of the East Anglian scene he evokes are 1947 (*16 March*) and 1998 (*10 April*):

'Outside it has begun to rain ... Not heavily, not torrentially, but with a steadiness, a determination that Fenlanders have come to know cannot be ignored. All over the country of the Ouse and the Leem that morning they are watching water-levels, fuelling auxiliary pumps, tending sluices and flood-gates ... And the rain increases. Moreover, if it was raining that day in the Fens, it was raining also over those upland regions to the south and the west whence the rivers descend for which the Fens are a basin ...

'The rain doesn't stop. It doesn't stop for two days and two nights ... but thoughts of divine weeping and so forth are soon put to one side as the flood takes hold. The folk of Gildsey know from long observation that however brown, swirling and threatful their old Ouse becomes, they have little to fear from a flood confined to that river alone ... But if the Leem floods simultaneously ... then the effect of the torrents discharged by the former into the latter will be like a liquid dam causing the Ouse to flow back on itself and spill out in every direction ...

'The waters rise. At first with a steady increase, and then with a sudden rush which signals that the Leem has indeed thrown in its forces.

'The waters rise. They creep up the slopes of Water Street. The lower buildings are, as everyone expected, inundated ... They creep further. They dump a ton or more of mud into the hastily cleared cellar of the Jolly Bargeman ...

6.7C Legnathie 1907 ❄

'From the air (though there are no helicopters in 1874 – no flickering newsreel shots of beleaguered rooftops and engulfed cars), Gildsey must look like a moated settlement drawn in on itself ... In the floods of 1874 eleven thousand acres of land are rendered uncroppable for a year. Twenty-nine people are drowned, eight missing, presumed so ... The damage to houses, highways, bridges, railways, drains and pumps is beyond clear reckoning.'

25.6C South Farnborough 1921 ☀

9 OCTOBER

7.6C min / 14.6C max

**LUTINE BELL /
GALE THAT
SAVED
SCOTLAND**

In a 'heavy gale from the NNW', HMS *Lutine* ('the tormentress') sinks off the island of Terschelling near Texel in the North Sea today in 1799. The ship is carrying £1,200,000 in gold and silver bullion to the banks of Hamburg: the loss precipitates just the stock market crash this cargo is designed to prevent. The gold is insured by Lloyd's of London, the marine insurers, who settle the claim. Repeated attempts over the following century to salvage the bullion, owned by Lloyd's under so-called 'rights of subrogation', are hampered by the silting up of the wreck, strong currents, shoals and appalling weather. Some £1 million remains at the bottom of the sea today. The bell, however, is recovered in 1858 and – some would say it's poor compensation for the bullion – is rehung, from the rostrum of the Lloyd's Underwriting Room. This is the famous Lutine Bell struck when news of an overdue ship comes in – once for bad news, twice for good news.

> 'Yesterday was startling for its beauty and warmth. We lunched outside off a picnic table and found it too warm to be wearing cardigans. "Off, off, you lendings!" Today promises to offer another Indian summer.'
> *Alec Guinness, actor, 1995*

> 'Poor Brown's legs ... dreadfully cut by the edge of his wet kilt ... just at the back of the knee, and he said nothing about it; but ... one became so inflamed, and swelled so much, that he could hardly move.'
> *Queen Victoria (aka Mrs Brown), Balmoral, 1865*

So violent is the storm, with 'hail and tempest', tearing up the west coast of Scotland today in 1263, that some longships in the 200-strong fleet of Haakon 'the Old' drop seven anchors. In others the crew cut down the masts. Several ships are still dragged on to the beach at Largs (⚔) where the Scots await with sharpened swords. There is no detailed account of the action – it may have been just a series of skirmishes, later given flavour by Scottish chroniclers – but the political consequences of the Battle of Largs are huge. The empire of Haakon, King of Norway, includes a large chunk of what is now Scotland – the Hebrides and Kintyre – and he is at Largs because the locals are getting restless. When the 'gale that saved Scotland' finally subsides, the Norwegian fleet is as battered as a Mars bar in a Glasgow chip shop. Haakon dies at Kirkwall in the Orkneys, sailing home. Negotiations for the return of the territories begin immediately. Scotland becomes whole again in 1266.

-3.3C Newton Rigg 1919 ❄

27.8C Kensington Palace 1921 ☀

10 OCTOBER

7.7C min / 14.5C max

YACHTING INVENTED

'The King lost it going, the wind being Contrary, but sav'd stakes returning' is John Evelyn's economical summary of the invention of yacht-racing today in 1661. The course is Greenwich to Gravesend (♥) and back. The competitors: Charles II versus his brother James. The wager: £100. The strange new word *jacht* ('hunt') only enters the English language the year before, when the Dutch present the newly restored Charles II with the *Mary*, one of the light, fast sailing vessels which the Dutch navy use for chasing pirates around the shallow waters of the Low Countries. To modern notions of pleasure yachts, of course, the *Mary* could hardly appear less sporty – she's a sort of mini-galleon designed for eight guns and thirty crew. However, compared with the great square-rigged ocean-going ships of the time, she's a racing machine. Crucially, with fore and aft rigging she can sail closer to the wind – making her much better suited to racing.

'The grayest, most silent days I ever saw: my Besom, as I sweep up the withered leaves, might be heard at a furlong's distance.'
Thomas Carlyle, historian, Craigenputtock, Dumfries and Galloway, 1830

'I have never seen such continued and heavy rain . . . The thunder was louder than falling bombs, the lightning as bright as day and the water pouring down as if a bucket was being emptied on each square foot every minute.'
Tony Benn, politician, 1944

Delighted with his new toy, Charles promptly orders an English version from his shipbuilder, Peter Pett. His brother James orders one from Pett's brother. Then another yacht arrives from Holland to swell the first British yacht squadron to four, and suddenly yachting is all the rage. It will be 165 years, however, before the sport (spurred, once again, by royal enthusiasm) finds its 'official' home at Cowes (*11 August*).

Tonight in 1795 it's raining with 'great violence' as proto-feminist Mary Wollstonecraft plans her suicide. Although she is distraught at being rejected by American adventurer Gilbert Imlay, what follows has become famous as 'one of the calmest acts of reason'. After she has rowed to Putney, the rain 'suggested to her the idea of walking up and down the bridge, till her clothes were thoroughly drenched and heavy with the wet'. Only when she is convinced her clothes are heavy enough to sink does she hurl herself off the bridge. Her deliberation, however, comes to nothing when two watermen drag her, unconscious, from the river and she is, as she later puts it, 'inhumanely brought back to life and misery'.

-5.8C Kingussie 1902

25.6C Kensington Palace 1921

WINDSCALE / NAVAL TACTICS IN THE AGE OF SAIL

Which way is the wind blowing in Cumbria this Friday in 1957? It's important because the Windscale plutonium plant, producing fuel for nuclear weapons, is ablaze and radioactive iodine is pouring from one of its chimneys. Initial reports that easterly winds are blowing the radioactive cloud out to sea, towards the Isle of Man and Ireland, are, at a 1974 inquiry, officially denied. Allegations of high-level government cover-ups follow. After visiting the Met Office archives, the Low Level Radiation Campaign (LLRC) alleges 'cooking of the books': 'Record sheets for 1957 had been removed from the Met Office's Windscale station volume and replaced with new sheets of a slightly different colour. The pages for 1957 read: NO RECORD – MAST DISMANTLED. The mast "reappeared" in November.' Wind direction, of course, has immense significance because of the long-term health implications of radioactivity. Half a century later, controversy still rages around this, the first of the great civilian nuclear disasters.

Today in 1797 Admiral Adam Duncan uses the brisk westerly wind to win an important victory over the Dutch fleet at the Battle of Camperdown. His daring tactic – using the wind to plunge through the enemy line to engage them on their leeward or downwind side – highlights the tactical dilemmas facing admirals dependent on the wind. Under sail, being upwind or to the windward – in naval-speak, 'holding the weather gauge' – confers key advantages during engagements. First, the tactical initiative. You can force battle or refuse it (by remaining upwind). A downwind admiral can avoid battle by withdrawing, but cannot force action. Secondly, boarding (a favourite Spanish tactic) is only possible from the windward side. Thirdly, retreating downwind opens vessels up to major risks. It exposes your stern, the most vulnerable part of the ship, to devastating damage by close-quarter cannon fire. Also, sailing in the lee of an enemy, as your ship heels to leeward, it exposes part of the ship's bottom. Get 'holed' in this area (known as being hulled 'between wind and water') and you will sink on the opposite tack (sailing the other way). Finally, cannon smoke drifts leeward, worsening the already considerable difficulties of communication during battle. Admirals in the days of sail often spend days manoeuvring to achieve the weather gauge – making it all the more unexpected when, as at Camperdown, an admiral deliberately relinquishes it.

-6.7C Braemar 1946
208.3mm Loch Quoich 1916

25.2C Greenwich 1978

12 OCTOBER

6.7C min / 14.2C max

FOG, FLYING
AND THE
MEANING OF
FEAR

A murky day in Farnborough (🌧), 1913, gives the young aviation pioneer Geoffrey de Havilland 'the most unpleasant experience I had in nearly fifty years of flying'. In these early days, the 'extreme dangers' of flying in fog still go unrecognized. De Havilland, due to test fly a new plane, describes in his autobiography, *Sky Fever*, how he takes off 'to see if the weather was good enough':

'Directly the wheels left the ground I could see nothing . . . and knew it would be impossible to return to the aerodrome. The great airship shed was straight ahead, so I turned left to avoid hitting it, and flew on until a tree loomed up very close in front of the nose. I just had time to pull the stick back and jump over it. Chimney pots on houses suddenly appeared in the same way and had to be "jumped". After about ten minutes of this nightmare I realised it was far too risky, so I started to climb steadily. By this time I understood, as never before, the meaning of fear.'

Unfortunately for de Havilland, his adventure is just beginning. 'There was no question of stopping to think things out, and I knew I just had to go on, concentrating all my attention on the crucial necessity of maintaining a safe speed above stalling point at the correct angle of ascent . . . As I climbed through the fog it gradually became more luminous and I emerged at three thousand feet into brilliant sunlight and an azure sky, with a level sea of dazzling white cloud extending to all horizons. But this was a false paradise, for I was a prisoner in this wide and lovely world above the fog-bound earth. I did what most of us instinctively do in times of great stress; I prayed hard . . .

-8.5C Lagganlia 1973 ❄

'Climbing steadily for about fifteen minutes, I occupied myself with calculations as to how long the fuel would last and just how unpleasant the result of engine failure would be. A long and anxious period passed before I suddenly spotted in the far distance what appeared to be a very small, dark patch. As I drew nearer I saw that it was a tiny break in the cloud, and with a sudden excitement that was like a shock I dived my 'plane towards it. There far below, as dimly perceived as the base of a deep well, was a small piece of good earth. I thrust down through it and landed in the first field I saw.'

25.2C Martyr Worthy 1978 ☀

13 OCTOBER

THE YEAR
WITHOUT A
SUMMER

Imagine global cooling. Imagine a layer collecting in the upper atmosphere that screens out the sun's energy, while doing little to prevent outgoing warmth escaping. Imagine the summer sun disappearing, average temperatures plummeting – by several degrees – and rainfall increasing dramatically, flattening unripened corn and causing widespread floods. This is what happens in 1816, in what's known as 'the year without a summer'. In Kent, one of the warmer parts of the British Isles, an abysmal wheat harvest is finally brought in on this dismal, chilly day – a month and a half late.

'The bright sun was extinguish'd and the stars / Did wander darkling in the eternal space / Rayless and pathless, and the icy earth / Swung blind and blackening in the moonless air; / Morn came and went – and came, and brought no day ...'
Lord Byron, 'Darkness', about the year without a summer

What's going on? These apocalyptic developments are caused by an event on the other side of the world eighteen months earlier. Mount Tambora, in eastern Java, explodes in one of the most powerful volcanic eruptions ever, blasting millions of tons of volcanic ash into the upper atmosphere. This spreads out to form an aerosol veil, shutting out incoming solar radiation (*13 June, 18 July, 26 September*). Only as the ash gradually clears from the atmosphere does equilibrium, and the even pace of the seasons, return – around 1820.

The ramifications on the ground are immense. The 1816 harvest is the worst for the next forty-two years, prompting soaring grain prices. Although Britain has some stored grain to fall back on, the army still has to be called out to control riots in East Anglia and Dundee. In southern Germany, according to the historian Carl von Clausewitz, 'true famine' follows. In France the wine harvest fails. In Ireland, 65,000 starve. In Iceland, the poor eat moss and cats. On both sides of the Atlantic, the effects ripple on for years. In Europe, 1816–20 is the last widespread famine of the Western world.

-8.3C Inverdruie 1971

The year has one curious artistic legacy. The dramatic colours of the volcanically induced sunsets can still be seen in the glowing, technicolour skies of an English painter whom they particularly captivate – Turner (*6 September, 25 September, 13 November*).

24.8C Pen-y-Ffridd 1990

14 OCTOBER

6.9C min / 13.6C max

BLACK FRIDAY AND OPIUM SUNDAY

Following a North Sea storm, the villagers of Eyemouth (✠) count their dead this morning in 1881. This, Scotland's worst-ever fishing disaster, devastates what was until today a prosperous east coast fishing village as well as claiming numerous ships off Ireland. 'I went to sea, taking three fishermen with me ... One of them, poor fellow, had lost three brothers ... even now I can hardly get his sad, wistful look out of my mind, ever scanning the face of the sea, with an eager, yearning look. Alas! All in vain ... the terrible state of distress at Eyemouth baffles description ... In one block of 12 houses there are eight widows, to say nothing of the orphans,' reports a letter to *The Times*. 'Black Friday', as it becomes known, claims nearly two hundred from Eyemouth alone. Many drown within 50 yards of the harbour, under the eyes of their families, prompting renewed calls for non-tidal 'harbours of refuge' along the coast (*19 August*).

HOW TO SPEND A RAINY SUNDAY AFTERNOON IN LONDON

'It was a Sunday afternoon, wet and cheerless: and a duller spectacle this earth of ours has not to show than a rainy Sunday in London. My road home-wards lay through Oxford Street; and ... I saw a druggist's shop. The druggist – as if in sympathy with the rainy Sunday, looked dull and stupid, just as any mortal druggist might be expected to look on a Sunday: and, when I asked for the tincture of opium, he gave it to me as any other man might do ... Never-theless, in spite of such indications of humanity, he has ever since existed in my mind as the beatific vision of an immortal druggist.

'Arrived at my lodgings, it may be supposed that I lost not a moment in taking the quantity prescribed. I was necessarily ignorant of the whole art and mystery of opium-taking: and, what I took, I took under every disadvantage [de Quincey was suffering from toothache]. But I took it: – and in an hour, oh! heavens! ... What an apocalypse of the world within me! That my pains had vanished, was now a trifle in my eyes: – this negative effect was swallowed up in the immensity of those positive effects which had suddenly opened before me – in the abyss of divine enjoyment thus revealed. Here was a panacea ... for all human woes; here was the secret of happiness, about which philosophers had disputed for so many ages.'
Thomas de Quincey, Confessions of an English Opium Eater, *1804*

-8.3C Crawfordjohn 1971 ❄

23.0C Brooksby Hall 1990 ☀

15 OCTOBER

6.7C min / 13.3C max

THE GREAT STORM OF '87

'A woman rang to say she heard a hurricane was on the way. Well, don't worry, there isn't,' says Michael Fish in the weather report following tonight's BBC News in 1987. A couple of hours later, as we all know, the great 'hurricane' of 1987 arrives. It's the most dramatic meteorological event in recent British history, ploughing the worst trail of havoc and devastation across London and the South-East since the Great Storm of 1703 (*7 December*). The hurricane-force winds cause eighteen deaths, result in transport mayhem (*16 October*) and run up an estimated bill for damages of £100 million. It's trees that suffer most, however. With disastrous timing, the gale arrives while the ground is soft after days of rain and the mild autumn has left many leaves on the branches to catch the wind. In a matter of hours, 15 million trees are flattened – Sevenoaks in Kent becomes 'Threeoaks', and a third of Kew's irreplaceable collection go (staff who tend them weep on tomorrow's news).

> 'Lovely day ... I am sorry to leave such a lovely day.'
> *Sir Arthur Sullivan, composer – notably with librettist W. S. Gilbert – makes his final diary entry, 1900*

> 'I woke to more mist flooding the land, floating the church towers and trees. Then it evaporated into a brilliant, shining blue, and I swam just outside Hambridge in one of the long, straight drains crossing the flat grazing meadows on West Moor like tall mirrors.'
> *Roger Deakin, nature writer, Somerset Levels,* Waterlog, *1997*

It isn't a hurricane, though (*25 August*): it's a Force 11 gale with 'hurricane-force' winds. Nor, compared to the great gales of history, is it that bad – a storm of such severity is expected about once every 150 years. Also, it's highly localized. It just happens to be localized over London and the South-East. In Scotland, storms of such severity occur every twenty years, but with fewer trees to uproot (now you know the reason) and fewer people to affect, there's less fuss.

-7.2C Balmoral 1919 ❄

22.8C Geldeston 1908 ☼

As for Michael Fish, he's soundly stitched up. His words are lifted out of context by a gleeful media. In fact, his fatal remark refers to conditions in Florida, a link to a news story about storm damage in the Caribbean, before he goes on to warn viewers to 'Batten down the hatches, there's some really stormy weather on the way.' That bit, however, is never repeated. Still, it's not a great day for the Met Office, especially when amateur forecaster Bill Foggett (*26 December*) announces that he correctly predicted the storm – when he saw his neighbour's cat go berserk.

16 OCTOBER

6.6C min / 13.1C max

'87: THE STORM AND THE CRASH

'There was trouble at the airport, the direct flight was cancelled, and ... it was nearly midnight before I arrived in London, thoroughly drunk. I slept deeply, waking briefly in the middle of the night because there seemed to be a rather strong wind. Next morning I set out for Earl's Court station and was surprised to see a tree lying across Redcliffe Square. I turned the corner to see two more. The streets were strangely empty. It was like *On the Beach*. I got a train – also eerily empty – and finally arrived at my agent Peggy Ramsay's office at eleven ... The deserted streets of the West End were shocking; it was as if the wind of God had passed through.'

It's no wonder that the actor Simon Callow, whose words these are, finds London empty this Friday morning. It's the morning after the 'Storm of '87' (*15 October*) – thousands of commuters from the Home Counties, especially Kent, cannot get to work due to fallen trees and power lines blocking roads and railways. Monday will be 'Black Monday', the worst ever single-day fall in share prices in the City. The question is: are these two events connected? A 2005 Channel Four documentary, *The Explosive 80s: The Storm and The Crash*, contends that they are.

'Today has been lovely – a stray day of summer that somehow got entangled in the red net of autumn.'
Oscar Wilde, poet and dramatist, 1887

The suggestion is that, had everyone got to work on Friday as normal, and had computers not been down due to power failure, then the downturn in global markets that begins on Wall Street the night before could have been averted, or at least slowed. This may be true, though a more significant factor is likely to have been the 'stop loss' computers on trading floors automatically selling as prices sink below predetermined levels – plunging the market into a downward vortex. Socio-economic factors play their part too: it's the senior decision-makers, living in their rural rectories and manor houses, who are least likely to make it to work, leaving markets in the hands of juniors, more prone to panic. Whatever, the result is fiscal Armageddon and the end of the City's longest lunch – the Loadsamoney eighties and Thatcherite yuppiedom are over. Tens of thousands of jobs will go and the market will take two years to recover. As one trader says: 'It felt like the end of capitalism.'

-9.2C Shap 1993

23.2C Valley 1977

WINDMILLS   On a breezy day in 1921, a time when 350 windmills are still in use across Britain, a young Londoner-turned-farm-apprentice on a Suffolk farm, Adrian Bell, is sent to take corn for milling. It's his first visit to a working windmill (he promptly betrays his ignorance by nearly coming within range of the sails 'without realising my danger'). His description of climbing the 'many narrow steps' to find the miller at the top of the mill beautifully evokes what it's like to be inside a working windmill:

THE MILLER'S SCALE OF WIND

Calm	–
Light breeze	Turns the cap so that the mill always faces the wind, preventing damage when the wind increases (i.e. prevents 'tail-winding')
Gentle to moderate breeze (12 mph/19 kph)	Slow milling possible if sails fully shuttered
Fresh breeze (20–25 mph/ 32–40 kph)	Perfect grinding
Strong breeze (25–30 mph/ 40–48 kph)	Milling possible with shutters open. Care required
Near gale/gale	Start, and you can't stop . . .
Storm	Batten sails. Check for damage

'There were three storeys . . . and in the floor between each were holes, through which ran chains . . . The chains were polished brighter than household silver by friction with the wood, and the edges of the holes were lustrous, rubbed, and rounded. The mill interior was something like that of an old ship. It was built round a tree, a stout, straight trunk which was covered with carved initials and dates. Reckonings in coombs and bushels were scrawled upon the curving walls. The top storey was a small chamber with a casement window. The mill creaked and trembled as [the sails] revolved.

    The miller was a small, rosy man with a powder of meal on his face . . . When next a sail broke he would not have it repaired, he said. "Why not?" I asked. "A new sail costs a hundred pounds, and for that I could get an oil engine which would grind all day and every day, whether the wind blew or not."

    So I pondered this paradox that the wind, which costs nothing, is more expensive than engine power to-day.

    I caught sight of what looked like a bunch of white sausages. "What are they?"

    He laughed. "You don't see many of them in London now, I'll be bound. They are tallow candles."

    Here I found them employed in their last use – greasing

-9.9C Carnwath 1993

22.8C Lowestoft 1954

the works of the mill. Oil would have run and fouled the meal.

This man's father, from what he told me . . . used to enjoy being up in the mill in a storm, and would let the sails revolve in dangerously high winds. He sang, he laughed, he loved wild weather. His son said, "It seemed to get into his blood."'

Post Mill   *Smock Mill*   *Tower Mill*

'ONLY A MILLER'S DAUGHTER MARRIES A MILLER'

So the saying went – because only she knows how wretched life is as a slave to the wind. Water mills were *always* preferable to windmills (you could mill when you wanted), even though Britain is one of the windiest countries in Europe. Exposed hills and coasts (for sea and land breezes) tend to be 'well-winded'. In winter, windmills might work most days, as winter is not only windier than summer but the wind goes further – cold air is denser. In an average day, 2–3 tons of corn might be milled, though windmills were also used for drainage pumping.

The sails of a windmill have to remain, at all times, square to the wind – every miller's fear was the disastrous damage done by 'tail-winding'. Windmills came in three types: the post mill where the whole building revolves and which probably appeared as early as the seventh century, and the later smock mill and tower mill on both of which only the cap (bearing the mechanism) revolves. Turning the mill into the wind – known as 'luffing' or 'winding the mill' – was therefore crucial. Fully automatic luffing, which vastly increased efficiency, arrives in 1745 with Edmund Lee's patented 'fantail' mechanism (shown on the tower mill above).

18 OCTOBER

INDIAN SUMMER / RED SKY AT NIGHT

A period of dry, calm weather, often with hazy skies and night frosts, beginning today, St Luke's Day, is traditionally called St Luke's little summer. The more contemporary term Indian summer – US in origin, it relates to the seasonal behaviour of Native Americans – is generic and refers to a spell with similar characteristics any time in mid to late autumn. Prior to the 1960s and the development of modern farming, a warm period at this time of year was important, especially after a wet summer, as it could make the difference between a poor harvest and a reasonable one.

Snippets of seasonal weather lore like St Luke's little summer, St Swithun's Day (*15 July*) and the Festival of the Ice Saints (*10 May*) indicate the tendency for particular types of weather at particular times of year. Memory and cumulative experience have distilled old nuggets of country wisdom, over centuries, into these sayings. Many are attached to Christian saints' days simply because it makes them easier to remember. While hardly infallible, they do provide surprisingly accurate rules of thumb (unlike the more random collection that constitutes animal weather lore). This ancient form of forecasting may be easily derided, but only 150 years ago, when a far greater percentage of the population was dependent on the weather, it was all our forbears had to go on.

ANIMAL WEATHER LORE
● Cows huddle together with tails to the wind (in fact, pretty much anything a cow does) - *rain* ● Eels are easy to catch - *thunderstorm* ● Robins singing from the treetops - *fine day* ● The cock goes crowing to bed - *rain* ● The cat sneezes - *rain* ● The old donkey blows his horn - *rain* ● Sheep feeding uphill in the morning - *fine weather* ● Rooks tumbling in the sky, as if shot - *rain* ● Cockles have more gravel sticking to their shells - *tempest* ● Porpoises in harbour - *imminent storm* ● Pigs can see the wind

Many local, short-term weather sayings also remain sound today. So when your web-linked BlackBerry fails in the hills, you can still look up and say, 'Mackerel sky and mare's tails, Make tall ships carry low sails', or 'Red sky at night, shepherd's delight; Red sky in the morning, shepherd's warning,' and you might even be right.

25.9C Nantmor 1997

-7.2C Usk 1926

19 OCTOBER

6.5C min / 13C max

SHIFTING SANDS / CROWN JEWELS LOST

An intense north-westerly storm today in 1694 is creating the British Isles' most implausible geographical feature – a Scottish desert. Throughout the second half of October the wind creates a moving sea of sand at Culbin, an estate of prime farming country covering more than 5 square miles (13 square km) at the mouth of the Findhorn River (♥). Coastal dunes, built up by prevailing south-west winds and possibly destabilized by the extraction of marram grass for use as thatch, are on the move. 'At first only fields were invaded,' a contemporary account says. 'The drift then advanced upon the village, engulfing cottages and the laird's mansion. The storm continued through the night, and next morning some of the cottars had to break through the backs of their houses to get out.'

> 'Midnight, sitting up in my bed, which I had drawn alongside the fire, with my head to the great window and the foot to the bookcase, my candle on the green table close by me – as I was reading – a flash of lightning came so vivid as for the moment to extinguish in appearance both the candle and the bright fire. It was followed by a clap of thunder that made the window belly in as in a violent gust of wind.'
> **Samuel Taylor Coleridge, poet, Lake District, 1802**

Trouble with shifting sands is not new to the north of Scotland. Forvie on the Aberdeenshire coast is buried, according to legend, by a nine-day storm in 1413. Nairn is threatened by sand drifts in 1663. The village of Udal on North Uist is finally buried in 1697 (*2 October*). The burying of the Culbin estate, though, is 'one of the greatest wind-borne deposits . . . in recent geological time,' according to Herbert Edlin, the twentieth-century natural history author. Sixteen farmhouses, huge expanses of farmland and the manse are engulfed. The laird, Alexander Kinnaird, is ruined and he petitions the Scottish Parliament for protection from his creditors. Parliament enacts legislation the following year prohibiting the extraction of marram grass. The Culbin estate remains a wandering desert until a pine forest is planted there in the 1920s.

24.6C Lairg 1899

In 1216, heavy rain today floods the Washes of East Anglia and the hapless King John loses his war chest: not just cash but the Crown Jewels themselves. A river in spate and a sharply rising tide combine to sweep the royal baggage train away. The treasure is never found – not that people aren't still looking.

-7.8C Bromyard 1926

20 OCTOBER

6.3C min / 12.6C max

SCOTLAND'S WORST AIR CRASH

Prestwick airport's (✈) exceptionally good weather record is the reason that the KLM New York flight from Amsterdam is stopping-over, tonight, in 1948. Sandwiched between two seaside resorts on the Ayrshire coast, Prestwick's almost guaranteed fog-free status is why it is the *only* Scottish airport granted a prestigious transatlantic link in these earliest days of civil aviation. Tonight's incoming Lockheed Constellation (the first pressurized civilian airliner, which brings flying to the masses) is captained by K. D. Parmentier, KLM's chief pilot – widely regarded as one of the great flyers of the era.

'My 30th birthday ... Sunshine on the Bassenthwaite window while rain and hail was scourging the Newlands window. The whole vale shadow and sunshine.' *Samuel Taylor Coleridge, poet, Lake District, 1802*

For once, however, Prestwick's celebrated micro-climate is not so ideal. The weather has steadily deteriorated all evening and as the KLM flight approaches, the airport is under drizzle, the cloud-base almost solid at 600 feet (200 metres). Parmentier, unfortunately, doesn't know this – a delayed take-off from Amsterdam means he misses the radio message – or he would certainly have diverted to Shannon in Ireland. Nor does he receive Prestwick's Morse weather warning on approach, because he has now switched to voice contact. Descending to land, with the main runway lights visible, Parmentier is, in any case, more concerned with cross-wind than visibility. He negotiates with air traffic control to use an alternative runway, one that KLM guidelines – drafted by Parmentier himself – expressly forbid using in low cloud. Parmentier, of course, in the dark, believes there is no low cloud.

The errors now come thick and fast. Over-flying the original runway, Parmentier climbs ready for his new approach and enters cloud. Assuming it's just an isolated patch, and expecting to see the runway any moment, he takes none of the usual cloud-flying precautions. As he heads, blind, for higher ground, the lowered landing gear of the Constellation strikes the main phase conductor of the 132,000-volt cables of the Scottish National Grid. Parmentier radios: 'I have hit something, going on fire, attempting to climb.' It's his last message. The aircraft crashes and explodes. All thirty-four passengers and crew are killed. Parmentier is found, still strapped in his seat, many yards from the aircraft. The flight is, excluding Lockerbie in December 1988, Scotland's worst air disaster.

-11.4C Braemar 1880 ❄

21.7C Heathrow 1969 ☀

21 OCTOBER

6.2C min / 12.6C max

ABERFAN /
CLOUD-
WATCHING

At 9.15 this wet and foggy morning in 1966, the children of Pantglas Junior School near Merthyr Tydfil (☞) have just finished singing 'All Things Bright and Beautiful' in morning assembly. As they head for their classrooms there is a sudden terrific, thunderous roar – the last thing many of them ever hear. Half a million tons of coal waste crashes down the Welsh valley to bury the school; 144 die, 116 of them children, mainly under the age of ten. It's one of the blackest days in the history of Wales. In a few seconds a closely knit mining community loses most of a generation. As news of the disaster filters out, it touches the heart of the world.

Two days of continuous heavy rain has loosened the slag tip from the Merthyr Vale Colliery, but that is not the real cause here. An underground spring beneath the tip, well known to everyone locally and now swollen by the rain, is largely to blame. The National Coal Board refuses to admit knowledge of it, then refuses to accept full financial responsibility for the disaster, forcing the Aberfan Disaster Fund – raised from donations – to contribute £150,000 for the removal of the remaining tips above the village. Not until 1969 does the Mines and Quarries (Tips) Act finally charge quarry owners with the securing and safety of slag heaps. The £150,000 isn't repaid by the government to the Disaster Fund until 1997.

OCTOBER SKY – A FAVOURITE FOR CLOUD-SPOTTERS

Meteorologists won't allow it, but cloud-watchers know that this is the best time of year. The weakened energy of the lowering sun, combined with the seasonal shift in the air masses affecting the British Isles, means, in the right conditions, a dramatic range of vapour effects. It's no accident that the artist John Constable chooses the 'noble clouds & effects' of autumn when painting his celebrated cloud studies (*23 October*). In the normally cloudless 100–800 feet (30–240 metre) zone, straggles of stray cumulus ('dragon's breath') often hang or lazily drift over fields like detached, leaky barrage balloons. High above, high-altitude ice particles of cirrus, up with the contrails, draw the eye past any combination of loose-knitted, mid-level strato- and alto-cumulus, like carelessly pulled cotton pleat. The mixture may be supplemented with every kind of thread, veil or haze. Lit by the low, orange-yellow autumn sun, the combination means one thing: nubilous delight.

188.2mm Glen Cassley 1971

Sun 'dazzle' causes worst ever
motorway pile-up, 1985

22.0C Chester 1893

-9.4C Rickmansworth 1935

*THE HOUND
OF THE
BASKERVILLES*

'Over the great Grimpen Mire there hung a dense, white fog. It was drifting slowly in our direction, and banked itself up like a wall on that side of us, low, but thick and well-defined. The moon shone on it, and it looked like a great shimmering icefield, with the heads of the distant tors as rocks borne upon its surface. Holmes's face was turned towards it, and he muttered impatiently as he watched its sluggish drift.

"It's moving towards us, Watson."

"Is that serious?"

"Very serious, indeed . . ."'

Yes, this evening, around the turn of the twentieth century, marks the fog-shrouded denouement for one of the greatest sleuth's greatest crimes. If one story has made Dartmoor (♀) and its notorious fogs famous the world over, it's Arthur Conan Doyle's *The Hound of the Baskervilles* (1902). Despite the fact that he visited Dartmoor only once, and then not in fog, Conan Doyle knows a local weather feature with dramatic potential when he sees one. Milking the fogs for the brooding sense of place they convey, he ratchets the tension: 'Every minute that white woolly plain which covered one-half of the moor was drifting closer and closer to the house. Already the first thin wisps of it were curling across the golden square of the lighted window . . . As we watched it the fog-wreaths came crawling round both corners . . .' And ratchets the tension: '. . . the steps grew louder, and through the fog, as through a curtain, there stepped the man whom we were awaiting . . .' And ratchets the tension: '. . . there was a thin, crisp, continuous patter from somewhere in the heart of that crawling bank. The cloud was within fifty yards of where we lay, and we glared at it . . . uncertain what horror was about to break from the heart of it . . .'

22.2C Cape Wrath 1965 ☀

-8.6C Braemar 1880 ❄

The scene culminates in the magnificent cliff-hanger with which the eighth instalment of the original magazine serialization ends: 'A hound it was, an enormous coal-black hound, but not such a hound as mortal eyes have ever seen. Fire burst from its open mouth, its eyes glowed with a smouldering glare, its muzzle and hackles and dewlap were outlined in flickering flame. Never in the delirious dream of a disordered brain could anything more savage, more appalling, more hellish, be conceived than that dark form and savage face which broke upon us out of the wall of fog.'

♀

23 OCTOBER

5.9C min / 12.2C max

CONSTABLE'S
CLOUDS

'I have done a good deal of sky-ing,' writes the landscape painter John Constable to a friend today in 1821. 'That Landscape painter who does not make his skies a very material part of his composition . . . neglects to avail himself of one of his greatest aids.' Constable may be famous for his big Suffolk landscapes such as *The Hay Wain*, but his revolutionary idea, the one that the art brigade in the last few years have been mad for, is displayed in his much smaller cloud studies. 'It will be difficult to name a class of Landscape in which the sky is not the "*key note*", the *standard of "Scale"* and the chief "*Organ of Sentiment*" . . . The sky is the source of light in nature – and governs every thing.'

'The wind was loud, the rain was heavy, and the whistling of the blast, the fall of the shower, the rush of the cataracts, and the roar of the torrent, made a nobler chorus of the rough musick of nature than it had ever been my chance to hear before.'
Dr Samuel Johnson, riding through a storm from Mull to Inverary, A Journey to the Western Isles of Scotland, 1773

'The late October sunshine, bowled underarm down the lanes, threw my pedaling shadow many yards ahead.'
Roger Deakin, Waterlog, 1999

'A wet day and all its luxuries.'
Caroline Fox, diarist, 1848

There remain almost a hundred laptop-size, oil 'sketches' on paper, painted on Hampstead Hill in the late summer and autumn of 1821 and 1822 (Constable specifically notes his preference for this time of year – *21 October*). On the back, he records exact notes of time and conditions, such as wind 'very brisk', or clouds 'running very fast'. Constable is the first painter to do anything so eccentric as painting *just* clouds. For him, of course, his studies are mere jottings, never intended for exhibition or sale but just part of his continuing quest to understand nature – his, as he calls it, 'natural history . . . of the skies'. Following Luke Howard's classification of clouds in 1802 (*18 July*), this is a time of general cloud fever, and Constable is aware of these developments. These paintings are now regarded as being among landscape painting's supreme achievements. Constable takes the credit for starting a new relationship between weather and art, culminating with Turner (*13 November*) and Monet (*23 February*).

-9.1C Braemar 1880 ❄

21.4C Red Wharf Bay 1996 ☼

Today in 1091, the earliest recorded tornado in Britain rips up central London killing two, destroying six hundred houses and damaging several churches. At St Mary le Bow, 26-foot roof timbers (a little under 8 metres) from the church are driven into the ground so deep that only 4 feet (just over a metre) remains visible.

24 OCTOBER

5.5C min / 12.1C max

ELIZABETHAN ENGLAND SAVED ... AGAIN

The weather saves Elizabethan England from a certain Spanish invasion – for the third time in a decade – today in 1597. The Third Armada – a fleet of 60 fighting ships, 8000 soldiers, horses, artillery and fortifications that is smaller than but just as purposeful as the great Armada of 1588 (*17 August*) – is lying-to, bow to wind, as little as 30 miles (48 km) off the Lizard (⊕) in Cornwall when a roaring and violent north-easterly gale strikes up and rages relentlessly for three days. One by one the Spanish galleons surrender to the storm until the fleet is completely broken up. Finally Martin de Padilla, the Adelantado of Castile and commander of the fleet, puts his flagship, *St Paul*, before the wind and runs for home.

As the British admiral, Sir William Monson, notes, 'The Spaniards never had so dangerous an enterprise on us.' It was all so unnervingly close.

'A fine day and all its liabilities.'
Caroline Fox, diarist, 1848

The Earl of Essex and Walter Raleigh, with almost the entire English fleet of seaworthy fighting ships, are on an ill-conceived mission in the Azores. Coastal look-outs have been disbanded. In fact, England is almost undefended. After a decade of constant vigilance, the island nation would have been taken completely by surprise had the Spanish ships reached the coast.

But the storm keeps them away. The air of complacency is only punctured when, eight days later, a messenger finally reaches London with news of Spanish ships off the Isles of Scilly. When a copy of the Adelantado's original plan – to invade at Falmouth – is seized, panic breaks out. Parliament is suspended. English troops in France are recalled. Queen Elizabeth is furious with Essex. The alarm does pass when a further report, of the storm-scattered Spanish fleet limping south, finally arrives. Everyone knows that the weather – that Protestant wind again – has delivered England from the Catholic grip of the gout-ridden King of Spain.

-9.4C Kingussie 1885 ❄

21.7C Prestatyn 1996 ☀

In 1933, the sun above London today turns a strange shade of yellow, then red, before disappearing altogether. A high fog has settled over the city to cause 'midnight at midday' on account of a temperature inversion which has trapped fog and smoke beneath it, just as it will to such disastrous effect in 1952 (*6 December*).

⊕

25 OCTOBER

5.4C min / 11.8C max

STORM
WARNINGS /
FIRST FROST

The storm of the century shatters the temerity of Victorian naval engineering today in 1859, when the *Royal Charter*, a 2719-ton, steam-driven iron clipper that supposedly represents the future of ocean-going, is tossed like a toy on to rocks on the Anglesey coast off North Wales (⌘). Some 450 drown. The gale rages across the British Isles for days, leaving a trail of death and devastation. At least two hundred ships are lost, but it's the destruction of the *Royal Charter* that shakes the nation and gives the storm its name.

In the aftermath Rear Admiral Robert Fitzroy (*1 August*), statist at the Meteorological Department of the Board of Trade, examines the course of the storm from carefully compiled observations: it develops in the Azores, then moves over Brittany before advancing up the Irish Sea and across Scotland. This convinces him that the route of the storm could have been predicted in enough time to save ships and lives. He visualizes, for the first time, the idea of foretelling weather by means of a synoptic chart. More significantly, the inquiry into the *Royal Charter* storm recommends that Fitzroy establish a network of forty British ports issued with barometers that will report daily weather observations via the new telegraph. Thus 'storm warnings' (*6 February*) can be issued, using a code of 'cautionary signals' involving drums, cones and, at night, lights suspended at ports (see opposite). It's the birth of weather forecasting.

FIRST FROST DUE TODAY

Global warming is messing with our frosts, but today is, on average since about 2002, the day when the first frost may be expected: the first true sign that winter is on the way. There are two kinds of frost: the severer 'air frost' and more common 'ground frost' (see *Definitions*). On average, there are two or three times as many ground frosts, between twenty and fifty days a year in inland, lowland areas, generally burning off by mid-morning. A full day's frost or 'ice day' is, these days, almost unheard of below 1000 feet (300 metres), or outside a 'frost hollow' (*27 January*). Thick 'white frost' or 'hoar frost' – usually caused by freezing dew – creates the 'winter wonderland' effect. Invisible 'black frost', equally rare, occurs in very dry air. Frosts play a vital role in the natural annual cycle, breaking down ploughed soil, checking or killing many bugs and plant diseases and promoting leaf coloration and fall. So don't complain about having to scrape the car windscreen; take the day off work and celebrate a disappearing phenomenon.

150.5mm Kinloch 1987
-8.9C Kingussie 1896 ❄

20.2C Jersey 1949

25 OCTOBER

FITZROY'S WARNING SIGNALS

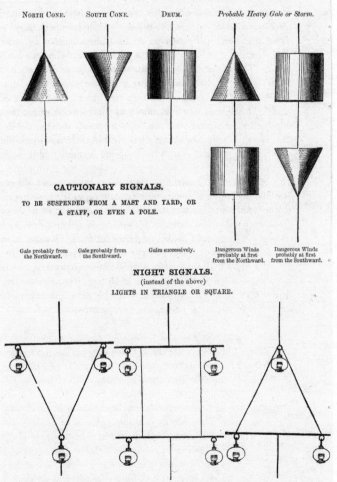

NORTH CONE.   SOUTH CONE.   DRUM.   *Probable Heavy Gale or Storm.*

CAUTIONARY SIGNALS.

TO BE SUSPENDED FROM A MAST AND YARD, OR
A STAFF, OR EVEN A POLE.

Gale probably from   Gale probably from   Gales successively.   Dangerous Winds   Dangerous Winds
the Northward.   the Southward.   probably at first   probably at first
                                                       from the Northward.   from the Southward.

NIGHT SIGNALS.
(instead of the above)

LIGHTS IN TRIANGLE OR SQUARE.

Four lanterns and two yards, each not less than four feet long, will be sufficient — as only one signal will be used at night.
These signals may be made with any lanterns, showing either white, or any colour, but *alike*.
Red is most eligible. Lamps are preferable to candles. The halyards should be good rope, and protected from chafing.
The lanterns should hang *at least* three feet apart.

26 OCTOBER

5.1C min / 11.6C max

Amongst the countless ships fighting for survival in the *Royal Charter* storm (*25 October*) this morning in 1859 is the largest ship in the world. Brunel's unfinished SS *Great Eastern* is berthed at Holyhead (♥) following first sea trials seven weeks earlier. As gale-force winds nudge hurricane-force, wind howls through the rigging 'with indescribable fury,' says *The Times*. Snow and hail plaster the decks. Cabin skylights are sucked open, then slammed shut as showers of glass and spray deluge the luxury saloons. The two anchor chains rise from the water 'like bars of solid metal'. Then Holyhead's breakwater, a massive stone cob, collapses, the masonry 'melting away into the angry sea like a mere fretwork of sugar-candy'. Throughout this maelstrom Captain Harrison nurses the great ship with indefatigable vigilance, his waterproofs sliced to shreds by the hail. He alternately hauls in and lets out her anchors, continually adjusting her position – until her propeller becomes fouled by debris.

This monument to Brunel's technical vision is way ahead of her time. Not just the largest ship ever built (remaining so until the twentieth century), she's also the first double-skinned, iron-hull vessel. With screw-propeller, paddles and six masts for sails, she's designed to carry four thousand passengers to Sydney and back, at up to 13 knots, without refuelling. Her four engines generate 8000 horsepower and burn 300 tons of coal a day. Of all the examples of Victorian engineering flamboyance she is the showpiece, the Concorde of her age. It's the project to which Brunel most completely devotes himself; his iconic portrait, in front of those massive chains, is aboard the *Great Eastern*.

Yet there is something odd with this ship. She has a reputation for being jinxed; unlucky. Never christened, the *Great Eastern* has her name changed from *Leviathan* when her first backers go bankrupt. Two are killed in the difficult sideways launch, six more in a boiler explosion. Missing riveters are reputed to be sealed inside her double skin. Just six weeks before today Brunel collapses on deck, dying ten days later. *The Times* devotes four columns of its dense print to Harrison's efforts during the storm – a measure of the place this ship occupies in Victorian sensibilities. And so, while all about her are dashed to pieces, somehow the *Great Eastern* survives – a rare piece of luck for a ship which never sheds her reputation or fulfils her promise. Soon she is sold for £25,000 (having cost £500,000) and converted for cable-laying.

-9.4C Braemar 1931 ❄

22.8C Southsea 1969 ☀

5.2C min / 11.7C max

DEADLIEST TORNADO / BUCKINGHAM'S WATER

'Roaring like a train in the Severn Tunnel,' is how a witness describes the 660-feet-wide (200 metres) tornado that snakes from east Devon to Lancashire today in 1913. The destruction is greatest in Edwardsville (♀) near Merthyr Tydfil in South Wales, where a miner is carried 1300 feet (400 metres), dashed on the ground and killed; another five die – the highest-ever death toll for a British tornado – and dozens are injured. Brick chimneys are skittled into the sky and a hayrick weighing several tons is carried over a mile.

'The glowing sun had retreated below a bank of rising mist, throwing upon the twilit sky a pinky primrose light which was translucent and pure. Against this backcloth a row of elms, their feet in the swirling river of mist, their arms drooping against their sides, looked tranquil and serene as though they must endure for ever in this setting, a kind of promise that England would remain as I had best loved it.'
James Lees-Milne, writer, Gloucestershire, 1973

Back in 1483 a great flood known as the Duke of Buckingham's Water ends a premature attempt to seize the crown of England by Henry Tudor, Earl of Richmond. The Duke of Buckingham, formerly a close ally of the King, Richard III, raises Henry's standard today in Brecon (♀). The plan – to cross the River Severn, join forces and march on London – goes awry when days of monsoon-like rain cause the Wye and Severn to break their banks and flood the surrounding land. As Richard's soldiers hold the fords, Buckingham is stuck in Wales. The whole thing is a disaster. With his fleet scattered in the Channel by a storm, the aspirant king limps back to Brittany. Buckingham pays with his head. Henry will return in August 1485 – an altogether much better time of year for invasion purposes – to defeat Richard III. Tudor England can begin.

In 2002 an 'extra-tropical storm' (*25 August*), the remnants of Hurricane Jeanette, blasts in off the Atlantic, crossing Ireland and northern England, before heading for Sweden. Gusts nudge 100 mph (160 kph). Seven die, mainly due to falling trees, three hundred thousand are left without power, and passengers on the ferry to Lerwick in the Shetlands endure eighty hours at sea before they finally enter port.

-10.6C Braemar 1926 ❄

20.3C Old Street 1888 ☼

'My birthday began with the water –
Birds and the birds of the winged trees flying my name
Above the farms and the white horses
And I rose
In rainy autumn
And walked abroad in a shower of all my days.'
Dylan Thomas, 'Poem in October'

28 OCTOBER

5.4C min / 11.7C max

A NIGHT IN A LIFEBOAT IN A HURRICANE

In a south-westerly gale this evening in 1927 the 'pulling and sailing' lifeboat from Moelfre (⚓) in Anglesey (where the *Royal Charter* and *Great Eastern* so recently ran into trouble – *25* and *26 October*) finally locates the sailing ketch *Excel*. She's wallowing and on the point of sinking. With the seas far too heavy to get alongside, the lifeboat crew hatch a daring plan: to sail directly over her amidships, hoping a big wave will carry the lifeboat aboard. It works. But as the lifeboat is 'grounded' on deck, she strikes the main hatch coamings of the sinking *Excel* with terrific force. Quickly the three crew members are hauled aboard, but when the water-logged ketch drops into a trough and the two vessels slide clear the lifeboat is badly holed. Captain Richard England, who was supposed to sail aboard *Excel* but missed her, records the rest of that grim night in *Schoonerman* (1927):

'The gale increased to hurricane force and it was hopeless to attempt sailing back to Moelfre. Throughout a long, dark night, the damaged lifeboat beat about in the raging seas. She was full of water and had lost most of her buoyancy. The waves broke over her continuously. Rescuers and rescued alike had to hold on grimly to the lifelines to save themselves from being washed overboard.

'The 65-year-old lifeboat Second Coxwain, William Roberts, who like his mates had been tumbled aft time and time again by the seas, had injured his head and was getting very feeble. He was washed right out of the boat on one occasion but still had the presence of mind to retain his grip on a lifeline. Two of his mates hauled him back aboard but he became very weak. The incessant battering of the waves, as the small craft plunged through, rather than over them; the blinding, choking brine in eyes, throat and nostrils and the stupefying effects of constant immersion in the icy water was as much as the strongest man could bear. Will Roberts died during the night. He talked to his friend Tom Williams right to the end. His last words were a request for a chew of tobacco.'

-11.7C Dalwhinnie 1948 ❄

20.6C Huddersfield 1958 ⚓ ☀

108mph Rhoose 1989, highest low level gust in Wales ⏱🌀🌀

'This empty year is fading into a dull grey mournful darkness: so slow-footed and yet so swift and evanescent. What of the new year and the spring? I wonder.' *J. R. R. Tolkien, writing* The Lord of the Rings, *1944*

29 OCTOBER

5.1C min / 11.3C max

ARCTIC FOOTBALL   In sleet, snow, thick mud and a stinging Arctic wind today in 1932, Blackburn Rovers play Sheffield United in the League (⚽). The game – 'an ordeal for players and spectators alike' – is suspended for ten minutes in the second half 'to give distressed players a respite' from the elements, the *Manchester Guardian* reports. The men are revived in the dressing rooms, but when the game resumes the referee, Mr Caseley of Wolverhampton, loses consciousness. A linesman – clearly made of sterner stuff – takes charge, but three Sheffield players refuse to come back on to the pitch. Down the road at Blackpool, five Chelsea players leave the field suffering from exposure.

> 'Wonderful downpouring leaf: when the morning sun began to melt the frost they fell at one touch and in a few minutes a whole tree was flung of them; they lay masking and papering the ground at its foot.'
> *Gerard Manley Hopkins, poet, Surrey, 1873*

Today in 1688 a south-westerly gale in the North Sea veers north-west and blows the Protestant Prince William of Orange and his fleet back into port on the Dutch coast. Married to Mary, daughter of James II, William has been 'invited' to take over the throne of England. He hastily repairs the damaged ships, and prepares for his next attempt (*15 November*).

In the same storm, the sands around the village of Santon Downham near Thetford in Norfolk are on the move again. A combination of rabbits and deteriorating climate has set several thousand tons of sand 'wandering' into the village over five decades. Storm by storm, and inch by inch, the village has been engulfed. 'The travelling sands . . . rouling from place to place, & like the sands in the Deserts of Libya, quite overwhelmed some gentlemen's whole estates,' the diarist John Evelyn notes on a visit. A century later, William Gilpin is amazed to discover 'absolute desert almost in the heart of England'. The area is afforested in the mid-nineteenth century, though some dunes still remain nearby.

-9.4C Balmoral 1911 ❄

19.2C Coltishall 1984 ☀

30 OCTOBER

5.2C min / 11.5C max

MOTOR
LIFEBOATS /
BRYSON
PICNIC

'Fog that you might cut with a knife all the way from London to Newbury. This fog does not *wet* things. It is rather a *smoke* than a fog.' So William Cobbett begins his seminal travel journal, *Rural Rides* (1830) today in 1821. A gale – that other dominating seasonal weather feature – this morning drives the hospital ship *Rohilla*, en route to collect wounded from the Western Front in 1916, on to a reef off Whitby in north-east Yorkshire (⊕). After rescuing thirty-four, the Whitby rowing and sailing lifeboat crew is finally defeated, unable to make any headway at all. A new motor lifeboat is sent for, from 44 miles (70 km) up the coast. With the power to make headway even in the massive swell – admittedly helped by pouring gallons of oil to calm the waters – the lifeboatmen save fifty more. The era of the 'pulling and sailing' lifeboat is closing.

PICNIC ON BOW FELL (☻)

'Walkers ahead of us formed well-spaced dots of slow-moving colour leading to an impossibly remote summit, lost in cloud. As ever, I was quietly astounded to find that so many people had been seized with the notion that struggling up a mountainside on a damp Saturday on the winter end of October was fun ...

As we pressed on the weather severely worsened. The air filled with swirling particles of ice that hit the skin like razor nicks. By the time we neared Three Tarns the weather was truly menacing, with thick fog joining the jagged sleet. Ferocious gusts of wind buffeted the hillside and reduced our progress to a creeping plod. The fog cut visibility to a few yards. Once or twice we briefly lost the path, which alarmed me as I didn't particularly want to die up here ...

We made it to the top without incident. I counted thirty-three people there ahead of us, huddled among the fog-whitened boulders with sandwiches, flasks and madly fluttering maps, and tried to imagine how I would explain this to a foreign onlooker – the idea of three dozen English people having a picnic on a mountain top in an ice storm – and realized there was no way you could explain it. We trudged over to a rock, where a couple kindly moved their rucksacks and shrank their picnic space to make room for us. We sat and delved among our brown bags in the piercing wind, cracking open hard-boiled eggs with numbed fingers, sipping warm pop, eating floppy cheese-and-pickle sandwiches, and staring into an impenetrable murk that we had spent three hours climbing through to get here, and I thought, I seriously thought: God, I love this country'.

Bill Bryson, Notes from a Small Island, *1994*

-10.0C West Linton 1926 ❄

182.4mm Thirlmere 1977 💧 ⚽

19.4C Reading 1927 ☼

31 OCTOBER

THE WICKER MAN

Snow, driving rain and Dumfries-shire winds are making the filming of *The Wicker Man* interesting today in 1972 (♀). The cult thriller is set around the neo-pagan festivities and rites of spring on a fictional Hebridean island, Summerisle, as every film buff knows. However, financial problems mean the cameras don't start rolling until autumn. Artificial leaves and blossom have to be brought in and glued to bare branches. Plastic fake apple trees are shipped in from Asia. Blow heaters roar. Stars (Christopher Lee, Edward Woodward and Britt Ekland) suck on ice cubes, so their breath isn't misty on the chilly autumnal mornings. In the final scene, Woodward, as Sergeant Howie, is set alight inside the huge wicker statue as the onlookers sing the folk song, 'Summer is i-cumen in'. He later recalls, 'I've never been so cold.'

During a thunderstorm today in 1638, a fireball (*12 July*) hits and enters the Parish Church at Widecombe (♀) on Dartmoor. 'There fell in time of Divine Service a strange darkness . . . a mighty thundering was heard, like unto the sound and report of many great Cannons; and terrible lightning therewith,' a witness records in a pamphlet entitled 'A True Relation of those most strange and lamentable Accidents':

'Some said they saw at first a great fiery ball come in at the window and pass through the Church which so affrighted the whole Congregation . . . with a great cry of burning and scalding . . . the lightning seized upon [the Minister's] poor wife, fired her ruff and linen burning many parts of her body . . . Master Hill, a Gentleman of good account in the Parish . . . had his head suddenly smitten against the wall, through the violence whereof he died . . . One more man, his head was cloven, his skull rent into three pieces, and his brains thrown upon the ground whole, and the hair of his head, through the violence of the blow at first given him, did stick fast unto the pillar . . . But it pleased God, yet in the midst of judgement to remember mercy, sparing some and not destroying all.'

-10.6C Wolfelee 1926

158.9mm Tollymore Forest 1968

A severe storm today in 1823 smashes the lights of three lighthouses on the Casquets, a group of rocks north-west of Alderney in the Channel Islands. And in 1976 it's still raining – the wettest October for seventy years has removed all memories of the century's longest summer drought.

19.4C Margate 1968

NOVEMBER

1 NOVEMBER

FERRYBRIDGE POWER STATION BLOWN DOWN

The photograph overleaf is the best testament to the power of the gale that blows across Yorkshire today in 1965. The first 377-foot high (115-metre), concrete cooling tower at Ferrybridge 'C' Power Station, near Knottingley (🖤), collapses at 10.30 a.m. Ten minutes later, a second crumples to the ground. One employee, though ready with his camera, forgets in the furore to take the lens cap off. Fortune smiles, however. Forty minutes later, a third tower ripples 'like a sail caught aback in the wind' and, with a huge report, sinks majestically inwards.

'Continuous rain for the last three days, and, in consequence, the novel progressing well and myself not sleeping.' *Evelyn Waugh, writing* Brideshead Revisited, *1939*

'No sun – no moon!
No morn – no noon –
No dawn – no dusk – no proper
time of day –
No sky – no earthly view –
No distance looking blue –
'No warmth, no cheerfulness,
no healthful ease,
No comfortable feel in any member –
No shade, no shine, no butterflies,
no bees,
No fruits, no flowers, no leaves,
no birds –
November!'
Thomas Hood, 'No!', 1844

The disaster is due to a substantive misunderstanding between civil engineering and meteorology. The towers, whose construction is begun in 1961, have been built to the standard allowance for wind speeds averaged over one minute – approximately 75 mph (120 kph), as they are during this gale. However, gusts today reach 103 mph (165 kph), and the force which wind exerts is proportional to the square of the speed. An inquiry into the disaster reports, also, that the proximity of the eight towers, arranged in two parallel, staggered rows, leads to increased wind velocity and turbulence between them by constricting the flow (the Venturi effect).

Dating from much earlier times, today traditionally marks the end of the 'lighted half' of the year, when livestock are brought off the hill pastures for winter. The Iron Age inhabitants of the British Isles lived by what has been called a 'Cattle Clock'. As there is little or no grass on upland pastures after this time – it stops growing when temperatures dip below 5C (*21 March*) – cattle and sheep are moved back to the shelter of permanent habitations in the lowlands.

-12.8C Braemar 1926 ❄

19.7C Hawarden Bridge 1971 ☀

1 NOVEMBER

Familiar November weather is fog and rain. It is the wettest month of the year on average (though *very* wet days are unusual). It's also the foggiest month, with a smattering of early frosts. All in all, a month to be endured. As the nineteenth-century French proverb says: 'In October de Englishman shoot de pheasant: in November he shoot himself.'

After rippling 'like a sail caught aback in the wind', the third concrete tower at Ferrybridge 'C' Power Station collapses, 1965.

2 NOVEMBER

5.4C min / 11.2C max

SIR CLOUDESLEY SHOVELL AND LONGITUDE

Gales from several directions cause the worst maritime disaster in British waters today in 1707. The naval fleet of twenty-one warships is returning from the Mediterranean. For days, without a glimpse of the sun, they have been navigating by 'dead reckoning'. Approaching the English Channel, the sailing masters know they are hopelessly lost, but they blunder on, north-eastward, until at 8 p.m. today the flagship HMS *Association* smashes into the notorious Gilstone Ledges south-west of the Scilly Isles (✪). She sinks in four minutes. Three more of Her Majesty's ships, *Eagle*, *Romney* and *Firebrand*, are wrecked on the same rocks and disintegrate in a welter of smashed timbers and swollen seas. Of the 1673 men aboard, 26 survive.

'Terrible rain today, but it cleared up at night enough to save my twelvepence coming home.'
Jonathan Swift, writer and satirist, London, 1711

'I really begin to doubt whether England is a beautiful country. Today it simply looks cold and dull.'
George Bernard Shaw, dramatist, on the Great Northern train from London to Bradford, 1896

'A hurricane blowing and bitter cold. Midwinter is no season to prospect for a house.'
Evelyn Waugh, author, 1946

The body of the commander-in-chief of her Majesty's fleet, the splendidly named Admiral Sir Cloudesley Shovell, 'fabricator of his own fortune' and an original British hero, is washed ashore next day 7 miles (11 km) away at Porthellick Cove. (There is a story that Shovell is found alive and promptly murdered by a Scillonian fishwife for his emerald ring, but it's probably not true.) The disaster shocks the nation and there is public outcry about the state of navigation in the Royal Navy. And though the error is really one of latitude, it precipitates the famed Longitude Act, by which the Board of the Admiralty offers £20,000 to anyone who can find a way of determining longitude practically, and accurately, at sea. The prize goes to Edward Harrison, a humble clockmaker who eventually perfects his chronometer in 1773, changing ocean navigation for ever.

-10.6C Glenlee 1893 ❄

19.4C Prestatyn 1948 ☀

It's the lack of wind, on the other hand, off the Lizard (⚽) this morning in 1805 that has Lieutenant Lapenotiere chafing with frustration. For he is a man in a hurry. He has news to deliver ahead of Captain John Sykes – vital, if bittersweet, news from Trafalgar for the Admiralty, for king and for country. He will finally make London tomorrow with his message: 'Sir, we have gained a great victory. But we have lost Nelson.'

3 NOVEMBER

5.3C min / 11C max

AUTUMN LEAF COLOUR: PHYTOGERONTOLOGY

In average weather, autumn leaf colour is at its neon best right now. Although leaves start turning as early as the end of July, most are still on the trees, these days, on Bonfire Night. The peak of colour, always much later than everyone thinks, is getting later every year with climate change.

Ideal weather conditions for autumn glory? The summer is as important as the autumn. The vintage show of 2001 came after a long hot summer, which builds up starch and sugars in the leaves ready for the long autumn. Then some sharp, cold weather – ideally an early frost – to 'flick the switch' that tells trees summer is over. Leaves turn because, as the sun gets lower and its rays feebler, as day length shortens and temperatures fall, the profits of photosynthesis (the life-sustaining process by which green plants convert sunlight into energy) get less and less for a tree until, eventually, running a leaf costs more than the leaf returns in energy. The tree then stops making chlorophyll (the pigment that makes leaves green) and the pigments of autumn colour are revealed: the yellow carotenes, orange xanthophylls and, most dramatically, red anthocyanins.

> 'The autumn always gets me badly, as it breaks into colours. I want to go south, where there is no autumn, where the cold doesn't crunch over one like a snow leopard waiting to pounce.'
> *D. H. Lawrence, writer*

> 'Misling rain all day.'
> *Reverend Gilbert White, 1770*

Phytogerontologists (yes, even leaf fall scientists have a name) have little idea what these pigments are for, other than making us 'ooh' and 'aah'. Possibly they allow trees to capitalize on the weaker sun. Carotenes, it's known, are responsible for leaf browning and fall. But the red anthocyanins? Anybody's guess. Anti-oxidants? Insect repellents? Whatever they do, the key ingredient for a visual feast is colour *variety*, which depends on varying rates of chlorophyll degradation. Finally, for a vintage year, September and October must be not too wet or windy, so that leaves remain on the trees. Too much wind, and treetops will be bare and unsightly.

Meet these conditions, add some fine, high-pressure weather to set the colours against blue skies, maybe a nip of frost or some hanging mist and – KERPOW!!!

20.6C Aber 1946

-12.1C Santon Downham 1985

184.2mm Trecastle 1931

4 NOVEMBER

THE RESULTS OF LOWERED HOUSE-BUILDING STANDARDS

At 4 a.m. today, in 1957, a resident of a house in a new development near Hatfield in Hertfordshire (𝕻) is woken by a gale. It seems to him, half-asleep as he is, that his ceiling is moving. Suddenly, with a roar, in a shower of dust the roof peels away and the resident is peering at the stars. A moment later, with a blinding flash, his neighbour's roof goes too, tearing the electrical wiring out. In the next few minutes, twenty-four more of the shallow-pitched roofs – acting in the wind like aeroplane wings – are sucked upwards for take-off. Next morning, buckled aluminium sheets from fifty houses litter the garden and trees.

Demand for new, low-cost housing after the Second World War, fuelled by the baby boom, prompts a house-building revolution, and a break with centuries of architectural tradition. Gone, in the scramble to build, are vernacular styles and materials, adapted over centuries to combat regional quirks of wind, damp or driving rain – traditions such as steep-pitched roofs, small windows, lime mortar, heavy slates or tiles and careful siting with regard to aspect and shelter. In come cheaper materials and quicker, if unproven, techniques. The consequence is Hatfield.

Wind causes, on average, three-quarters of the UK's insurance losses due to natural disasters. Above 70 mph (112 kph; Force 11), damage is certain. The Building Research Establishment has complex formulae for assessing wind stresses at different heights and hill positions, and the design code for most post-war building in south-east England allows for a 'wind pressure' tolerance of 10 per cent. Average winter wind speeds are predicted by the UK Climate Impacts Programme to increase by 10 per cent by 2080 – an increase in wind pressure of 20 per cent. The consequences of a windstorm as severe as, say, those of 1987 (*15 October*) or 1990 (*25 January*), tracking directly up the Thames valley and across London, would be disastrous.

'It is raining in torrents. The light is greenish & unnatural, objects being as if seen through water . . . A silver fringe hangs from the eaves of the house to the ground. A flash. Thunder.' *Thomas Hardy describes the weather on which he bases his storm in* Far from the Madding Crowd *(30 August), 1873*

'Came back to London . . . in bright sunshine, which duly dwindled away . . . and disappeared in impenetrable fog as I reached Belgravia.' *Matthew Arnold, poet, 1858*

'Ten days of torrential rain punctuated with blizzards, hurricanes, whirlwinds, earth-quakes, and like upheavals. My only comfort is indulgence in a gruelling, man-of-the-elements walk through the blinding rain.' *Martin Amis, novelist, aged eighteen, writing home from Sussex Tutors, Brighton, 1967*

-10.0C Braemar 1942

21.7C Prestatyn 1946

5 NOVEMBER

4.9C min / 11C max

BONFIRE NIGHT

Still, misty conditions today in 2006 (as in 2001 and 1997) lead to strikingly high air pollution readings. The reason: fireworks – most notably in London and Lewes, Sussex. Made by combining gunpowder with other chemicals that create the colours and effects, fireworks release carbon dioxide, sulphur compounds, metal oxide and other particulates as they go off, all of which affect air quality. These pollutants are most toxic when the air is still, with hanging mist, because they linger. Local authority guidelines for Guy Fawkes Night advise against letting them off in unsuitable weather. And you thought it was only dangerous to play catch with a lit banger.

'It being an extraordinary wett morning, & I indisposed by a very greate rheume, I could not go to Church this day, to my greate sorrow, it being the first Gunpouder conspiracy Anniversary, that had ben kept now this 80 yeares, under a Prince of the Roman Religion: Bonfires forbidden &c: What dos this portend?'
John Evelyn, diarist, 1685

'In the evening we went round to the Cummings', to have a few fireworks. It began to rain, and I thought it rather dull. One of my squibs would not go off, and Gowing said: "Hit it on your boot, boy; it will go off then." I gave it a few knocks on the end of my boot, and it went off with one loud explosion, and burnt my fingers rather badly. I gave the rest of the squibs to the little Cummings' boy to let off.'
George and Weedon Grossmith, Diary of a Nobody, 1888

'The wind, indeed, seemed made for the scene, as the scene seemed made for the hour ... Gusts in innumerable series followed each other from the north-west, and when each one of them raced past the sound of its progress resolved into three. Treble, tenor, and bass notes were to be found therein. The general ricochet of the whole over pits and prominences had the gravest pitch of the chime. Next there could be heard the baritone buzz of a holly tree. Below these in force, above them in pitch, a dwindled voice strove hard at a husky tune, which was the peculiar local sound alluded to. Thinner and less immediately traceable than the other two, it was far more impressive than either. In it lay what may be called the linguistic peculiarity of the heath.' *Thomas Hardy, The Return of the Native, 1878. The novel opens on Egdon Heath today in 1842*

-11.1C Braemar 1968

21.1C Chelmsford 1938

6 NOVEMBER

THE FLOODS OF 2000: 'PUT A SANDBAG IN THE TOILET BOWL'

The rain stops, fleetingly, today in 2000 to reveal . . . a new Atlantis. It's been the wettest October, and what will turn out to be the wettest November, for a century. March 2001 will confirm it's been the wettest twelve months since records begin (*3 April*). But as the clouds break, a new realization dawns – ten thousand houses are significantly flood-damaged, to the tune of £1.5 billion. 'Floodline', the Environment Agency's new helpline, becomes the country's most dialled phone number after 999, with nearly a million calls.

'There was a most violent Gale of Wind this Morn' early about 3. o'clock, continued More than an Hour. It waked me. It also shook the House. It greatly frightened our Maids in the Garrett. Some limbs of Trees blown down in my Garden. Many Windmills blown down . . . Dinner today a boiled Chicken and a Pigs Face and some beef Steaks.'
Reverend James Woodforde, diarist, Weston Longeville, Norfolk, 1795

The repercussions of all this rain are enormous. There is a 10 per cent hike in home cover for properties in flood plains (rising to 60 per cent in high-risk areas). Black-listed homes lose up to 80 per cent of their value. The Environment Agency starts a flood-risk scoring system to prioritize areas for flood defences – controversial, says the National Flood Forum, because the scoring system 'ends up excluding poorer areas'. It also issues two-hour flood warnings, along with scary instructions about how to protect your home. ('Put plugs into sinks and weigh them down with something heavy.' 'Put a sandbag in the toilet bowl to prevent backflow.') 'Thirty times as costly as getting burgled, this thief takes everything in its path,' the website warns, to clarify that this is *your* problem. The point is made: with rising sea levels and a wetter climate, if yours is one of the half a million houses on a flood plain, or a new-build scheduled for one, start filling those sandbags.

-7.2C Grantown-on-Spey 1981

19.8C Aber 1972

Back in 1697, ball lightning (*12 July*) scores a direct hit on the armoury of Athlone Castle in central Ireland (☞), causing a huge explosion: most of the town has to be rebuilt. In 1889, after eight years, the Forth Rail Bridge is finished, but there's a hitch: the unseasonably cold weather means the two central aprons don't meet. Only when the weather warms, and the metal expands, can the final rivets be driven home. (The bridge isn't officially open until the Prince of Wales, later Edward VII, drives home the final, gold-plated, rivet next March.)

Rivers and coastal flood-risk areas, England and Wales, 2007

7 NOVEMBER

4.2C min / 10.3C max

THE SEVEN ILL YEARS

After two weeks of north-westerly storms (*19 October*), the weather turns cold in Scotland today in 1694, and a legendary winter begins. Severe frosts and heavy snow will still cover the country in late April when the diarist Anthony à Wood writes, 'A most unnatural season, many die, no appearance of spring.'

'A sharp frost; and this had set the poor creatures to digging up their little plots of potatoes. In my whole life I never saw human wretchedness equal to this; no, not even among the free negroes in America ... And, this is "prosperity," is it? These, Oh, Pitt! are the fruits of thy hellish system!'
William Cobbett, radical journalist and MP, Wiltshire, 1821

'Day opened wet but became lovely – warm & sunny. Trees glowing like burning bushes. Beeches beacons of red & gold ... Egyptian advance continues.'
Violet Bonham-Carter, Liberal campaigner, Mottisfont Abbey, Hampshire, 1942

This winter is one of a spectacular series during the 1690s, a decade of meteorological menace – the summers are also lousy – which marks one of the coldest phases of the Little Ice Age (*3 March*). The period 1693 (when the harvest first fails) to 1700 becomes known in Scotland as the 'Seven Ill Years' or the 'William Ill years' (for the reputed part the Protestant king plays in the misery).

Up to a fifth of the population die from starvation and related diseases, or emigrate. Whole villages are abandoned in the Highlands and Aberdeenshire, as starving people pour into refugee camps in the towns. Even the profitable oat crops on the east coast fail repeatedly. The Edinburgh Parliament establishes the Bank of Scotland, increases poor relief and promotes commercial textiles and fishing, but all to little avail. Is it this decade, as some say, which so emasculates the nation that the Scottish Parliament submits to the Act of Union with England early in 1707? Whatever, the kingdom of Scotland ceases to exist. Had the 1690s been 'years of plenty', there might still be a Stuart monarch in Scotland today.

-10.6C Glenlivet 1949 ❄

Today in 1980 it's exceptionally cold, with snow on the ground in the Home Counties. This, the coldest early November of the twentieth century in southern England, is not blamed on climate change though: it's thought that the millions of tons of ash deposited into the upper atmosphere by the catastrophic eruption of Mount St Helens in the USA in May is the culprit.

17.8C Eastwick Lodge 1966 ☀

8 NOVEMBER

4.3C min / 10.2C max

MIGRATORY
BOGS / JANE
AUSTEN'S
STORM

The rain is falling in buckets again in northern England in 1771. After several days of continuous downpour the landscape is saturated and swollen, so much so that it begins to wobble – then, in a few days, to move. A huge upland peat bog known as Solway Moss, north of Carlisle (♥), blows its solid earth cap, sending tons of liquid peat, glutinous mud and wet sphagnum moss pouring into the valley, forming a mile-long (1.6 km) quagmire of black liquid which suffocates livestock, overturns buildings, fills cottages and smothers hundreds of acres of farmland. Thomas Pennant, the naturalist and travel writer, records how the 'smiling valley' becomes a 'dismal waste'. Some farmers, he notes, 'received no other advice than what this Stygian tide gave them: some by its noise, many by its entrance into their houses; and I have been assured that some were surprised with it even in their beds.'

'We have had a dreadful storm of wind in the forepart of this day, which has done a great deal of mischief among our trees. I was sitting alone in the dining room, when an odd kind of crash startled me – in a moment afterwards it was repeated; I then went to the window, which I reached just in time to see the last of our two highly valued Elms descend into the Sweep!!!!! ... I am happy to add however that no greater Evil than the loss of Trees has been the consequence of the Storm in this place, or in our immediate neighbourhood. We grieve therefore in some comfort.'
Jane Austen, novelist, Steventon Rectory, Hampshire, 1800

The 'migration' of peat bogs, caused by heavy winter rains, is not uncommon before the twentieth century. In 1853, a bog with a mile (1.6 km) circumference in County Clare is set loose by days of rain. And Leland, Henry VIII's librarian who tours Britain in 1533, reports how Chat Moss in Lancashire (☞) 'brast up . . . first corrupting with stinkinge water Glasebrooke, and so Glasebrooke carried stingkinge water and mosse into Mersey water, and Mersey carried the roulling mosse, part to the shores of Wales, part to the Isle of Man.'

'A grand frosty forming, and as soon as the sun came, he came in red, tinting the thick hoar saffron, between the melted patches where the sheep had been lying down.'
T. H. White, author, Wiltshire, 1934

19.1C Alston 1881

-8.9C Lenton Fields 1923

4.4C min / 10C max

GIRONA: END OF THE ARMADA

In a terrific gale, the galleass *Girona* is driven on to Lacada Point near the Giant's Causeway in Ireland today in 1588 (✠). Of the 1300 men crammed aboard, including the flower of Spanish chivalry, only 5 survive. It's the tragic finale of the Spanish Armada, in which far more ships have been lost to gales than to English guns.

Storms that 'blow with most terrible fury' harry the Spanish fleet from the moment the Duke of Medina Sidonia makes his critical decision to sail for home via north-west Scotland (*17 August*). Despite his warning,

SWIFT'S SLOW DEPARTURE
A swift – the bird that usually leads the migration to Africa, departing before the end of July – is seen over West Sussex today in 2006. It's so warm, he's forgotten to leave.

'take great heed lest you fall upon the Island of Ireland', many ships, missing anchors and leaking badly, are unable to avoid the notorious west coast. Galleons and converted merchantmen are wrecked in Glenagivney Bay and Loghros Bay in Donegal, off Tyrawley in Mayo, at Loop Head in Clare, and near Valentia Island. In Blasket Sound, off the Dingle

Peninsula, 'There sprang up so great a storm on our beam with a sea up to the heavens so that the cables could not hold nor the sails serve us and we were driven ashore,' a Spanish captain records. Three more ships are lost the same day. The entire western seaboard of Ireland is a graveyard of broken driftwood and bloated corpses in the surf.

The fate of the survivors who make landfall is little better. The panicky English administration in Dublin issues orders to execute all Spaniards who can't be ransomed. Life becomes a dangerous game of hiding from both the weather and English soldiers. Don Alonso de Leiva has already survived two wrecks on the Irish coast when he finds the deserted oared fighting ship *Girona* in Killybegs harbour, Donegal. With his company of young aristocratic adventurers, he repairs the ship and sets sail for Scotland. Sailing round the coast of Antrim, he encounters his final storm. A gold ring, found three centuries later, bears the inscription: '*No tengo mas que dar te*' – I have nothing more to give you. An epitaph for Don Alonso? Of the 130 or more ships that set sail as part of La Felicissima Armada in May, fewer than 80 return to Spain. At least 20 sink in storms off the Irish coast. King Philip II goes into mourning at the loss of a generation of noblemen.

-11.7C Welshpool 1921
18.5C Hereford 1977

10 NOVEMBER

THE GREAT
PLAGUE

The Great Plague rages in London throughout the unusually warm, long summer of 1666, peaking at the end of September with four thousand deaths per week. Only when autumn finally kicks in at the beginning of November, and cold weather starts to kill off the fleas which transmit the bacillus *Yersinia pestis* from rats to humans, does the plague abate. Samuel Pepys records in his diary today: 'Merrily . . . after some fears that the plague would have increased again this week, I hear for certain that there is above 400 decrease.'

'Terrible weather and very high seas, which gave me a bad night, aided by a broken shutter banging so much that I thought it would break the window. Went out, muffled up in top boots, leather coat, etc., and grappled with the bloody thing in the rain and spray. I finally broke it from its hinges, hurting my thumb considerably in the process.'
Noël Coward, writer and composer, at his new home on the south coast, under what he called 'the greyish cliffs of Dover', 1945

The Great Plague is the last major epidemic in the British Isles. Though small-scale compared with the fourteenth-century Black Death (*30 June*), it kills nearly a hundred thousand in England, including a sixth of London's population. The outbreak begins in the overcrowded parish of St Giles-in-the-Field during the 1664–5 winter, but the intensity of the cold seems to control it initially. A warm, dry spring, however – ideal conditions for breeding rats – sets the disease loose on the squalid city. By early summer there is a stampede of merchants, peers and professionals, led by King Charles II and his court, to leave the capital. In the East End, people take to barges on the Thames.

The Lord Mayor and the aldermen of the City remain, instituting a number of fruitless public health schemes: dogs and cats are destroyed (leaving rats with no predator), fires are kept burning night and day to cleanse the air, and everyone, including children, is encouraged to smoke tobacco. In the end, it's the cold weather that subdues the disease (though it takes the Great Fire to kill it off completely – *12 September*). Towards the end of November, Pepys writes, 'the plague is come very low . . . and great hopes of a further decrease, because of . . . a very exceeding hard frost'. The King returns to the capital in February.

18.3C Aber 1938

-12.2C wokingham 1908

11 NOVEMBER

4.2C min / 10.2C max

EAT BATCHELOR'S PEAS

The wind drops and clouds part over Manchester (☂) at mid-morning today in 1938 as Captain Michelmore takes to the skies in his biplane, trailing a banner. The captain is under contract with mushy pea manufacturer Batchelor's to advertise their product for an agreed number of hours. Unfortunately, high winds and rain have kept him grounded for weeks. Michelmore is also contractually bound to call his employer before each flight. But phones are unreliable in 1938. Today, buoyed by the sunshine, he takes to the air.

Alas, he's forgotten one small detail. At the first stroke of eleven, the crowd in Salford's main square bow their heads and fall silent, to commemorate the Armistice and remember the dead of the First World War. At precisely this moment Captain Michelmore arrives overhead, engine roaring, trailing his banner exhorting the public to 'Eat Batchelor's Peas'. The two minutes' silence are scarcely up before the telephones at Batchelor's head office are jammed with complaints. An action for defamation and breach of contract follows. The case sets a precedent for recoverability of damages in breach of contract cases, which remains good law today.

In 1570, a gale and sea flood threaten the east coast of England from the Humber estuary to the Straits of Dover. It's estimated that storm activity increases by 85 per cent in the second half of the sixteenth century. Holland gets the worst of the 'All Saints' Flood', however: one report states that five-sixths of the country is under water.

'Our weather, for this fortnight past, is chequered, a fair and a rainy day; this was very fine, and I have walked four miles; wish MD would do so, lazy sluttikins.'
Jonathan Swift, writer and satirist, London, 1711. 'MD' is Esther Johnson, with whom Swift had a mysterious and controversial relationship.

'London does strike a blow at the heart, I must say: tonight, in a black rain out of doors, and a tube full of spectral, decayed people.'
D. H. Lawrence, writer, 1915

'It is noon and I am writing by candlelight ... But I like fogs; they leave the imagination so wholly to herself, or just giving her a jog every now and then. I shall go out into the park [Hyde Park] by and by, to lose myself in this natural poesy of London which makes the familiar strange.'
James Russel Lowell, American Poet, critic and diplomat, 1888

-11.7C Balmoral 1966 ❄

18.9C Hodsock Priory 1915

211.1mm Lluest Wen Reservoir 1929

12 NOVEMBER

4.2C min / 9.9C max

'NUDGER' NEEDHAM

'The bitter cold wind and sleet pierced one, numbing muscle and brain. Men on both sides succumbed and were carried away to hot baths and stimulants. I left the field before the finish and by doing so probably saved my life.' So recalls Ernest 'Nudger' Needham of Sheffield United (♀) (and future England captain) of a League Division One football match today in 1894. The weather is so bad that several Aston Villa players don greatcoats, while the winger briefly uses an umbrella to keep the sleet off. United's goalkeeper, William 'Fatty' Foulke (who weighs over 20 stone and at whom, it's said, the chant 'Who Ate All the Pies?' was originally directed), seems the least bothered by the cold. Seven thousand hardy fans endure the game.

Today in 1991, pop pickers curse the skies. A hurricane-force storm – the remnants of the north-easter that is the subject of Sebastian Junger's best-seller *The Perfect Storm* – drags the *Ross Revenge*, the broadcast ship of the offshore 'pirate' station Radio Caroline, off its anchor in the Channel (✛) and on to the Goodwin Sands (*27 November*). The crew are rescued, but any chance of the station recommencing broadcasting (it went off air in 1990) goes down with the ship. The offshore radio era is over.

USING RAIN TO ASSIST ROMANCE

'"Can I have the carriage?" said Jane.

'"No, my dear, you had better go on horseback, because it seems likely to rain; and then you must stay all night."

'"I had much rather go in the coach."

'"But, my dear, your father cannot spare the horses, I am sure. They are wanted in the farm, Mr Bennet, are not they?"

'"They are wanted in the farm much oftener than I can get them."

'"But if you have got them to-day," said Elizabeth, "my mother's purpose will be answered."

She did at last extort from her father an acknowledgement that the horses were engaged; Jane was therefore obliged to go on horseback, and her mother attended her to the door with many cheerful prognostics of a bad day. Her hopes were answered; Jane had not been gone long before it rained hard. Her sisters were uneasy for her, but her mother was delighted. The rain continued the whole evening without intermission; Jane certainly could not come back.' *Jane Austen, Pride and Prejudice, 1813. Mrs Bennet's ruse works: Jane gets soaked, catches a chill and remains at Netherfield*

-12.8C Braemar 1927 ❄

204.0mm Seathwaite 1897 ☂

19.1C Aber 1989 ☀

TURNER'S
'SNOWSTORM'

The most famous painting of a storm is Turner's 1842 depiction of a steamer in a blizzard – a single, vast swirling vortex of water, wind and spray with, somewhere in the middle, scarcely discernible, a ship. The painting has a story attached: 'The author was in this Storm on the Night the Ariel left Harwich.' Aged 67, Turner tells his friend, the Reverend W. Kingsley, 'I got the sailors to lash me to the mast to observe it; I was lashed for four hours and did not expect to escape but I felt bound to record it if I did.'

Bravo! The only hitch to this appealing anecdote is that no ship called *Ariel* ever leaves Harwich. According to Turner's biographer, Anthony Bailey, however, a vessel called *Fairy* does set out from Harwich in 1840 (☼). We know this, because it's one of numerous ships lost in a great storm that devastates the south-east coast today. 'Turner's wonderfully associative mind could easily have exchanged *Fairy* for *Ariel*,' suggests Bailey. It seems plausible. If Turner had painted the *Fairy*, it might well make artistic sense to him to change her name to that of the spirit whom Prospero invokes in Shakespeare's *The Tempest*. Besides, Turner doesn't say he was actually aboard the *Ariel*, just that he was in the storm – perhaps he was observing from another boat.

True or not (and the French painter Claude Vernet makes the same claim for his *Storm on the Coast*, 1754), there's no doubt that *Snow Storm* is revolutionary. 'There is nothing else remotely like it in European art,' says Lord (Kenneth) Clark, describing its stormy chaos as 'portrayed as accurately as if it were a bunch of flowers'. Ruskin, Turner's great champion, calls it 'the grandest statement of sea motion, mist and light that has ever been put on canvas'. Contemporary critics call it 'soapsuds and whitewash', to which Turner replies: 'I wonder what they think the sea's like? I wish they'd been in it.' And, of course, that's the nub. Whether he was lashed to the mast or not, *Snow Storm* represents, above all, the sensation of being in it. It is less an observation of reality than the *experience* of it – and with this leap of the imagination, Turner pre-empts the Impressionists by half a century.

-12.8C Braemar 1919 ❄

18.9C Nantmor 1989 ☀

14 NOVEMBER

3.3C min / 9.1C max

'Fog continuing' is the gist of the first-ever broadcast weather forecast – not to be confused with the first newspaper forecast (*1 August*) or the first live television forecast (*16 May*) – when radio transmission in Britain officially begins today in 1922. The first of the two reports is read out when the station opens, the second between 9 and 10 p.m. As all broadcasting is live, and there is no effective way of recording it, the exact wording of the text supplied by the Meteorological Department of the Air Ministry to the BBC does not survive. *The Times*, however (which then got its weather information from the same source), predicts 'settled conditions generally . . . fog and night frost are likely'. Visibility will be bad nearly everywhere, 'flying generally impracticable'.

And they're right! Fortunately, with a big anticyclone parked directly over Britain, predicting more of the same is a banker – though it turns out it's even foggier and colder than forecast. Outlying districts of London, reports *The Times* next day, 'which had hitherto escaped with white autumnal mists, were enshrouded throughout the day by a dense yellow canopy, making the use of artificial light necessary . . . in the late afternoon it was impossible to see more than five or ten yards . . . bitterly cold.' So, all in all, this important landmark is a great success. From September of the following year, the forecast will be 'stabilized' at 7 a.m. and 9.30 p.m, with the News.

WHY GIRLS SHOULDN'T ROLL THEIR SLEEVES UP

'It was one of the last autumn days when the leaves were falling in little gusts. They fell on the children who were thankful for this excuse to wriggle and for the allowable movements in brushing the leaves from their hair and laps.
"Season of mists and mellow fruitfulness. I was engaged to a young man at the beginning of the War but he fell on Flanders' Field," said Miss Brodie. "Are you thinking, Sandy, of doing a day's washing?"
"No, Miss Brodie."
"Because you have got your sleeves rolled up. I won't have to do with girls who roll the sleeves of their blouses, however fine the weather. Roll them down at once, we are civilized beings . . ."'
Muriel Spark, The Prime of Miss Jean Brodie, *1961, set in 1930s Edinburgh*

-23.3C Braemar 1919 ❄

17.7C Cwmbargoed 1989 ☼

15 NOVEMBER

PROTESTANT V POPISH WIND

'Only the finger of God could have brought William through the storms and high seas safely to anchor here,' the chaplain, Gilbert Burnet, declares from the flagship of William of Orange in Brixham harbour (♥) in Devon today in 1688. His words mirror the anxious thoughts of the nation. The last successful invasion of England has begun – and the weather helps it happen.

The ambitious Protestant William, Prince of Orange and Chief Magistrate of the Dutch Netherlands, has been 'invited' to replace King James II, who's trying to establish a Catholic government. From the moment William starts assembling a large fleet on the Dutch coast, the wind intervenes. During October, he's pegged back by persistent westerly ('Popish') winds. His first attempt to invade (*29 October*) is driven back by a 'crewell storme'. Twelve days later, the wind finally veers easterly ('Protestant') and the fleet of several hundred ships, carrying 28,000 troops, fills the Straits of Dover (where it's viewed by thousands) and sweeps through the Channel unchallenged. The same wind, crucially, keeps the English Navy (still loyal to James II) pinned near the mouth of the Thames estuary. 'Heard the news of the Prince . . . passing through the Channel with so favourable a wind that our navy could not intercept or molest them,' notes the diarist John Evelyn. 'This put the King and Court into great consternation.'

THE DEVELOPMENT OF LIFEJACKETS
When, in a gale today in 1928, the lifeboat from Rye in Sussex (⛑) capsizes, drowning seventeen crew, the subsequent inquiry questions the buoyancy of the crew's kapok life-jackets and recommends that all jackets should now be made with waterproof covers. Kapok – a vegetable fibre with honeycombed air cells – replaces the cork and canvas lifebelt (invented by Captain Ward, an RNLI inspector, in 1854) in 1904. The collar is added in 1917 to keep an unconscious man afloat. The Second World War sees the advent of the inflatable jacket. And with the development of synthetic foam in the 1960s, the shape and styles of lifejackets broaden into the extensive range we know today.

-22.8C Braemar 1919 ❄

18.5C Aber 1997 ☼

When William's fleet sails past Torbay (⛵) this morning, because of a navigational error, there's a crisis. But the wind, now faithfully Protestant, veers round and the expedition sails comfortably into Brixham. William leads a triumphal procession on London. There is almost no bloodshed. James II abdicates. In February 1689, William and his wife Mary (James's daughter) are crowned. The ballad 'Lilliburlero', which includes the line 'Ho! by my shoul, 'tis a Protestant wind', becomes the nation's most popular song. Henry Purcell's jaunty march, borrowed from an Irish jig, is still played as the signature tune of the BBC World Service.

The 'wind dragon' – the weathervane atop the steeple of St Mary-le-Bow church on Cheapside, the central marketplace in the City – is watched anxiously by thousands of Londoners, in the weeks preceding William's invasion, for a change in direction from westerly ('Popish') to easterly ('Protestant'). While it is westerly, people habitually 'curse the dragon'.

3.1C min / 8.6C max

VET IN A BLIZZARD

'This was my second winter in Darrowby so I didn't feel the same sense of shock when it started to be really rough in November. When they were getting a drizzle of rain down there on the plain the high country was covered in a few hours by a white blanket which filled in the roads, smoothed out familiar landmarks, transformed our world into something strange and new. This was what they meant on the radio when they talked about "snow on high ground".'

'A lot of piano playing and jolly games. Retired to bed exhausted in the coldest room I have ever encountered . . . woke frozen. Shaving sheer agony and glacial bathroom with a skylight that would not shut. The loo like an icebox.'
Noël Coward, writer and composer, at a house-party with Vivien Leigh, Laurence Olivier and Princess Margaret, 1952

The November snow that James Herriot describes in his bestselling story of life as a rural vet in the Yorkshire Dales, *It Shouldn't Happen to a Vet* (1972), is recalled with some artistic licence. Herriot's second November in Darrowby (Thirsk ♀), in 1943, cannot, in fact, have been anywhere near this snowy. There's little November snow anywhere between 1930 and 1950. Almost certainly Herriot, who started writing in 1966, has today in 1962 in mind. Not that this should detract from his delightful anecdote following a call from Mr Clayton, a farmer high in the remote Dales.

'"What's the road like?" I asked.

"Road? Road?" Mr Clayton's reaction was typically airy. Farmers in the less accessible places always brushed aside such queries. "Road's right enough."'

Needless to say, Herriot has to abandon his car and stumble through waist-deep snow. Falling into holes and crawling out, he loses all sense of time and is tempted 'to sit down and rest, even sleep; there was something hypnotic in the way the big, soft flakes brushed noiselessly across my skin'. Eventually he stumbles his way to a building he thinks must be the farm, 'mouth gaping, chest heaving agonisingly. My immense relief must have bordered on hysteria.' A little later, Mr Clayton rubs the pane of the single small window. 'Picking his teeth with his thumb-nail, he peered out at the howling blizzard. "Aye," he said, and belched pleasurably. "It's a plain sort o' day."'

18.8C Aultbea 1997

-18.3C West Linton 1919

162.1C Seathwaite 1897

17 NOVEMBER

2.9C min / 8.4C max

BAROMETERS — A storm accompanies what is probably the lowest barometric pressure ever measured in London (about 931 mbar) today in 1665, and the Anglo-Irish gentleman scientist Robert Boyle realizes the fundamental connection between air pressure and weather. The reading is recorded on one of the earliest mercury barometers – the Italian physicist Evangelista Torricelli only discovers the principle of an instrument that measures atmospheric pressure in 1643. In addition to his crucial discovery, Boyle is also credited with naming the instrument, 'barometer' coming from the Greek words for 'weight' and 'measure'. Robert Hooke (*2 September*), Boyle's assistant and the most prolific inventor of scientific instruments in the seventeenth century, modifies Torricelli's concept to produce a usable domestic instrument with a dial, the wheel barometer, around the same time.

The simple mercury barometer, which rises and falls with variations in atmospheric pressure against a set scale, transforms the science of meteorology from the eighteenth century. Today, readings are often taken from electronic sensors, but the principle – that rising pressure usually indicates fair weather, and falling pressure heralds unsettled weather – remains.

WHY BAROMETERS ARE NO GOOD

The barometer is useless; it is as misleading as the weather forecast. There was one hanging up in a hotel at Oxford at which I was staying last spring, and, when I got there, it was pointing to "set fair". It was simply pouring with rain outside, and had been all day; and I couldn't quite make matters out. I tapped the barometer, and it jumped up and pointed to "very dry" ... I tapped it again the next morning, and it went up still higher, and the rain came down faster than ever. On Wednesday I went and hit it again, and the pointer went round towards "set fair", "very dry", and "much heat", until it was stopped by the peg, and couldn't go any further. It tried its best, but the instrument was built so that it couldn't prophesy the weather any harder than it did without breaking itself. It evidently wanted to go on, and prognosticate drought, and water famine, and sunstroke, and simooms, and such things, but the peg prevented it, and it had to be content with pointing to the mere commonplace "very dry". Meanwhile, the rain came down in a steady torrent, and the lower part of the town was under water, owing to the river having overflowed. *Jerome K. Jerome*, Three Men in a Boat, *1889*

-15.6C Crathes 1909

20.7C Aber 1997

18 NOVEMBER

2.9C min / 8.5C max

THE MOST DANGEROUS COAST IN BRITAIN

Gales spring up in the Channel today, in 1795, as Rear-Admiral Christian sets out from Spithead (♀) with a fleet of two hundred heavily laden troop, ordnance and merchant transports. (He's bound for the West Indies, where there's a spot of bother with the French.) Disaster strikes within forty-eight hours. In a south-westerly wind, they are fighting every sailor's worst nightmare: the dreaded 'lee shore'. This is a shore towards which a sailing vessel is being irresistibly blown, making shipwreck inevitable. Thus a gale from the south or east converts the whole of the south and Cornish coasts into a lee shore. Small wonder, then, that with only Falmouth, Fowey or Plymouth Sound providing shelter, this is the most treacherous and wreck-strewn coast in the British Isles.

'After heavy rain the hill sides are slippery, and I saw a neighbour's cow tobogganing as if she had been shot out of a gun – she flew down hill sitting on her tail. If she had not kept all her legs in front of her, she would have broken her neck, but she finished on a flat piece of grass, sitting down like a cat, just before she reached the river.'

Beatrix Potter, Lake District, 1927

Once a sailing ship became caught in a bay, people would crowd the cliff-tops to watch the hapless vessel beating back and forth, trying to claw her way seaward to safety, often for days. Sometimes, to avert disaster, a ship's captain might 'strike' the mast to reduce wind resistance, and anchor. But before heavy iron anchor chains, ropes (up to 10 inches/25 cm in diameter) could snap under the strain – and the ship would drive ashore. Accepting the inevitable, a captain might even pick a safe stretch and beach his ship, in the hope of saving life and reducing damage (*4 January*).

In this instance, Christian orders his ships to make for the safety of Torbay (♈). It's a lousy decision. Several of the transports are already unable to stand out to sea sufficiently to clear Portland Bill and Chesil Bank (⊕). At daybreak, six smash to pieces, drowning three hundred. 'The Chesil Bank was strewn . . . with the dead bodies of men and animals, with pieces of wreck, and piles of plundered goods, which groups of people were at work to carry away, regardless of the . . . drowned bodies,' says one witness.

17.5C Aultbea 1997
-14.9C Braemar 1905

19 NOVEMBER

3.1C min / 8.6C max

THE SIEGE OF
BASING HOUSE

Icy rain dampens both the gunpowder and the ardour of Sir William Waller's Parliamentarian soldiers today in 1643, and he abandons his siege of the Royalist stronghold, Basing House, at Basingstoke (✂). The years 1643 and 1644 are extremely wet, and rain frequently affects military operations during the English Civil War. One of the biggest problems is dragging cannons weighing up to 3 tons along roads reduced to muddy quagmires.

Waller's troops retreat to their damp, cheerless billets, of which one lieutenant notes, 'Our lodging and our service did not agree, the one being so hot, the other so cold.' In a few days the assault is renewed, but Waller's single piece of artillery makes no impression on the fortifications and his men 'hazarded themselves on the very muzzles of the enemy's muskets', to no avail. When the rain turns to sleet and snow, the sodden army retreats again. The epic resistance of Basing House, the mansion of the Marquis of Winchester and an isolated Royalist fortress in the Parliamentarian-dominated south, continues until October 1645, when Cromwell himself turns up, in dry weather, with a Cannon Royal, the heaviest siege gun then known.

'This afternoon the weather turning suddenly very warm produced an unusual appearance; for the dew on the windows was on the outside of the glass, the air being warmer abroad than within.'
Gilbert White, Selborne, Hampshire, 1776

'Fearfully cold. Landscape trees upon my window-panes.'
John Everett Millais, Pre-Raphaelite painter, Surrey, 1851

'Tis very warm weather when one's in bed.' *Jonathan Swift, writer and satirist, London, 1710*

Today in 1093, after days of torrential rain, Malcolm III, King of Scotland, and his eldest son are killed at the Battle of Alnwick in Northumberland (✂). The Aln valley is flooded, preventing retreat by the Scots who are slaughtered by the Norman army. As the chronicler Simeon of Durham records: 'Those who escaped the sword were carried away by the inundations of the rivers, more than usually swollen by the winter rains.'

-17.2C Dalwhinnie 1947 ❄

Persistent fog over much of England in 1921 causes major disruption to transport: the Calais mail steamer collides with another vessel in the Channel and a 'private motor-car' crashes into a shop in Acton in west London. And in 1877 Valentine Salvage petitions his employer, South Eastern Railways, against its 'no moustache' rule. 'The wearing of moustachios,' Salvage argues, 'is a protection against the inclemency of the weather.'

17.7C Tivington 1994 ☀

20 NOVEMBER

FIDO   It's foggy early this morning over the hundreds of low-lying airbases across central and eastern England in 1943 – as it's been on so many mornings recently. During an intense autumn, Allied planes, now equipped with radar, are making hundreds of successful night-time bombing sorties. But now almost as many aircraft are being lost in murky landing conditions as to German flak.

So, this morning, at RAF Graveley, near Huntingdon (♠), an experiment is being tried. The idea of burning oil to disperse fog – raising the temperature so that the air can hold more moisture – has been considered before and rejected, largely because of the cost. But tests in late 1942 convince Churchill to authorize the Fog Investigation and Dispersal Operation (FIDO). And the experiment works, albeit at the cost of burning some 22,700 gallons (320,000 litres) of petrol per hour. Four pilots of 35 Squadron land their Halifax bombers safely.

Fifteen airfields across England become FIDO-equipped and some 2500 Allied aircraft are guided down this way before the end of the war. Pilots, however, are not enthusiastic. The petrol, forced in jets through rows of perforated horizontal pipes parallel with the landing strip, shoots sheets of flames 15 feet (5 metres) into the air, providing roaring updraughts of rising air just where the planes are trying to descend. As one pilot says, 'It was a bit like landing in Dante's *Inferno.*'

> 'A thorough storm, so that it was hard to walk or stand, the wind being ready to take us off our feet. It drove one of the boats, which were on the strand, from its moorings out to sea. Three men were in it, who looked for nothing every moment but to be swallowed up.'
> *John Wesley, founder of Methodism, Lowestoft, 1776*

> 'Truly wintry and miserable. They are all tearing about the windy streets with their over coats huddled around them; the English are not very happy ... We are far from the proud days of 'We – Victoria, Queen of Great Britain, Empress of India ...' etc etc. and it's beginning to dawn on everyone that we are just a rather small, foggy little island that must either make things and sell them, or die.'
> *Kenneth Williams, actor, 1967*

−17.1C Braemar 1880 ❄

17.2C Dunoon 1947 ☀

Today in 1556, in 'tempest torn seas' the *Edward Bonaventure* sinks near Fraserburgh off the north-east coast of Scotland (✛). The English explorer Richard Chancellor saves the life of Russia's first ambassador to the Tudor court, but not his own. The chest of gifts for Queen Mary Tudor is never recovered, though rumours of locals suddenly taking early retirement abound.

21 NOVEMBER

3C min / 8.3C max

THREE FOGGY STORIES

'Unreal city, under the brown fog of a winter dawn' is how T. S. Eliot describes London in *The Waste Land* (1922), describing the seventy-two-hour fog beneath which England is immersed today, in 1921, seizing all transport. In 1979, the England v. Bulgaria European Championship qualifier football game at Wembley is postponed due to fog. The game goes ahead next day, but without Kevin Keegan, who has to return to his club in Hamburg. In his place, a new player makes his debut: Glenn Hoddle – and a star is born.

A HILL-FARMER LOSES HIS VIRGINITY

'The valley was lost in the fog. Spider's webs, wavering white with dew, were stretched over the dead grass; and all she could see, down the line of the hedge, were the grey receding shapes of oak trees.

" . . . we might get lost in the fog, like."

"Well, you're not afraid of getting lost?"

"No, mam!"

"So there! Besides, it'll be sunny on the tops. Just you wait!"

He handed her the reins of the grey. She cocked her leg and swung into the saddle. She led and he followed. The hawthorns made a tunnel over their heads; the branches ripped against her hat, and showered her with crystal drops . . . They stopped at the gate that leads on to the mountain. She opened the latch with her riding-crop . . .

The path was muddy and the gorse brushed their boots. She leaned forwards, rubbing herself against the pommel. The damp mountain air filled her lungs. They saw a buzzard. On ahead it was already looking lighter.

"Look! What did I tell you? The sun!"

Then they cantered on into the sunlight with the clouds spread out below, on and on, for miles it seemed . . . In a hollow, out of the wind, there were three Scots pines.

"I love Scots pines," she said. "And when I'm very old I'd like to look like one. Know what I mean?"'

He was breathing beside her, hot under his mackintosh. She clawed at the bark, a flake of which came away in her hand. An earwig scuttled for safety. Judging that the moment had come, she transferred her lacquered fingers from the tree-trunk to his face . . .

It was dark when she pushed through the door of the cottage and Nigel was drowsing by the fire. She banged her riding-crop on the table. There were moss-stains on her breeches: "You lost the bet, duckie. You owe a bottle of Gordon's."

"You had him?"

"Under an ancient pine! Very romantic! Rather damp!"'
Bruce Chatwin, On the Black Hill (♀), *1982, referring to events in 1938*

-17.2 Lauder 1880 ❄

18.3C Llandudno 1947 ☀

22 NOVEMBER

THE CAIRNGORM TRAGEDY

Stumbling, frost-bitten and incoherent, after nineteen hours in a Highland blizzard and waist-deep snow: that's the state the teacher from Ainslie Park School in Edinburgh is in when the RAF Mountain Rescue team finally find her this Monday in 1971. She is just able to say where she's left the others – another teacher, and six teenage schoolchildren, on an outdoor adventure 'survival' weekend. Sadly, this is not a story with a happy ending. All the children except one are dead, as is the other teacher. Only by blind mountain flying through low cloud does a helicopter pilot manage to pluck the one surviving child to safety.

The story that unfolds is described by one climbing website as 'one of the most cautionary ever known in the Scottish hills'. The two women teachers, trainee adventure instructors aged eighteen and twenty-one, set off with the teenagers on a survival training weekend on 4084-feet (1244-metre) Cairn Gorm (♥). They head for the remote Curran mountain hut, one of the basic shelters scattered across the Highlands. Although the weather starts clear, the instructors are aware that heavy snow is forecast – in fact, the subsequent inquiry is told, that's the point. Everyone has waterproofs, spare clothes, 'survival bags' and emergency food rations. Unfortunately, when the snow begins the wind increases, and it becomes a classic 'white-out'. All sense of direction, and reference points, disappear. After three hours knee-deep in soft snow, realizing they will never make the hut, the group prepares to bivouac. But the snow obliterates every feature that might provide shelter. Also, wearing jeans or thin cords, they are getting cold (*14 March*). Following a grim night, on Sunday morning one instructor goes for help. After eight hours stumbling in the snow, the group's absence has yet to be even noticed.

It's the worst weather-related incident in British mountaineering history (*1 January*), and the tragedy raises questions about numerous aspects of accepted mountain craft. Notably, should mountain huts exist? Do these supposed havens kill more people than they save, as tired walkers press on into bad weather towards a mirage of safety? Calls for certification for mountain instructors follow, plus a requirement for mandatory radio contact where children are involved. As part of a new onus on carrying your own survival equipment, the Curran bothy, and two others, are demolished.

-18.0C Kingussie 1904

18.3C Hawarden 1947

23 NOVEMBER

3.1C min / 8.7C max

THE OUTRAGE The combined gale and storm surge that strikes the Dorset coast today in 1824 – a freakish confluence of spring tide and dying Atlantic hurricane (*25 August*) – is the weather event of the nineteenth century: one of the two greatest storms ever to strike the region. In the 'Great Gale', or 'The Outrage', as it becomes known, high tide arrives five hours early, and then the water just keeps on rising. It rises up the East Fleet valley 'as fast as a horse can gallop', carrying with it a hay-stack and debris from the fields. The mighty Chesil shingle bank (**P**) is breached, destroying eighty houses behind it and drowning between fifty and sixty. Hurst Spit, the other great south coast shingle bank, on the West Solent (**P**), is described by Sir Charles Lyell (Darwin's geology tutor) as 'moved bodily forward for 40 yards'. The Plymouth breakwater is ruined, as is the Cobb at Lyme Regis. Weymouth wakes to find most of its handsome Esplanade demolished and the seafront basements full of sand and water. As for shipping, wrecks litter the coast of Hampshire, Dorset, Devon and Cornwall. 'In the midst of this sublime and terrible storm,' records the Reverend Sydney Smith, 'Dame Partington, who lived upon the beach, was seen at the door of her house with mop and pattens, trundling her mop, squeezing out the sea-water, and vigorously pushing away the Atlantic Ocean.'

RECORD 'TORNADO SWARM'
Today, in 1981, the largest-ever number of tornadoes is recorded in a single day in Britain - 105 in just over five hours as a cold front moves from Anglesey to Norfolk. It's a vintage period: 152 tornadoes are recorded in 12 days. British tornadoes may seem apologetic alongside the American Mid-West's twisters, but we still have, relative to our land area, more reported tornadoes than any other country (chiefly, of course, because there are more of us to report them) - an average of around 33 per year. They can occur at any time, though are commonest in autumn, especially November, in eastern England. They are rare in Northern Ireland and Scotland. No one seems to know how or why they form, other than that they tend to be associated with humid, unstable air trying to rise, with colder air above – thunderstorm cells on cold fronts. They seem to be on the increase – *21 May, 28 July, 2 August, 18 August, 22 September, 26 September, 23 October, 27 October, 7 December.*

20.0C Lairg 1906

-12.2C Balmoral 1910

24 NOVEMBER

THE HARSHEST
ENVIRONMENT
IN THE WORLD

A gale-driven 'night of fear' (*The Times*) in the North Sea in 1981 ends this morning when forty-eight oil workers are finally air-lifted to safety off the Transworld 58 platform. In 100 mph (160 kph) winds and 60-foot (18-metre) waves, eight of the twelve massive anchor chains on the platform in the Argyll field (⊕) snap, casting it adrift at the mercy of a wild sea. Eight tugs eventually regain control, after it has drifted 10 miles (16 km). Near by, in another near-catastrophe, a 28,000-ton Norwegian 'service platform' vessel, which has also dragged its anchors, almost collides with the neighbouring drilling platform.

Apart from the 1988 Piper Alpha explosion, all the worst disasters in North Sea drilling are weather-related. The difficulties and expense of operating in this, the harshest environment for major engineering works outside space, become most extreme after the mid-1970s. Soaring prices, prompted by the oil crisis, make it economically viable to exploit fields down to 1000 feet (300 metres) below sea level in the West Shetland Basin – the most storm-battered region of the Atlantic (𝇋). 'They told me you don't have hurricanes here,' a hard-bitten American oil man is supposed to have said, shortly after his arrival in Scotland's oil capital, Aberdeen. 'Well you darn well do, but you jus' call 'em depressions.'

FOG, MIST AND SNOW – CLEARED BY CAR BATTERY

'It was one of those iron days when the frost piled thickly on the windscreen blotting out everything within minutes. But this morning I was triumphant. I had just bought a wonderful new invention – a couple of strands of wire mounted on a strip of Bakelite and fastened to the windscreen with rubber suckers. It worked from the car batteries and cleared a small space of vision.' So writes James Herriot, a vet in the Yorkshire Dales at the end of the 1930s. The first electrically heated windscreen, the 'Mistproof Plate', appears in 1922 – a 16-inch-wide (40 cm) glass panel with 'practically invisible' resistance wires across its surface. Unlike most automotive inventions, which are American, it, along with the fog lamp, is proudly British. Say no more. 'No more did I have to climb out wearily and scrub and scratch at the frozen glass every half mile or so. I sat peering delight-edly through a flawlessly clear semicircle about eight inches wide at the countryside unwinding before me like a film show ... I was enjoying it so much that I hardly noticed the ache in my toes.'

-14.8C Grantown-on-Spey 1993 ❄

⊕

17.2C Prestatyn 1970 ☼

25 NOVEMBER

3.1C min / 8.5C max

MUD AND FOOTBALL Under a quilt of grey cloud, England loses its first-ever home football game to an overseas team (Hungary) at Wembley today in 1953. But the reason has more to do with weather than just grey clouds. The defeat emphasizes the yawning skill gap which has developed between doughty Brits used to heavy, mud-covered pitches, and fancy foreigners used to drier, harder, faster surfaces.

THE STORM THAT LEADS TO THE PLIMSOLL LINE

The storm that the forty-year-old Samuel Plimsoll endures tonight, in 1864, aboard a passenger steamer inbound to Redcar in Yorkshire (♥), is responsible for saving countless numbers of merchant seamen's lives. In Victorian England, hundreds of merchant sailors drown every year because grossly overladen ships sink while their indifferent shipowners profit from the insurance. But so alarmed is Samuel Plimsoll tonight that, later, as a Liberal MP, he tirelessly campaigns against un-seaworthy ships and overloading, giving tonight's voyage as the reason 'why he troubled himself so much on this subject'. The furore he stirs up – the 'Plimsoll Sensation' – culminates, in 1876, in the Plimsoll Line: a compulsory line that is still marked on every cargo ship's hull today, showing the level of maximum submergence.

From the moment we invent modern football (as opposed to the unregulated hooliganism of the medieval sport – *15 February*), our game develops as a winter pursuit of manly Victorian values. It is supposed to be played on a wind-driven mire, in ankle-high, block-toed boots with a heavy leather ball, which doubles in weight when soaked. In these circumstances, key elements of the game emerge – the sliding tackle, the long ball over the top, the Desperate Dan-style centre forward. A little wizardry is expected from the wingers, but they play on the only part of the pitch likely to have grass all season. In midwinter, when even Football League pitches look like Flanders fields, the most skilful players are often dropped. Strength is the key to success. Finesse is irrelevant.

In fact, mud is even occasionally employed as a secret weapon. A swamp helps England beat Italy in 1934 at the 'Battle of Highbury'. It's decades before professional clubs use scientifically modified grasses and artificial pitches (*17 January*) to nullify the part weather plays, and before the culture shifts emphatically from attrition in the muddied morass to the delicacy of 'the beautiful game'. But the revolution begins today, with our 6–3 humiliation at the fleet feet of the 'Magical Magyars'.

-15.0C Dalwhinnie 1952

185.3mm Honister Pass 1979

17.0C Chivenor 1970

26 NOVEMBER

MANCHESTER, FOG AND COTTONOPOLIS

For the eighth consecutive day, Manchester is blanketed in fog in 1936. It's the most prolonged fog since records begin and the city is besieged: with visibility down to a few metres, people have to navigate by following tramlines. Buildings are coated in a thick, black soot-slime, and today, acid rain (pH 3) begins to fall. As Mae West says in her autobiography: 'I looked out of the train window and all I could see was rain and fog. "I know I'm going to love Manchester," I told Jim, "if I can only see it."'

'You cannot figure the stillness of these moors in a November drizzle: nevertheless I walk often under cloud of night ... conversing with the void of heaven, in the most pleasant fashion.'
Thomas Carlyle, historian and writer, Nithsdale, Dumfries and Galloway, 1828

The smog – an acrid mix of smoke and fog – is, of course, a by-product of industrial pollution. In sixteen years' time, Manchester will become the first municipality to address this problem when a 'black smoke tax' creates a 'smokeless zone' in 100 acres (42 hectares) of the city centre. The same climate that smothers the city today, however, plays a vital role in creating it. In the Middle Ages when Lancashire first attracts textile workers, it is the powerful Pennine streams, swollen with rainwater, that turn the waterwheels of the emerging cotton industry. When Manchester (nicknamed 'Cottonopolis') emerges at the beginning of the Industrial Revolution as the 'Workshop of the World' and the centre of worldwide cotton manufacturing, it's the damp air – perfect for maintaining the moisture in cotton yarn while it is spun, thus ensuring it doesn't snap – that gives it the edge over the older English cloth centres in eastern England.

It's the 'sad, damp sky', too, that Manchester's greatest artist, L. S. Lowry, so loves about the city. 'You'll never see the sun in one of my pictures,' he once says – and you don't. His 'matchstalk men and matchstalk cats and dogs' (as Brian and Michael's 1978 No. 1 hit calls them) are invariably set against white, damp, Mancunian overcast skies.

-13.7C Braemar 1904

It is probably a combination of fog and dampness, Lowry's skies and Old Trafford cricket ground's record of rain-affected matches (*26 July*) that underpins Manchester's dismal weather reputation. Yet, it's all a myth. Glasgow, Bradford, Plymouth, Cardiff and Penzance are wetter.

18.1C Shipston on Stour 1979

SECRETS OF
THE GOODWIN
SANDS

No light shines from where the South Goodwin lightship should be, off the Kent coast (✪), as a gale blows through the early hours this morning in 1954. While the mist and rain might hide the other two lightships – the North Goodwin and East Goodwin – that warn sailors off the treacherous sandbanks, the South vessel should always be visible. The gales have been raging for days, but no distress signals have been seen or heard. There's no murmur from her radio-telephone. Then, at 1.27 a.m., the East Goodwin lightship reports the South Goodwin vessel drifting north-west of her. The wind has wrenched her from her two giant sea anchors.

The gale is so strong that when the Ramsgate lifeboat station fires its warning maroons, they go unheard; the crew have to be alerted by motorbike. After a long search in the dark, at dawn the sad truth is revealed: the red-painted lightship responsible for saving so many lies on her beam end on the sands. Seven crew are dead; one survivor, after eight hours is clinging grimly to the rigging. Britain's deadliest shipping hazard has struck again.

The Goodwin Sands are a notorious mariners' graveyard, lying adjacent to one of the busiest shipping routes in the world. They are created when a giant storm inundates Earl Godwin's island off Kent in the late eleventh century. 'A very dangerous flat and fatal, where the carcasses of many a tall ship lie buried,' says Shakespeare in *The Merchant of Venice*. Beguilingly pretty in fine weather, only a tenth of the 12 miles (19 km) of ever-shifting golden sands are ever revealed. An annual cricket match used to be played on them every summer. But in fog, or gales, combined with the strong currents around the North Foreland, it's a different story. More than two thousand ships have been driven or pulled onto the sands. The Great Storm of 1703 (*7 December*) is the worst single night: thirteen men o'war and forty merchantmen are wrecked – 268 lives lost. Once a ship is aground, the tide sweeps the sand from under bow or stern or both, and most vessels break their backs. Famous wrecks abound, from the *Rooswijk* of 1739, laden with Spanish silver bullion, to Radio Caroline's *Ross Revenge* in 1991 (*12 November*). The masts of two are still visible, even at high tide. 'The dead of mankind . . . they tell no tales,' writes Herman Melville in *Moby Dick*, 'though containing more secrets than the Goodwin Sands.'

-14.4C Appleby 1904 ❄

16.3C Hartland Point 1970 ☼   ✪

28 NOVEMBER

2.7C min / 8.1C max

THE SALTIRE A white cross of cloud hanging in a perfect, azure sky: that's the divine inspiration, according to legend, that the Pictish King Angus mac Fergus receives when he's surrounded by a great force of Saxons in East Lothian (✕) late in the autumn of AD 832. Angus duly gains the victory, and the saltire (a diagonal cross – upon which St Andrew, the patron saint of Scotland, was crucified – in white on a dark blue background) is adopted as the national flag of Scotland. Of course, the opposite – two slim streaks of blue in a cloud-covered sky – might have been a truer representation of Scottish weather, but you can't hold a legend back.

'A glorious day, still autumnal, and not wintry ... On our way to the park the view from Richmond Hill had a delicate blue mist over it, that seemed to hang like a veil before the sober brownish-yellow of the distant elms. As we came home, the sun was setting on a fog-bank, and we saw him into that purple ocean – the orange and gold passing, into green above the fog-bank, the gold and orange reflected in the river in more sombre tints.'
George Eliot, novelist, 1857

'The weather is really most curious, snow in the morning, rain in the afternoon and thunderstorm in the evening. Apparently some American professor puts it down to the approaching end of the world by collision with some fragment of the sun.'
Evelyn Waugh, author, aged 16, 1919

'Only one other occasion ... stands out as an exciting emotional experience from all my flying before the 1914 War, and that was my first attempt to probe into the clouds. Very few people attempted to do this before the days of instruments, for it was inevitably accompanied by some degree of risk. But the attainment of sheer height was something I always strived for, and one overcast day ... I climbed at maximum speed and allowed myself to be swallowed by the low cloud, attempting to maintain only by instinct and a sense of balance the same angle and attitude as before. Some tense minutes passed, and then as I broke clear of the damp darkness that had enveloped me during my ascent, I suddenly found myself in a blinding world of billowing whiteness that stretched in every direction, magnificent and vast and thrilling. Besides the wonderful peace and beauty of the scene, there was also a complete absence of the motion and turbulence that always accompanied flying, particularly in small light aircraft, beneath or in cloud. This was a flying experience I can never forget.' *Geoffrey de Havilland, Sky Fever, 1979*

-12.3C Lagganlia 1977 ❄

17.0C Ruthin 1913 ☀

29 NOVEMBER

2.7C min / 8C max

'MOTORWAY MADNESS'
Most of the early morning fog has burnt off, revealing brilliant sunshine, on the M1 this morning in 1971. Most, but not quite all – it's a lethal situation. Drivers speeding north and south suddenly enter a solid white bank between Luton and Dunstable in Bedfordshire as fog drifts on to the motorway. Blinded, they slam on their brakes. The vehicles behind plough into them . . . the next do the same . . . and the next. A lorry jack-knifes. And suddenly, in seconds, there's carnage. One driver who's jumped clear runs down the road shouting 'No, no, no . . .' It makes no difference. When the fog clears, ten minutes later, the scene is like a scrapyard. Some cars, according to *The Times*, are 'unrecognisable as cars'. The scene is so gruesome, the screams so piercing, that one uninjured man faints. Fifty vehicles are piled up on the south carriageway, twenty on the north. The final toll is eight dead, forty-five injured – the worst fog pile-up until 1991.

Unbelievably, just ten weeks earlier a series of multiple pile-ups on the M6 in Cheshire kill 11 and injure 60. Following this crash, the tabloids coin a phrase which catches, and is still repeated now in newspapers: 'MOTORWAY MADNESS'. The phrase has a specific meaning, referring to the altered state of mind, the false sense of security, that makes sane people drive insanely fast in fog. 'Is it sheer recklessness . . . aggressive fixity to speed . . . semi-somnolence? Or does the condition arise from . . . the distortions fog introduces of speed and distance?' wonders a *Times* leader.

Five weeks later, another fog pile-up kills a lorry driver and injures fourteen on the M1 near Nottingham. The truth is that drivers just aren't very motorway-savvy yet. In 1971, less than half the present 2200 miles (3500 km) of motorway exist. Most of that (apart from the M1 and M6, where all the accidents happen) is in unconnected sections. There's no 'network', as there is today, when 70 per cent of long journeys are by motorway. Motorways are still a novelty. The pile-ups prompt calls for rear red fog lamps (mandatory from 1978), though, ironically, bad as fog sounds, the worst British motorway pile-up ever, killing 13 on the M6 in 1985, is also blamed on the weather – but not the fog or the wet. The conditions? Dazzling sunshine (*21 October*).

-17.8C Braemar 1912 ❄

158.0mm Honister Pass 💧

18.3C Ashburton 1913 ☀

30 NOVEMBER

FIRST METEOROLOGICAL BALLOON EXPEDITION

A light north-westerly breeze blows on an otherwise murky afternoon in Grosvenor Square, London, today in 1784 as an eccentric ballooning party departs. It's just seventeen months since the Montgolfier brothers demonstrated the possibilities of hot air balloons on the Continent, and the world's mad for them. This is the third such flight in Britain, instigated by an American physician, Dr John Jeffries. A Frenchman, Jean-Pierre Blanchard, 'in consideration of 100 guineas' is taking the good doctor, plus his dog – early balloon trips invariably involve a dog – up for the purpose of 'examining the atmosphere'. Accordingly, they carry an assembly of thermometers, barometers and other meteorological kit.

'A most comfortable gale day. The potato failure has been much exaggerated, the disease is by no means so far spread as was supposed and the crop so over abundant that the partial failure will be the less felt, particularly as the corn harvest was excellent.'
Elizabeth Smith, County Wicklow, 1845. The 'partial failure' turns into total failure – and catastrophe (8 September)

The departure isn't entirely dignified. Released by their ground crew, they alight briefly on the roof of the adjoining stables. Heaving out sand ballast, they become airborne again, only for the wind to blow them into some chimneys where they dislodge a number of earthenware funnels. At 2.38 p.m., Blanchard loftily records, 'we rose above the reach of any further terrestrial obstructions'.

The trip offers dramatic new experiences. As they rise above the fog and cloud, Jeffries records St Paul's coming into view, its 'dome suspended', and describes ships on the Thames 'like canoes in a foggy creek'. By ten past three, having 'lost sight of sun and earth' in cloud, they are over 6000 feet (1830 metres) up, the temperature falls below freezing, the dog begins 'to shake and shed tears' and Jeffries pulls on a fur cap 'which in great measure relieved me'. Ten minutes later, at around 9000 feet (2740 metres), they are in 'beautiful azure sky', by which time the distended state of the balloon alarms them so much that they start releasing the gas. They freak themselves out by dropping 4000 feet (1220 metres) in five minutes. 'We now refreshed ourselves with cold chicken,' reports Jeffries. 'And drank a few glasses of wine.' They eventually land near Dartford in Kent (🌂). Quite how much scientific discovery is achieved by the trip is open to question, but a paper about the flight is duly read to the Royal Society.

-20.9C Kinbrace 1985 ❄

15.8C Winfrith 1979 ☀

DECEMBER

1 DECEMBER

3C min / 8C max

THE CRYSTAL PALACE BURNS

Today is the official start of winter. If Advent calendars and Christmas cards rooted in Dickensian images of snow and iron-hard freeze-ups seem like figments of distant memory, that's because they are. December hasn't been like that since Dickens was growing up (*24 December*). Over the last century, December is the coldest month only half as often as January or February. There have been only two properly snowy Decembers – 1950 and 1981 – and the average, in central England, is two days of lying snow. December is more accurately – if less picturesquely – characterized by rain, wind and short days. Having the shortest days does not make for the coldest month, because the sea surrounding us still contains residual warmth. Having said this, it's been getting steadily warmer since 1970, making December now essentially an extension of autumn and scarcely distinguishable from November.

> 'An utterly dull day. A steady downpour of rain all day made Clubs impossible so we had ridiculous PT games in the Great School. Slackness is ever so much more tiring than energy and I feel quite washed out.'
> *Evelyn Waugh, author, aged 16, 1919*

Otherwise, today's weather events are London-centric. A 'high wind' in 1937 is instrumental in incinerating the Crystal Palace, Joseph Paxton's vast iron and glass masterpiece re-erected after the Great Exhibition of 1851 in Sydenham, South London: 'In virtue of its construction, a natural flue for any flame,' says *The Times*. The wind, as with the Great Fire of London (*12, 14 September*), gets up at just the wrong moment, and from the wrong direction: a north-easterly 'strong enough to carry large pieces of glass to the shore of the lake 150 yards away'. The fire rips through the building: within half an hour of the outbreak yesterday evening, Crystal Palace is 'a vast bonfire', visible for miles. Thousands gather to watch on Parliament Hill in Hampstead, and the glow in the clouds can be seen from near Brighton. It's a severe loss. First, London is deprived of its most capacious venue, which seats forty thousand. Then it turns out that this proud emblem of Victorian innovation has been chronically under-insured, so there's no question of rebuilding. 'The end of an age,' says Winston Churchill, viewing the glow on his way home from the Commons. The site remains a vacant lot.

-17.5C Fyvie Castle 1973 ❄

16.0C Harrogate 1985 ☀

WHITE SHIP
DISASTER

'Agitated by dreadful tempests,' the *White Ship* strikes a rock in heavy seas and sinks in the Channel today in 1120. Prince William Adelin, the only legitimate son and heir of Henry I, is aboard. Along with all hands, he drowns, leading to a succession crisis and civil war in England.

Because the Anglo-Norman nobility own estates in both France and England, cross-Channel sailings are almost as routine in the twelfth century as today. And though this is undoubtedly late in the year, the *White Ship* is both 'excellently fitted out, ready for royal service',

'We are living here with open windows, and complaining of the heat.'
Sydney Smith, clergyman and wit, Taunton, Devon, 1833

according to the contemporary chronicler Orderic Vitalis, and captained by a well-known sailor whose father bore William the Conqueror across the water in 1066 (*18 September, 3 October*). At the port of Barfleur, 'while a south wind blew, the earls embarked in the first watch of night, hoisted their sails to catch the wind, and put to sea'. There are several barrels of wine aboard the *White Ship*, and the disaster is probably the combination of bad weather and alcohol: you might say the crew are, both literally and metaphorically, three sheets to the wind. As a contemporary poem notes: 'the raging waves / Sweep the high decks – the royal seed goes down / Lost in the barren sea the world's renown.'

Henry I fails to produce another male heir and is forced to name his daughter, Matilda, as successor. In the confusion following the King's death, his nephew Stephen of Blois (who should have been aboard the *White Ship*, but is ill) seizes the English throne in a coup in 1135. England is plunged into nineteen years of civil war, known as 'the Anarchy' (*13 December*), which neither side has the resources to win outright.

18.3C Achnashellach 1948

-21.1C Kelso 1879

Nine centuries later, in 1948, 18C (65F) is registered in the unlikely location of Achnashellach, a remote glen in Wester Ross in the Scottish Highlands. And in 1975, seven tornadoes whirl across East Anglia, sucking turnips out of the ground.

3 DECEMBER

2.8C min / 8C max

INVENTION OF THE CAT'S EYE

On a cold and foggy night in 1933, Percy Shaw, a blacksmith in Halifax (♥), is driving back to his home when he hits a perilous stretch of road with a sheer drop to the side. Before street lights and road markings, drivers along this stretch depend on the reflection of their headlights in tram tracks to find their way in fog. But tonight the tracks have been removed. As Shaw inches uneasily along, he is startled when his lights are brilliantly reflected by the eyes of a cat sitting beside the road. Shaw has a flash of inspiration – and perhaps the greatest contribution to road safety of the twentieth century is conceived.

At least, that's the romantic version of the invention of the cat's eye. The Eureka moment is probably more pedestrian. But Shaw is astute enough not to allow cold truth to de-mist a good story. He patents the cat's eye light-reflecting road stud in 1934 and develops its durability over several years, encasing the glass beads in a rubber mould which sinks in a cast-iron mount, cleaning the beads in rainwater. Sales are slow until the Second World War, when blackouts make night driving even more hazardous. Then, in 1947, a junior Transport Minister, Jim Callaghan, introduces the cat's eye nationwide. Shaw's company, Reflecting Roadstuds Ltd, manufactures and exports a million cat's eyes a year at its peak. Shaw becomes one of Britain's most eccentric millionaires, eschewing all luxuries (including carpets and curtains) except for a pair of Rolls-Royces and a cellar of Worthington's India Pale Ale. The cat's eye still regularly features in lists of the greatest-ever British designs – and coloured, reflective road markers continue to guide motorists safely home in fog.

> 'You cannot conceive how dark and hideous London is today, mouldering in a dank fog.'
> *D. H. Lawrence, writer, 1915*

'It's going to be a dirty night,' says the Isle of Man Steam Packet Company's agent to Captain Teare of the SS *Ellan Vannin*, shortly before his ship departs Ramsey harbour today in 1909. Sometimes instinct is more reliable than the barometer and 'storm warning cones' (*25 October*), which suggest all is fine. Just three hours after the ship sails for Liverpool in a north-westerly breeze, a Force 10 or 11 storm is tearing up the Irish Sea and 20-foot (6-metre) seas are battering the *Ellan Vannin*. The ship goes down before first light, with fourteen passengers and twenty-one crew (♻).

-26.7C Kelso 1879 ❄

16.7C Hawarden Bridge 1948 ☀

3 DECEMBER

Still regarded as a defining design classic, Percy Shaw sold over a million of his 'Catseyes' a year at the company's peak

2.8C min / 7.8C max

HORROR IN
THE FOG

'HORROR IN THE FOG,' read the headlines following today's Lewisham rail crash in south London in 1957, one of the three worst rail crashes in British history (*28 December*). The crash happens when the driver of the 4.56 p.m. steam train from Cannon Street to Ramsgate goes through a crucial red light, smashing into the back of the 5.18 p.m. electric train from Charing Cross to Hayes. Both trains are crowded with around two thousand Christmas shoppers and rush-hour commuters: both are running late because of the fog. The derailed steam train swings sideways on impact, striking a steel column of the fly-over viaduct, which another train is crossing. The viaduct falls on to two coaches of the steam train, crushing them to half their size.

Ironically, the Clean Air Acts that will eventually banish the 'killer smog' (*6 December*) – the disaster's chief cause – are passed the year before, but the legislation doesn't come into force until the early sixties. Meantime the 'grimy fog' is so dense that no one sees the crash. Even the sound is muffled. It's so dense that the stationmaster, whose house overlooks the scene, sees nothing from his upstairs windows – he just hears the cries and screams. 'The compartment toppled onto its side,' a woman on the electric train tells *The Times*, 'and we were one struggling mass of tangled arms and legs.' Some survivors grope their way, unaided, a mile along the track to the next station. The final death toll is 90, with 173 injured.

A spokesman from the Southern Region of British Railways robustly defends the 'colour light' signalling system: 'A little over 1000 trains pass through that area daily, and have done so with perfect safety since 1929.' The sixty-two-year-old driver, it turns out, has missed the two preceding signals because the fog is so thick, and is unable to stop in time on seeing the red light. Tried for manslaughter, but discharged, he dies a year later. The official inquiry concludes that the accident could have been prevented if an Automatic Warning System (AWS) had been installed – the disaster finally leads to its universal adoption.

-22.2C Lauder 1879

164.3mm Seathwaite 1864

In 1855 and 1879, intense frosts kill trees but provide fine skating, though 'large numbers of work people are suffering immensely through enforced idleness,' reports *The Times*.

17.2C Thetford 1953

THE UMBRELLA STORY

Today is a story of rain. For today, in 1758, is the first *documented* use, as a *rain shield*, in *Britain*, by a *man*, of an umbrella. You will surmise from these qualifications that the umbrella concept is not entirely new – four thousand years old, in fact. The umbrella begins life in the Middle East, as a sunshade (*umbra* is Latin for shade), arriving in Europe, waxed against rain, only around 1685 – and then exclusively for women. For the eighteenth-century man-about-town, carrying an umbrella is like wearing a dress. Not from today, however, when the physician, polemicist and political writer Dr John Shebbeare is standing pillory at Charing Cross – his punishment for seditious, anti-Hanoverian pamphleteering. As his popular cause happens to be shared by the sub-Sheriff of London, however, Shebbeare arrives at the pillory not in irons but by coach. There, in front of the large crowd, far from being confined and ridiculed he struts and poses with – Horace Walpole supplies the crucial detail – 'a footman holding an umbrella to keep off the rain'.

Even if Shebbeare is the first man to shelter publicly under one, however, he can't take credit for popularizing the umbrella. That honour falls to Jonas Hanway, eighteenth-century traveller, philanthropist and indefatigable campaigner against the evils of tea. Importing his umbrella-carrying ideas from Portugal, Hanway endures thirty years of ridicule from 1750 onwards. He is especially attacked by coachmen who see his portable shelter as threatening their trade. Unphased, however, this courageous pioneer persists. By the time of his death, in 1790, the carrying of 'Hanways' has caught on.

The story then becomes one of mere technical improvements. The early umbrellas are monsters: 13 lb (16 kg) of oiled silk with stretchers of cane or whalebone. The intractable engineering conundrum of the age – how to reduce the umbrella's weight while maintaining its strength – falls to the genius of Samuel Fox. Fox, inspired by the U-shaped girders of Telford's Menai Bridge (♀), makes ribs in the form of a trough with flat sides, by which greatest strength is obtained. Fox it is, in 1852, who develops the strong, lightweight 'paragon' frame, halving the umbrella's weight. Thus he mobilizes Britain during inclement weather – leading, as we all now know, to the heyday of our empire, victories in two world wars and the splitting of the atom.

-18.9C Alston 1879 ❄

17.2C Hampton 1979 ☀

THE GREAT SMOG

At first, the freezing fog which settles on the capital this Saturday morning in 1952 seems just like any other London fog. For days the weather's been cold, and Londoners have been loading coal on to their fires to keep warm (almost no one has central heating yet). But as the extra smoke pours into the air now, the yellow-grey fog just gets thicker – from a dozen yards, visibility dwindles to 12 inches (30 cm). And thicker – people report not being able to see their own feet. And thicker – fog enters buildings, so that concerts and film screenings have to be cancelled because the audience cannot see the stage or screen. Roads become blocked with abandoned cars. It's the start of the worst environmental disaster in British history: the four days known as the Great Smog.

Smog is not new. The word ('smoke' + 'fog') is coined by Dr H. A. des Vœux in a paper delivered to the Public Health Congress in 1905, to refer to the dirty, smoke-generated fogs found in great cities. London's smogs are notorious (*5 October, 16 December, 23 December, 23 February*). But, even by London's standards, today's is exceptional.

The first hints of disaster are a shortage of coffins and soaring flower sales. 'No one realized at the time that the number of deaths was increasing,' Dr Robert Waller of St Bartholomew's Hospital later recalls to the BBC. 'There weren't bodies lying around in the street.' Yet each day, it's been calculated, 1000 tons of soot and 370 tons of highly toxic sulphur dioxide are being pumped into London's sky – just as an anticyclone has settled a pool of cold, stagnant air above the Thames valley. Sealed by chalk hills either side, this air acts like a lid clamped over the capital. Three days later, when westerly winds finally clear the fog as quickly as it arrives, four thousand are dead – mainly the young, old or infirm, suffocated by bronchitis and pneumonia. The final figure rises to between eight and twelve thousand. The figures are so appalling that, as with the Big Stink in 1858 (*23 June*), the problem can no longer be ignored (although, initially, the government tries to attribute the death toll to a flu epidemic). In 1956 the Clean Air Acts are passed, designating the capital a 'Smokeless Zone'. It's a landmark in the modern environmental movement, though two more killer smogs will strike before the Acts actually take effect – 1957 (*4 December*) and 1962 (also today, killing 750).

-18.3C Buxton 1879 ❄

16.4C Guernsey 1979 ☀

7 DECEMBER

2C min / 7.5C max

THE GREAT STORM OF 1703

'The mercury sank lower than ever I observ'd . . . which made me suppose the tube had been handled . . . by the children,' Daniel Defoe notes today in 1703. It's not his children, though. It's a deep low-pressure system drawing an extra-tropical cyclone (*25 August*) of unusual ferocity across the Atlantic. It's the worst recorded storm in the history of the British Isles.

For two weeks leading up to this day, storms, squalls and violent rain showers have torn down chimneys, rattled windows and agitated trees. But nothing could prepare people for what follows. 'No pen could describe it, nor tongue express it, nor thought conceive it, unless by one in the extremity of it,' writes Defoe afterwards. Approaching midnight, the storm slams into the British Isles, carving a 300-mile (480-km) swathe through southern England and Wales. With winds reaching 120 mph (190 kph), it rages for six hours.

The damage is astonishing: 17,000 trees are uprooted in Kent alone; 123 people are killed on land (including the Bishop of Bath and Wells, in his bed); 400 windmills are destroyed, many igniting from the friction of the rotating blades; the lead roof of Westminster Abbey is 'rolled up like parchment'; hundreds of churches lose their spires; most of Bristol is flooded. The sea is 'swept clean of all shipping': the lower Thames, the Goodwin Sands, the Solent and Spithead become one crushed mass of wreckage. Some 8000 sailors drown and the Royal Navy loses 12 ships. The Eddystone Lighthouse (*3 January*) disappears without trace off Plymouth Sound.

Queen Anne orders a public fast since the storm is 'a token of the Divine displeasure'. In a public proclamation, she states: 'a Calamity so Dreadful and Astonishing, that the like hath not been Seen or Felt, in the memory of any Person Living in this Our Kingdom.' For once, it's not hyperbole. Daniel Defoe, author of *Robinson Crusoe*, is amazed by this single night of destruction: 'Not a house, not a Family that had anything to lose, but have lost something by this Storm, the Sea, the Land, the Houses, the Churches, the Corn, the Trees, the Rivers, all have felt the fury of the Winds.' Using 'exact' and 'true' first hand accounts of the event, he writes his first book, *The Storm*, published in July 1704. A bestseller on publication and still in print today, it fixes the event in the hierarchy of British weather disasters.

-21.0C Hodsock Priory 1879 ❄

London tornado 2006

16.1C Paignton 1921

8 DECEMBER

1.9C min / 7.3C max

EXCURSIONS TO LOOK AT SNOW

'The effects of the hurricane and the tempest of wind, rain and lightning thro' all the nation, especially London, were very dismal. Many houses demolish'd and people kill'd. As to my own losses . . . not to be parallel'd with any thing happening in our age. I am not able to describe it, but submit to the pleasure of Almighty God . . . Methinks I shall hear, and am sure feel the dismal Groans of our *Forests*, so many thousand of goodly *Oaks* subverted by that late dreadful *Hurricane*; prostrating the Trees, and crushing all that grew under them, lying in ghastly Postures, like whole *Regiments* fallen in *Battle*.' So records the diarist John Evelyn, author of the first book of forest management (*28 February*), after the 'Great Storm' yesterday in 1703 (*7 December*), when at least two thousand oak trees are blown down at his Surrey estate, Wotton (♀).

'Everything was white; and the hedges looked as though they had grown old in the night. Everything glistened and all was still; the whole country was like Sleeping Beauty's park. Now it has all gone, and there is only slush.'
Vita Sackville-West, writer, Sevenoaks, Kent, 1925

'Before I had been out long, I actually saw the sun looking red and rayless, much like the millionth magnification of a new half-penny . . . I went home by way of Holborn, and the fog was denser than ever, – very black, indeed, more like a distillation of mud than anything else; the ghost of mud, – the spiritualized medium of departed mud, through which the dead citizens of London probably tread, in the Hades whither they are translated. So heavy was the gloom, that gas was lighted in all the shop-windows; and the little charcoal-furnaces of the women and boys, roasting chestnuts, threw a ruddy, misty glow around them. And yet I liked it. This fog seems an atmosphere proper to huge, grimy London; as proper to London, as that light neither of the sun nor moon is to the New Jerusalem.'
Nathaniel Hawthorne, novelist and American Consul in Liverpool, 1855

Today in 1990, snowdrifts of 24 inches (60 cm) on the motorways around Birmingham cause chaos when, despite the severe weather warnings, the public go out and about as usual. The problem is greatly aggravated by the number of cars abandoned on entirely unnecessary journeys. To the consternation of the emergency services, people call in to report getting stuck – on excursions 'to look at the snow'.

-17.8C Cambridge 1879 ❄

9 DECEMBER

1.9C min / 7C max

SELF-RIGHTING LIFEBOAT

In a storm in 1849, twenty of the twenty-four crew of the South Shields lifeboat, *Providence*, are drowned when she capsizes (✠). Horrified by the tragedy, local landowner and First Lord of the Admiralty the Duke of Northumberland offers a prize of 100 guineas to design a new lifeboat – one that will right itself. James Beeching, of Great Yarmouth, wins the competition. His design is improved by James Peake, Master Shipwright of the Royal Naval Dockyard at Woolwich. High end-boxes fore and aft and a 7cwt (356 kg) iron keel make the vessel self-righting. Eight one-way, self-relieving drainage tubes make her self-baling. Beeching and Peake's boat is the forerunner of the Standard Self-Righting Lifeboat – the most important development in sea rescue since the first lifeboat (*15 March*).

'A very hard frost; which is news to us having none almost these three years.' *Samuel Pepys, diarist, 1662*

'Why do owls hoot during nights of frost? Does it mean he is hungry, sexy or merely happy? Perhaps, miserable? I must ask an ornithologist.' *James Lees-Milne, writer, 1973*

Thirty-seven years later, on this black day for lifeboatmen, a Force 7 gale in 1886 causes the worst disaster in the history of the RNLI (*6 October*). Among innumerable tragedies, the gale drives a German merchant vessel, *Mexico*, aground off Southport in Lancashire (✠). Forty-four lifeboatmen launch, in three open rowing lifeboats – two from Lytham St Anne's, one from Southport. The second Lytham lifeboat, with great difficulty, rescues all twelve sailors. The cost, however, is dear. In the treacherously shallow Ribble estuary, waves are biblical. When the Southport lifeboat capsizes, despite being self-righting, fourteen of her crew drown. The first St Anne's lifeboat is washed ashore next day – her entire crew of thirteen dead.

-16.6C Braemar 1972 ❄

The tragedy shocks the nation. A disaster fund is established to support the sixteen widows and fifty fatherless children. Assisted by Queen Victoria and the German Emperor, it raises £50,000. In 1891, Mancunian industrialist Sir Charles Macara organizes the world's first charity street collection, 'Lifeboat Saturday', to augment the fund. It's the template for the street collections by which the RNLI still raises funds. An insurance scheme is established for dependants of lost lifeboatmen. And the loss of two Standard Self-Righting boats prompts a new lifeboat design, by the naval architect George Lennox Watson, with a drop-keel and water ballast tanks for added stability.

16.7C Torquay 1921 ☀

10 DECEMBER

1.9C min / 7C max

Cold air sinks. This remarkable discovery is made this freezing day in 1784. Living in a temperate climate, we generally presume that air gets colder as it gets higher – hence we need warm clothes for hilltops, and mountain peaks are snow-capped. The discovery that this isn't necessarily the case is made accidentally by the observant rural clergyman Gilbert White (*27 January, 10 April*). In his notes about this evening's exceptionally intense frost in his Hampshire parish (**♀**), published in *The Natural History of Selborne* (1789), White describes his thermometer reading '*one degree below zero!*' – or -18C:

> 'A dark sky, an obscure horizon, inclined to moisture. Wore my new great coat for the first time.'
> *John Fitzgerald,*
> *Cork schoolteacher, 1793*

> 'Wind howling savagely.'
> *George Gissing, writer, 1891*

'This strange severity of the weather made me very desirous to know what degree of cold there might be in such an exalted and near station as *Newton*,' White writes, before requesting his neighbour, some 200 feet (61 metres) higher up, to hang out a thermometer. But, to White's astonishment, the reading is 10C lower at his own house. 'Disturbed at this unexpected reverse,' and assuming his neighbour's thermometer is at fault, he promptly despatches one of his own, only to find 'when the instruments came to be confronted, they went exactly together: so that, for one night at least, the cold at *Newton* was . . . less than at *Selborne* . . . and indeed, when we came to observe consequences, we could readily credit this; for all my laurustines, bays, ilexes, arbutuses, cypresses . . . and (which occasions more regret) my fine sloping laurel-hedge, were scorched up; while, at *Newton*, the same trees have not lost a leaf!'

White reports no further temperature experiments, and it's not until 1814 that Dr William Wells, FRS, identifies the process at work here: cold air, because it is heavier, flows downhill, collecting in hollows in the landscape. It is 150 years, however, and following a disastrously cold May in 1935, before fruit growers at East Malling Research Station in Kent begin to come to terms with the now well-understood phenomenon of 'frost hollows' or 'frost pockets'. Britain has many of these gardening disaster spots, for example at Rickmansworth in the south, Santon Downham in the east and Houghall in the north.

-14.4C Carnwath 1967 ❄

17.2C Salisbury 1915 ☼ ♀

10 DECEMBER

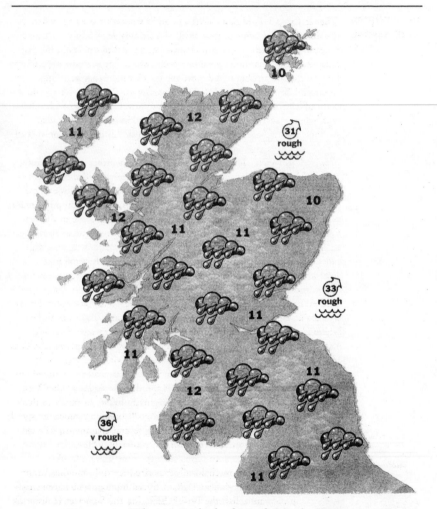

Welcome to Scotland – from today's Sunday Times
Scottish edition, 2006

11 DECEMBER

THE INVENTION OF SKATING

'Thence I to my Lord Sandwich's . . . and then over the Parke (where I first in my life, it being a great frost, did see people sliding with their Sckeates [sic] which is a very pretty art,' Samuel Pepys notes in his diary today in 1662. That other great chronicler of the late seventeenth century, John Evelyn, is present at the same event: 'Having seene the strange and wonderful dexterity of the sliders on the new canal in St James's Park, perform'd before their Maties [Majesties] by divers gentlemen and others with scheets [skates], after the manner of the Hollanders, with what swiftness they passe, how suddainely they stop in full carriere upon the ice.'

Charles II is widely credited with introducing the 'new' sport of skating to the British aristocracy. Certainly the Merry Monarch brought the innovative iron-bladed skate from Holland, where he spent time in exile. However, a more rudimentary form of racing on ice has been popular for centuries. William FitzStephen writes how, in the 1170s, when the marshes near London freeze, 'vast crowds of youngsters go out to play on the ice . . . some bind the shinbones of animals to their feet, and with poles furnished at their lower ends with sharp spikes, they skim away, swift as a darting bird, or the bolt from a crossbow.'

Medieval skates made from cattle and horse metatarsals have been found in marshes across the British Isles, particularly in the East Anglian Fens, where in the coldest winters skating was as much a means of getting around as a sport. Vermuyden, the Dutch engineer who reclaims the Fens in the early seventeenth century, brings workers with him and the locals adapt their skates into what become known as 'fen runners'. Throughout the eighteenth and nineteenth centuries the shallow waters of the Wash, and its lattice of rivers and dykes, regularly freeze, so much so that the village of Welney in Cambridgeshire (♀) briefly becomes the speed skating capital of the world, producing several champions. The last major Fenland skating event is in the winter of 1996–7.

-15.0C Carnwath 1981 ❄

17.7C Penkridge 1994 ☀

Although the first mechanically generated ice rink, the Glaciarium, is created in London in 1876, artificial outdoor winter rinks only proliferate across the British Isles after the Somerset House rink in London opens in December 2000. These refrigerated carnivals evoke memories of winters past: a shared romantic ideal of iron-hard frosts, cracked cheeks and visible breath, as the rasp of iron on ice merges with yells of delight.

12 DECEMBER

1.9C min / 7.1C max

THE ONE-MAN FOOTBALL TEAM

It's grim up north today in 1891. In a snowstorm at Turf Moor stadium (♀), Burnley are playing Blackburn Rovers in a First Division fixture. It says much about Victorian virility that the game kicks off at all. In the second half, however, the weather worsens and resolve weakens. Blackburn players revolt, walking off the pitch, numb and simply too cold to play. Only the Rovers goalkeeper, Herby Arthur, remains. The match descends into farce, to derisive cheers from the crowd, as Arthur successfully appeals for offside when one Burnley player passes to another. What can the referee do? With no one to pass to, Arthur engages in a lengthy period of time-wasting until the bemused ref is forced to abandon the game thirty minutes before time. Burnley win 3–0.

Down the road at Deepdale, Preston North End are playing Notts County. There's so much snow on the ground that 'the ball can scarcely be passed', and in the second half, as the light deteriorates and freezing rain starts to fall, six Notts County players also walk off. In similar conditions, Accrington are playing Aston Villa. Again, in the second half, after a 'council of war', according to the *Manchester Guardian*, the home side 'make tracks for the pavilion' in 'the best move of the day'.

Major snowstorms rarely strike before Christmas, but today in 1901 the whole of southern Britain and Ireland is blasted by snow and gales. In a blizzard, telegraph wires collapse and railway lines are blocked as Birmingham is temporarily, but completely, cut off. In 1946, freezing fog brings much of the country to a halt. Not everyone is stranded, however. The *Daily Telegraph* reports an enterprising gang of thieves using the thick veil that cloaks the south east to get away with £8000 worth of jewellery in Surrey.

'So unseasonably warm yesterday – 15 degrees – there were small flies about, and green shoots in the garden.'
Christine Evans, poet, Lleyn peninsula, 1997

'Up to catch the dawn over the Suffolk coast. A golden sun is rising slowly into a clear sky as we approach the low-lying, neat green shoreline with Harwich's old church and surrounding houses on their low headland and Felixstowe opposite. I didn't expect such a caricature of England ... We are home on a morning of glassy calm.'
Michael Palin, on day 79 of Around the World in 80 Days, 1989

17.4C Cape Wrath 1984

-22.6C Shawbury 1981

188.0mm Oakeley Quarry 1964

13 DECEMBER

2.1C min / 7.3C max

THE SNOW QUEEN DOWN THE PUB

In a blizzard this Sunday afternoon in 1981, there's a knock on the door of the Cross Hands pub in Old Sodbury, Gloucestershire (☞). 'I wish to inform you,' says a well-spoken stranger to landlord Roberto Cadei, 'that Her Majesty the Queen is outside. Can you accommodate her?' There's no sign of Jeremy Beadle. The guy seems to be for real.

Earlier in the day, Mr Cadei, assuming no one will want a room in such weather, has sent the staff home and settled down with his family to watch a TV rerun of the summer's royal wedding of Prince Charles and Lady Diana Spencer. Within an hour, more than a hundred stranded motorists are packing the bar, their cars abandoned in waist-deep, drifting snow. And now, the stranger explains, the royal entourage, on its way back to Windsor after visiting Princess Anne a few miles away, has met the same fate: two royal Range Rovers, containing two chauffeurs, two detectives, a lady-in-waiting, a staff aide and Her Majesty – are stuck.

> 'I do not remember such a series of North-Pole days ... the sky looks like ice; the earth is frozen; the wind is as keen as a two-edged blade.'
> *Charlotte Brontë, writer, Haworth, Yorkshire, 1846*

> 'Snow is a bloody nuisance in London. It ruins the uppers of your shoes – gets inside the stitching and you start letting water, or treading water or whatever, and you end up with the chill blains and frostbite and the ends dropping off.' *Kenneth Williams, actor, 1967*

And so the Queen comes to tea, on what turns out to be the coldest day of the year. The final vacant twin – Room 15 (complete with rosebud bedspreads and matching wallpaper, Victorian-style tassel lampshades, TV and hot beverage selection) – is hastily prepared, and the Cross Hands becomes the only hotel in England with a room that's been occupied by the reigning sovereign.

17.1C Cape Wrath 1984 ☀

-25.2C Shawbury 1981 ❄

Another queen is having an adventure in the snow 839 years earlier. In 1142, Matilda makes a daring escape from Oxford Castle where she's trapped under siege during a long civil war (*2 December*). Taking advantage of the snow, she wraps herself in a white cloak as camouflage and, with a handful of knights, slips out of the castle and through the middle of King Stephen's camp. The weather's so bad that no one troubles them. Having 'deceived the eyes of the besiegers, dazzled by the reflection of the snow', she crosses the frozen Thames to safety. Her escape ensures that the civil war continues, and it's ultimately won by her son, Henry II.

14 DECEMBER

2.1C min / 7.3C max

'Heavy snow, which raised the roads to the top of the hedges' is how
Laurie Lee recalls mid-December 1917 in *Cider With Rosie* (1959), after
he first arrives in the Cotswolds (🍐). Memory famously plays tricks with
the weather. Summers are sunnier, winters are snowier – extremes,
however fleeting, stick in the mind. The most evocative weather recollec-
tions often show scant regard for the facts (*16 November*). In fact,
December 1917 is cold, dry, sunny – and snowless. Given that Lee was
only 3 at the time, and publishes his memoir more than forty years later, it
seems likely that he telescopes together details of the subsequent run of
notably snowy winters into a single, vivid evocation. His account of a hard
winter remains a classic:

'One never remembered the journey towards it; one arrived, and winter
was here. The day came suddenly when all details were different and the
village had to be rediscovered. One's nose went dead so that it hurt to
breathe, and there were jigsaws of frost on the window. The light filled
the house with a green polar glow; while outside – in the invisible world
– there was a strange hard silence, or a metallic creaking, a faint throb-
bing of twigs and wires.

The kitchen that morning would be full of steam, billowing from
kettles and pots. The outside pump was frozen again, making a sound
like broken crockery, so that the girls tore icicles from the eaves for water
and we drank boiled ice in our tea.

"It's wicked," said Mother. "The poor, poor birds."

There was an iron-shod clatter down the garden path and the milkman
pushed open the door. The milk in his pail was frozen solid. He had to
break off lumps with a hammer.

"It's murder out," the milkman said. "Crows worryin' the sheep.
Swans froze in the lake. An' tits droppin' dead in mid-air . . ." He
drank his tea while his eyebrows melted, slapped Dorothy's
bottom, and left.

"The poor, poor birds," Mother said again.

Now the winter's day was set in motion and we rode through
its crystal kingdom . . . We saw trees lopped-off by their
burdens of ice, cow-tracks like pot-holes in rock . . . The
church clock had stopped and the weather-cock was frozen,
so that both time and the winds were stilled; and nothing,
we thought, could be more exciting than this; interference
by a hand unknown, the winter's No to routine and laws –
sinister, awesome, welcome.'

-20.7C Braemar 1882 ❄

16.8C Aber 1972 ☀

15 DECEMBER

2.1C min / 7.1C max

DEATH OF GLENN MILLER

'Even the birds are grounded. We'll never get off today,' Glenn Miller reputedly says, peering out of the car window into low cloud, heavy rain and patchy fog today in 1944. The big band leader is driving to RAF Twinwood Farm in Bedfordshire, to fly to recently liberated Paris where he's organizing the Allied troops' Christmas concert. 'The fog and rain could be cut with a knife,' says his manager, Don Haynes. Most airports across Britain are closed today; nevertheless the single-engine Noorduyn Norseman aircraft takes off, disappearing quickly into low cloud.

'The tiercel peregrine flew steeply up above the river, arching and shrugging his wings into the gale, dark on the grey clouds racing over. Wild peregrines love the wind, as otters love water. It is their element. Only within it do they truly live. All wild peregrines I have seen have flown longer and further and higher in a gale than at any other time. They avoid it only when bathing or sleeping. The tiercel glided at two hundred feet, spread his wings and tail upon the billowing air, and turned down wind in a long and sweeping curve. Quickly his circles stretched away to the east, blown out elliptically by the force of the wind.'
J. A. Baker, The Peregrine, 1967

Miller is never seen again. Nine days later, a terse BBC communiqué announces that he is missing. As befits the greatest pop star of his era, the conspiracy theories and rumours begin immediately – and sixty years later, they're still circulating. Some say he's captured and tortured by the Nazis, others say he dies in the arms of a Parisian prostitute. Another explanation is that Miller's the victim of 'friendly jettisoning': a fleet of 139 Lancaster bombers, returning from an aborted mission over Germany, unload their unused bombs in the Channel at about the time Miller is flying to France. Miller would have been flying on the Supreme Headquarters Allied Expeditionary Forces air route – and the unofficial 'jettison zone' for returning bombers is near the SHAEF path. It's not inconceivable that a bomb falls out of the heavy cloud on to the Norseman. The truth behind Miller's death, however, is probably more prosaic: an inexperienced pilot gets out of his depth crossing the Channel in bad weather.

-22.4C Braemar 1882 ❄

Also today, in 1981 (*13 December*), as the country is being flooded by a sharp thaw, mini-icebergs are floating in the Wash at King's Lynn in Norfolk (𝄢) – it's the last incidence of sea ice on the east coast of England.

16.3C Prestatyn 1982 ☼

103mph Gwennap Head 1979
highest ever low level gust in England

16 DECEMBER

2C min / 7.2C max

LONDON'S
MURKY PAST

Today, in 1890, is the middle of the foggiest month of the foggiest year in all London's foggy history. How can we be sure? Because, this December, out of a possible 242 hours, 40 minutes of sunshine, Westminster scores 0 – the only place ever to do so. It is also – partly due to this – the capital's coldest December on record. This fog, of course, is not the clean, white country fog that we know. It is yellow-brown, smells of rotten eggs, leaves a taste in the mouth and coats clothes and curtains with a greasy film. It is so opaque that, along the Thames, watermen have to bang drums so that barges can find their wharves.

There's nothing new in all this. In 1661 John Evelyn publishes *Fumifugium*, a pamphlet denoucing the 'Hellish and dismall Cloud of SEA-COALE' which, he says, makes the capital resemble 'the face of Mount Etna, the Court of Vulcan, or the suburbs of hell'. Around 1750, when the furnaces of the Industrial Revolution really begin to roar, density (and frequency) increase sharply. Gilbert White (*27 January, 10 December*) records smelling London's smoke in Hampshire. He refers to the city, for the first time, as 'the big smoke'.

Population growth is key. Already the largest city in the world, London sees its population double between 1800 and 1840 – when the 'pea-souper' is born – and then double again between 1840 and 1890. In 1853 Dickens refers, in *Bleak House* (*23 December*), to the 'London Particular'. Over the next century – until the Great Smog of 1952 (*6 December*) – the fog is at its thickest. Notable fogs strike in 1873, 1880, 1882, 1886, 1888, 1891 and 1892. They last anything from a few days up to, in 1879, four months. The densest fogs kill between five hundred and seven hundred a week from respiratory ailments.

-22.0C Braemar 1882 ❄

17.0C Aber 1972 ☀

Not everyone hates the fog, however (*23 February, 5 October*). Some visitors deliberately wait until November, to view 'the darkness'. The Savoy Hotel advertises rooms with 'a splendid view of the Thames in fog'. Other industrial cities, such as Manchester (*26 November*) and Leeds, have their fogs – Edinburgh is even called 'Auld Reekie' or 'old smoky'. But for London, until the Clean Air Acts take effect in the 1960s, one feature characterizes Britain's capital beyond all others: fog.

17 DECEMBER

BOMBER COMMAND'S BLACK THURSDAY

Imagine, for a moment, that it's 1943, the weather's grim and you are Flight Lieutenant Charles Owen of 97 Squadron, Bomber Command, stationed in Lincolnshire. Despite the adverse forecast, your squadron is ordered out as part of a massive five hundred-bomber raid on Berlin, a 1168-mile (1870-km) round trip. You have been at the controls of your Lancaster, a plane requiring not just concentration to fly, but brute strength, for nearly nine hours.

Somehow you have escaped the anti-aircraft guns and the enemy fighters that have claimed fifty other aircraft. Your wireless transmitter and radio navigational aids have packed up, but you have found your way back to your base following a radio beam. Cold and exhausted, you begin the descent. You finally break through the cloudbase 250 feet (76 metres) above the ground, to find the worst sight in the world: the airfield is blanketed in fog. At that moment your radio system packs up, so you cannot communicate with the tower and have to await visual signals for permission to land. After circling for ten minutes at 200 feet (61 metres), you cross your fingers and decide to try to land anyway, without permission. Somehow you get down in one piece.

> 'We have now a fine frost, and walk safe from the dirt; but it is like a life at court, very slippery.' *Jonathan Swift, writer and satirist, County Cavan, 1735*

> 'The constantly heavy-clouded, and often wet, weather tends to increase the depression. I am inwardly irritable and unvisited by good thoughts.' *George Eliot, novelist, 1862*

Three other crews, you discover, have landed elsewhere. Three crews have baled out after running out of fuel looking for somewhere to land. Four have crashed trying to land. One is still missing.

This disastrous morning becomes known as Black Thursday. In fact, it's not an untypical experience. The winters of the Second World War are notoriously severe, especially those between 1939 and 1942. But the greatest curse for air crews returning to the dozens of American and RAF airbases that litter East Anglia and Lincolnshire is not ice or snow. Low-lying, surrounded by sea, they are notoriously prone to fog. Blind landing equipment is but a dream for the future, and although trial schemes are tested for fog dispersal (*20 November*), crews frequently have to resort to the best available option – baling out.

-18.0C Braemar 1981
199.1mm Dalness 1966
17.8C Aber 1972

18 DECEMBER

2.1C min / 6.8C max

LORNA DOONE The fearsome frost, 'with its iron hand, and step of stone', which R. D. Blackmore describes in his historical Exmoor romance, *Lorna Doone* (1869), is in fact a real frost, beginning around now in 1683 (☞). In reimagining it, Blackmore brings a novelist's eye for detail: 'All the birds were set in one direction, steadily journeying westward, not with any heat of speed, neither flying far at once; but all (as if on business bound), partly running, partly flying, partly fluttering along; silently . . . This movement of the birds went on, even for a week or more . . . all we had in the snowy ditches were hares so tame that we could pat them . . . their great black eyes . . . seemed to look at any man, for mercy and for comfort.'

The winter that Blackmore describes is probably Britain's coldest ever (*19 January*), with the longest frost on record. Dover and Calais are joined for the first time since the last Ice Age. The Thames freezes for two months. In Somerset, the ground freezes to a depth of more than 3 feet (over a metre). (The grave-diggers burying Sir Ensor Doone 'broke three good pickaxes, ere they got through the hard brown sod'.) The winter forms part of a pronounced cool period in the Northern Hemisphere known as the Little Ice Age (*3 March*), the severest stage of which is between 1650 and 1850 – a period that Blackmore, who lives from 1825 to 1900, just catches.

Then, after the frost, one morning comes the snow: 'An odd white light was on the rafters, such as I never had seen before,' says narrator John Ridd. Puzzled, he goes to the first-floor window where normally he can 'see the yard, and the woodrick, and even the church beyond. But now, half the lattice was quite blocked up, as if plastered with grey lime . . . I spread the lattice open; and saw at once that . . . all the earth was flat with snow, all the air was thick with snow . . . for all the world was snowing.' The kitchen is 'darker than the cider-cellar', the windows have either fallen right in through the weight of snow against them; or bulged in, 'bent like an old bruised lanthorn'. That night . . . many men were killed, and cattle rigid in their head-ropes. Then I heard that fearful sound, which never I had heard before, neither since have heard . . . the sharp yet solemn sound of trees, burst open by the frost-blow.'

18.0C Aber 1972
-18.6C Lincoln 1981

**FOG AND THE
SERIAL KILLER**

Today in 1811, a day of fog so thick that 'two feet ahead of you, a man was entirely withdrawn from your power of identification', the Williamson family is found murdered in their home in London's East End. Not only is the attack horrific in its brutality, it's not the first. A few doors down, twelve days earlier, the Marr family are also murdered. These are the notorious Ratcliffe Highway Murders. They begin a grim association between fog, London's East End and serial killers – a romanticized mythology cemented, in 1888, by Jack the Ripper (*5 October*). In his essay 'On Murder Considered as One of the Fine Arts' (1827), Thomas de Quincey describes – stimulated by opium, some say – the public alarm, and the fog that allows the murderer to escape: 'Never . . . could there be a night . . . more propitious to an escaping criminal.' After forty false arrests John Williams, a lodger at a local public house, is arrested, but he hangs himself before committal proceedings. What truly shocks Victorian Londoners, however, is the thought that such killings can be committed *in their own homes*. And from this irreversible loss of confidence comes a lasting memento: latch chains on front doors.

A 'strange note . . . a screaming sort of noise' is how one witness describes the gale that brings tragedy of a very different kind to the tiny Cornish fishing village of Mousehole in 1981. Eight miles (13 km) off the south Cornish coast (✠), the cargo ship *Union Star* develops engine trouble. The captain refuses the offer of assistance by a tug, not wanting to pay the costs. As the weather worsens, and the gale drives the *Union Star* towards the rocky coast, he finally sends a distress signal. A helicopter from RNAS Culdrose is unable to rescue anyone, so the lifeboat *Solomon Browne* launches from Penlee Point. Coxswain Trevelyan Richards realizes it's going to be tough: from twelve volunteers, mainly fishermen, he chooses only one member from each family. In 60-foot (18-metre) seas, the lifeboat is hurled on to the *Union Star*'s deck twice before they manage to pull four crew off. On the final attempt, as the *Union Star* keels over, the radio goes silent – and the lights of the *Solomon Browne* disappear. Lifeboat tragedies always touch the nation's heart, but the Penlee disaster, devastating a single, tiny community, does so more than most. The official inquiry in 1982 blames no one, attributing the tragedy to the weather. But new legislation empowers coastguards to declare a Mayday and authorize salvage on behalf of a ship's captain.

-18.8C Carnwath 1981 ❄

15.5C Hemsby 1993 ☼

20 DECEMBER

1.4C min / 6.4C max

THOMAS
HARDY'S
DESOLATING
WIND

'A desolating wind wandered from the north over the hill ... The hill was covered on its northern side by an ancient and decaying plantation of beeches, whose upper verge formed a line over the crest, fringing its arched curve against the sky, like a mane. Tonight these trees sheltered the southern slope from the keenest blasts, which smote the wood and floundered through it with a sound as of grumbling, or gushed over its crowning boughs in a weakened moan. The dry leaves in the ditch simmered and boiled in the same breezes, a tongue of air occasionally ferreting out a few, and sending them spinning across the grass.'

So Thomas Hardy sets the scene on Gabriel Oak's farm at the beginning of *Far from the Madding Crowd* (1874). Hardy, an astute meteorological observer (*2 January, 4 November, 23 December*), uses the weather continually to reflect the emotions of his characters: Sergeant Troy woos Bathsheba in blazing sunshine. Fanny treks forlornly through the snow. Gabriel saves Bathsheba's harvest after a summer storm (*30 August*). The weather descriptions begin tonight, with the wind across Norcombe Hill:

'The thin grasses, more or less coating the hill, were touched by the wind in breezes of differing powers, and almost of differing natures – one rubbing the blades heavily, another raking them piercingly, another brushing them like a soft broom. The instinctive act of human-kind was to stand and listen, and learn how the trees on the right, and the trees on the left, wailed and chanted to each other in the regular antiphones of a cathedral choir; how hedges and other shapes to leeward then caught the note, lowering it to the tenderest sob, and how the hurrying gust then plunged it into the south, to be heard no more.'

On a frozen grass strip at Croydon aerodrome (♥) today in 1929 the young aviator Francis Chichester takes off in his De Havilland Gypsy Moth on a solo journey to Australia. It's an inauspicious start: on the brick-hard ground, he rips open a tyre and a tube. Then the moonlight, guiding him across fields covered in hoar frost, disappears behind cloud at the coast. 'I had expected to cross the Channel in about fifteen minutes,' he later writes, 'and when I was still over water after three-quarters of an hour I began to feel lost.' Nevertheless, he safely arrives in Australia forty-one days later.

-17.4C Drumlanrig 1886 ❄

16.7C Llandudno 1900 ☀

21 DECEMBER

1.4C min / 6.5C max

In 1796, a week of gales destroys a planned French invasion that might have ended English rule in Ireland. The French fleet – forty-three ships carrying fourteen thousand troops commanded by the brilliant young general Lazare Hoche – leaves Brittany for Ireland a week earlier in weather 'warm and bright as May'. With the English military in Ireland weak and ill prepared, the invaders have every prospect of success.

> 'The shortest day: a truly black, and dismal one.'
> *Gilbert White, naturalist and writer, Selborne, Hampshire, 1776*

The weather intervenes almost immediately, however. In the first storm the fleet encounters, Hoche's flagship is separated from the rest and not seen again. After five wretched days at sea, the coast is finally sighted. Wolfe Tone, the father of Irish republicanism and coordinator of this invasion and uprising, writes: 'We are under Cape Clear . . . so I have at all events once more seen my country.' The fleet edges round the south-west coast as far as the wide expanse of Bantry Bay (✂), and casts anchor.

Then tonight, another 'heavy gale from eastward with snow' sweeps half the fleet out into the Atlantic, leaving just sixteen ships and a small fraction of the troops. The enterprise is now doomed. 'Everything depends upon the promptitude and audacity of our first movement,' Tone writes, but the 'infernal wind continues without intermission.' In fact, the storm strengthens with every passing day. In complete disarray, the last ships cut their anchor cables on 27 December and retreat to France, 'in a perfect hurricane'. Had Hoche and Tone succeeded, Ireland might have been independent 125 years before its time. The English haven't had such an escape since the Spanish Armada (*17 August*).

Today (or tomorrow) is the Winter Solstice, the shortest day of the year (7 hours 50 minutes of daylight in London, 6 hours 20 minutes in Wick) and the time when the sun is weakest. It's a turning point in the year and the traditional moment for winter celebrations. Two of the most important Neolithic monuments in the British Isles – Newgrange (over five thousand years old) in County Meath (♀) and Maes Howe in the Orkneys (☌) – are aligned so that sunlight on the solstice illuminates the interior of a chambered cairn. Even Stonehenge is now thought to be aligned for the Winter, rather than the Summer, Solstice. Our prehistoric ancestors worshipped the winter sun. As do we, of course – in Majorca.

15.5C Mackworth 1971
-15.6C Mayfield 1909

FIXTURE CHAOS IN FOG

Fog plays havoc with the sporting fixture list today in 1962. Eighteen League football matches are postponed and eight are abandoned after kick-off when the referees can't see both goals though the murk. At Richmond, the rugby match with Harlequins goes ahead and *The Times* reports that 'only thuds, shouts and the whistle confirmed they were still playing'. At one point, the Harlequins' full-back is spotted on the touchline smoking a cigarette, and the referee has to appeal to both captains to keep their sides down to fifteen.

Otherwise, it's a day of wind and frost. In 1991 the great white turbines at Delabole, Cornwall (❀), Britain's first wind farm, start generating electricity. About time too, environmentalists say – the British Isles is, after all, the windiest place in Europe. On a 'cold, wintry' day in 1926 the pilot John Leeming embarks on a wacky attempt to popularize civil aviation: to the astonishment of a lone walker he lands his Avro Gosport two-seater plane on the top of Helvellyn (✈) in the Lake District, using the strong wind to stop the plane abruptly on the minute summit.

Tonight in 1643 a heavy frost comes to the aid of the Parliamentarian army during the English Civil War. Sir William Waller takes advantage of the iron-hard frozen lanes – usually thick with mud – to march his troops 10 miles (16 km) in double-quick time, allowing him to surprise and rout the Royalist soldiers at Alton, Hampshire (⚔) before dawn.

And the immortal literary Christmas at Dingley Dell, in *The Pickwick Papers* (1836) by Charles Dickens, begins today in 1827. On the journey to Kent, the party take 'ale and brandy to enable them to bid defiance to the frost that was binding up the earth in its iron fetters'. Though modern editions of *The Pickwick Papers* set the festivities in 1827, records show that this is a mild Christmas. In 1830, however – the year Dickens originally assigns to events at Mr Wardle's home – it is extremely cold and snowy. Not that the fictional frost is sufficiently intense for the pond ice to bear the substantial weight of the illustrious Pickwick. 'The sport was at its height ... when a sharp smart crack was heard ... Mr Pickwick's hat, gloves and handkerchief were floating on the surface; and this was all of Mr Pickwick that anybody could see.'

16.1C Hoylake 1910

-17.4C Stokesay 1890

23 DECEMBER

1.9C min / 6.5C max

BLEAK HOUSE 'Fog everywhere. Fog up the river, where it flows among . . . meadows; fog down the river, where it rolls defiled among the tiers of shipping and the waterside pollutions of a great (and dirty) city. Fog on the Essex marshes, fog on the Kentish heights. Fog creeping into the cabooses of collier-brigs, fog lying out on the yards, and hovering in the rigging of great ships; fog drooping on the gunwhales of barges and small boats. Fog in the eyes and throats of ancient Greenwich pensioners, wheezing by the firesides of their wards; fog in the stem and bowl of the afternoon pipe of the wrathful skipper, down in his close cabin; fog cruelly pinching the toes and fingers of his shivering little 'prentice boy on deck . . .

'Since the 21st, the fog has never once broken, . . . the room being in pure night effect, with three candles. As the clock struck 10, the gardeners could not see to work.'
John Ruskin, critic, Surrey, 1871

'Before day. A lavender curtain, with a pale crimson hem, covers the east & shuts out the dawn.'
Thomas Hardy, writer, 1873

'Sissinghurst is almost intolerably uncomfortable in winter.'
Harold Nicolson, diplomat and writer, after the pipes and lavatories freeze, 1945

'The raw afternoon is rawest, and the dense fog is densest, and the muddy streets are muddiest, near that leaden-headed old obstruction, appropriate ornament for the threshold of a leaden-headed old corporation: Temple Bar. And hard by Temple Bar, in Lincoln's Inn Hall, at the very heart of the fog, sits the Lord High Chancellor in his High Court of Chancery.

'Never can there come fog too thick, never can there come mud and mire too deep, to assort with the groping and floundering condition which this High Court of Chancery, most pestilent of hoary sinners, holds, this day, in the sight of heaven and earth.

'On such an afternoon, if ever, the Lord High Chancellor ought to be sitting here – as he is – with a foggy glory round his head, softly fenced in with crimson cloth and curtains, addressed by a large advocate with great whiskers, a little voice, and an interminable brief, and outwardly directing his contemplation to the lantern in the roof, where he can see nothing but fog . . .'

-15.6C Cambridge 1890 ❄
16.0C Innsworth 1977 ☼

Charles Dickens's opening to *Bleak House* (1853), datable to today, is the first great literary evocation of London's fog, a metaphor for the suffocating and obstructive procedures of the Chancery Division of the High Court that have, by now, been a national scandal for twenty-five years. The Chancery Court structure he describes is reformed in 1842, setting his story some time before this date.

24 DECEMBER

1.5C min / 6.8C max

SONNING
CUTTING
COLLAPSES

After days of rain, the embanked earth of the Sonning cutting (❂), the 2-mile (3-km) slice hewn out of Sonning Hill in Berkshire, on Brunel's Great Western Railway, gives way today in 1841. A train hurtles into the landslip. Eight die and dozens are injured, thrown from or crushed between the open-topped wagons. Following the accident, William Gladstone, President of the Board of Trade, introduces new legislation: all passenger carriages will have roofs from now on.

A couple of years later, it's 'cold, bleak biting weather . . . the fog came pouring in at every chink and keyhole . . . Piercing, searching, biting cold . . . gnawed and mumbled by the hungry cold as bones are gnawed by dogs . . .' Charles Dickens piles on the adjectives and similes in *A Christmas Carol* (1843). And Dickens should know: he was born in 1812, and the first decade of his life is the coldest period since the 1690s (*7 November*) – six Christmases white with snow or frost. It's his memory and literary interpretation of this that's done more than anything to influence our own yearning for white Christmases.

An Atlantic storm in 1946 is so bad that no doctor can get to Great Blasket, off County Kerry (❂), where a young man has a terrible headache. When he dies next day, it's a catalyst. The remaining islanders, representing the last vestiges of pure Irish culture, decide to leave Great Blasket for ever.

Back in 1811 a fierce North Sea storm wrecks five of His Majesty's ships – *St George* (formerly Nelson's flagship), *Defense* (a veteran of Trafalgar), *Hero*, *Grasshopper* and *Archimedes*. Britain is at war with Napoleon, and the loss of nearly 1900 sailors is devastating.

-15.6C Braemar 1935 ❄

And today in 1988, taking the pop industry by storm, A Tribe of Toffs reaches number 21 in the UK singles chart with the unlikely hit 'John Kettley is a Weatherman', named after the moustachioed, one-time 'housewives' favourite' BBC TV weather presenter. The song remains a call to arms in Student Union bars for years to come. All together now: 'John Kettley is a weatherman, / And so is Michael Fish.'

15.6C Rhyl 1910 ☀

25 DECEMBER

1.4C min / 6.5C max

CHANCES OF A WHITE CHRISTMAS

The fog creeps past the Chelsea goal and up the pitch at the Valley in south London today in 1937 as Sam Bartram of home team Charlton Athletic, 'the finest goalkeeper never to play for England', watches from the far end. Finally the curtain of murk envelops him. After twenty minutes, a phantom in blue emerges out of the pea-souper. Bartram, bouncing on the balls of his feet, readies himself for the Chelsea attack.

'It was a policeman,' he later writes. 'He gaped at me incredulously. "What on earth are you doing here?" he said. "The game was stopped a quarter of an hour ago."'

'Intense frost. I sat down in my bath upon a sheet of thick ice which broke in the middle into large pieces whilst sharp points and jagged edges stuck all round the sides of the tub like chevaux de frise, not particularly comforting to the naked thighs and loins.'
Francis Kilvert, diarist, Clyro, Wales, 1870

'A typical Christmas morning of the good old-fashioned sort; a slight misty haze, aided by a frost ... the trees and hedges in ... glittering, scintillating white. Scarcely a breath of wind ... Underfoot the grass crunched ... and all along the river's edge, clinging lovingly to the lowest twigs ... a thin skim of ice, that swayed and dipped with every swirl of the current ... I gloried in the beauties of a keen winter's morn, and considered a day's chubbing ... the very beau-ideal of a sportsman's life.'
J. W. Martin, angler (the 'Trent Otter') and author, 1906

Who's dreaming of a white Christmas? Not the bookmakers. Betting on snow on Christmas Day begins in the early 1970s and develops into a multi-million-pound business. The bet used to depend on a single snowflake falling on the London Weather Centre in the twenty-four hours of 25 December, with a Met Office official present throughout to adjudicate. In 2006, however, technology takes over, and electronic recordings are now taken so that people can bet on their own region being 'white'. Either way, it's a bet of passion over probability. The last proper white Christmases in London are 1906, 1916, 1927 (one of the most severe snowstorms of the century, with 23-foot [7-metre] drifts on Salisbury Plain), 1938 (snow falls daily from 18 to 26 December; people ski in the Chilterns), 1956, 1964, 1968, 1970, 1976, 1981 (more than half the country has snow – truly 'deep and crisp and even'), 1996 and 1999. In recent years, the chances of snow on Christmas Day have receded and receded. Strangely, few seem interested in betting on it being warm, wet and windy.

15.6C Leith 1896

108.0mm Foffany Reservoir 1914   -18.3C Gainford 1878

1.4C min / 6.7C max

FOGGITT'S
FORECAST

It's an exceptionally mild Boxing Day in 1946, and Bill Foggitt, a young Yorkshireman, is walking through the woods near Thirsk (♥) when his father notices some small crested birds called waxwings. He pronounces that a terrible winter is imminent. The rationale – that the birds have flown south to escape the bitter Scandinavian weather heading this way – impresses Foggitt junior, who sends the prediction to a national newspaper. Within days, one of the cruellest winters of the twentieth century has begun (*4 March*).

The forecast is no fluke. Successive generations of the Foggitt family have been observing, recording and forecasting the weather in North Yorkshire for 175 years, with the dogged, meticulous rigour that is a hallmark of the British amateur (*21 March*). No Foggitt finds fame like Bill, however. Using the family theory on cyclical weather (hot summers recur every twenty-two years, severe winters every fifteen) and observations of plant and animal behaviour (*18 October*), Bill rises to international notoriety over the years – with forecasting columns in the *Yorkshire Post* and the *Daily Mail*, a BT 'Foggitt Forecast' payline and innumerable television appearances. Foggitt's finest hour, however, comes in an Arctic spell in January 1985. He publicly refutes the Met Office prediction of a long, hard winter when he sees a mole stick his head through the snow. 'The cold snap is over,' he declares. The thaw duly commences next day.

'Windows opaque, painted over with frost-patterns, and outside ... a light jewelry of rime; every cobweb a little lace net ... spent the day ... making a fire the smoke of which rose in a still unmoving column straight up into the fog-roof.'
J. R. R. Tolkien, academic and writer, Oxford, 1944

'One of the most beautiful days I can remember in my life. The whitest-hardest frost – unimaginable. The beauty of Long Hedges frozen into white coral ... The whole world seemed the work of a master draughtsman.'
Violet Bonham Carter, Liberal campaigner, Stockton, Wiltshire, 1944

-17.7C Braemar 1981
16.0C Haddington 1983
103mph gust Prestwick airport, 1998
153.7mm Buttermere 1924

Of course, Foggitt is wrong just as often as he is right. But he provides a bridge to the weather lore of the Middle Ages. His popularity probably has more to do with our collective reluctance to have the weather explained away by science and supercomputers than it does with the behaviour of moles.

The snow starts falling today in 1962, heralding the beginning of one of the greatest winters of them all (*6 February*). Most of the country will be blanketed white for two months.

27 DECEMBER

1.7C min / 6.5C max

Is this the most weather-prone day of the year? In 1812 the Prince Regent (later George IV) gets lost in fog at the start of the greatest frost of the nineteenth century – the winter that is Napoleon's undoing in Russia. In 1836, headline writers are able to write for the first – and, presumably, last – time: 'AVALANCHE KILLS EIGHT IN SUSSEX.' A giant cornice of snow, sculpted by winds whipping over the lip of a 330-foot (100-metre) chalk precipice in Lewes (♀), thaws in the sunshine and breaks off, sending hundreds of tons of icy snow on to houses beneath. Compressed air pockets within the ice cause the blocks to explode as they hit the ground, lifting the houses before burying them.

'Weather more like April than ye end of December. Hedge-sparrow sings.'
Reverend Gilbert White, naturalist and writer, 1768

'When we came to try to enter Dover harbour, after a four-hour crossing, we had to wait nine hours at the entrance ... in what is known in seamen's terms as une tourmente, half tempest, half squall ... every one very ill.'
Alexandre de La Rochefoucauld, aged 16, travelling with his father, a French social reformer, 1783

In 1852, a 'terrific storm' drives the English sailing brig *Lily*, carrying 61 tons of gunpowder, on to rocks off the Isle of Man (✛). It explodes next day, killing thirty salvage men – all except one, whose only memory is the head of a companion flying through the air. In 1887, attending a political meeting in Dover, ex-Liberal Prime Minister William Gladstone is 'saluted with a shower of snowballs, one of which struck him on the shoulder,' reports *The Times*. In 1965, a north-westerly gale sinks Britain's first offshore drilling platform, the giant *Sea Gem* rig, in the North Sea (✛). Weakened by the wildly fluctuating water pressures caused by 20-foot (6-metre) waves, two of the drilling legs collapse and thirteen die (*27 March, 24 November*).

In 1972 (as in 2006) flights are severely disrupted by fog, with so many diversions from Heathrow (*17 February*) and Gatwick (including Prime Minister Ted Heath's flight from Canada) that Birmingham airport is packed nose-to-tail with aircraft. And in 1981, following a severe cold snap, the thaw introduces a horrifying new road hazard: blocks of melting ice falling from motorway bridges.

-21.4C Grantown-on-Spey 1995 ❄

16.7C Ashburton 1921 ☀

28 DECEMBER

1.6C min / 6.6C max

TAY BRIDGE DISASTER

The collapse of the Tay Rail Bridge in a gale tonight in 1879 isn't the worst British rail accident, but it's the most dramatic. The whole train, with around seventy-five passengers and crew, plunges from the central 'high girder' section of the bridge 100 feet (30 metres) into the icy estuary (♥). There are no survivors. These days the disaster is remembered chiefly because of William McGonagall's poems about it – often cited as the worst ever written – but the story behind the collapse makes even more painful reading.

'Twas about seven o'clock at night,
And the wind it blew with all its might,
And the rain came pouring down,
And the dark clouds seem'd to frown,
And the Demon of the air seem'd to say –
"I'll blow down the Bridge of Tay."
The Tay Bridge Disaster,
William McGonagall

Open for just six months, the bridge is the product of Victorian free market economics – namely a war between two railway companies vying to connect Glasgow and Edinburgh with Dundee and Aberdeen, even though the potential revenues will scarcely support a single line. A route across the Tay promises to slice an hour off the journey, so one company, the Northern British, starts building what has been called 'the longest and cheapest bridge in the world'. The engineer, Thomas Bouch, knows nothing about 'wind-loading'. He accepts unquestioningly the 10lb per square foot estimate advised by the (equally ignorant) Astronomer Royal. (Storm gusts exert anything from four to five times this figure.) It's a grave miscalculation. Before Queen Victoria even opens the bridge in 1878 – or knights its engineer – iron tie-bars are 'chattering' in the wind, a sign that the joints have loosened. Painting gangs report movement as trains cross, and passengers complain of a 'strange motion'. Rivets, nuts and bolts are found in the sand beneath the bridge, and the 25 mph (40 kph) speed restriction on the bridge is routinely flouted.

-23.7C Altnaharra 1995 ❄

16.8C Long Lawford 1974 ☼

'Badly designed, badly built and badly maintained,' concludes the official inquiry, ending Sir Thomas Bouch's career. Having taken the full blame, and with his son-in-law one of the dead passengers, he dies within a year. Following the disaster, wind loading is finally taken seriously. The Forth Rail Bridge (*6 November*) is over-engineered almost to the point of insanity. And via the 'Wind Force Committee' of the Royal Meteorological Society, the disaster leads in 1885 to William Henry Dines's Pressure Tube Anemometer – the first instrument to accurately measure wind speed, revolutionizing our understanding of 'wind pressure' (*4 November*).

1.5C min / 6.6C max

A STORM CREATES THE NORFOLK BROADS

After the storm, the calm. Who'd have guessed that the Norfolk Broads – the tranquil waterscape of winding rivers and shallow lakes (or 'broads'), described by one naturalist as a 'breathing space for the cure of souls' – are the result of a titanic storm around today in 1287? Indeed, it's assumed that the Broads are the remains of natural estuaries, until, in 1952, a study of the sediments reveals them to be man-made. In the thirteenth century, east Norfolk and Suffolk are the second most urbanized parts of the British Isles, where peat is burnt on a vast scale (the monks of Norwich Cathedral use four hundred thousand 'turves' a year for cooking alone). The huge, shallow pits left by digging out all this peat flood in 1287's cataclysmic, gale-driven storm surge.

> 'Snow falls in great flumps as the men shovel it off the roof. We are quite cut off except by walking.'
> *Vita Sackville-West, writer, Knole, Kent, 1927*

> 'Snowing; a thing rather unusual here ... London looks very curious; there is such a silence in it, the wheel-vehicles making no noise.'
> *Thomas Carlyle, historian and writer, 1836*

'This flood took more life than any sea inundation in memory,' writes the chronicler John of Oxenedes. 'Sleeping men and women from their beds, beasts and fresh water fish choked or drowned, houses were pulled down to their foundations into the sea.' This 'tempest of all tempests' drowns some eight hundred along eastern Britain. In the village of Hickling (), the flood waters rise to a foot above the high altar of the Priory church. While this single storm surge doesn't leave the Broads exactly as they are today, it's the event that begins the process which succeeding floods complete.

By the Middle Ages, however, the Broads are a forgotten part of the landscape. Only in the eighteenth and nineteenth centuries are they rediscovered. By this time the lock-free waters, with their drainage windmills, Norfolk wherries (trading barges), wooden bridges and brightly painted boatyards, have acquired a character all their own (even down to their own wind, the 'Roger's Blast' – *15 July*), making them popular for boating holidays. 'One does not go on the Broads for variety of scenery, since from the water one can see very little but reeds and green banks,' says the 1949 *Batsford Guide*. 'One goes on the Broads because . . . time does not matter – it's no use fussing about time when progress depends almost entirely on the state and direction of the wind.'

-24.2C Altnaharra 1995

170.0mm Glaspwll 1986

16.1C Usk 1925

30 DECEMBER

1.5C min / 6.4C max

CROSSING THE
IRISH SEA

Westerly winds confound the ambitions of the Tudor court today in 1566. Soldiers and provisions wait on the wharves at Chester, to embark on ships for Ireland. But the wind will not abate and this urgent journey to resupply the English garrisons, struggling to put down the rebellions in Ireland and establish a central administration, cannot take place.

Though you can see across the Irish Sea on a clear day – it's only 65 miles (just over 100 km) from Holyhead (⚓) to Dun Laoghaire (⚓) – wind, in the days of sail, means passages are plagued with uncertainty. Elizabeth Freke, travelling between Ireland and England in the late seventeenth century, records numerous dreadful voyages. In 1677, sailing to Ireland, she comes 'within a watch of reaching Watterford; butt on a sudden, about sunsett nightt, the wind changed, with the most hideous tempest of wind & raine, wch brought us backe againe next day att nightt to Lundy, wher we lay wth 4 ships more, despairing of life, and our mast all downe ... We lay roleing till next night, when, being In a despratt condition, we attempted to shoot the Bay of Barnstaple.' In 1696, she takes eighteen days, 'in a tempest and mists', to sail from Cork to Deal in Kent. After that, she never returns to Ireland.

> 'Strong freezing wind ... so inexpressibly sharp that some post boys who were forc'd to go into it were kill'd by it.'
> Thomas Barker, Lyndon Hall, Rutland, 1739

> 'This is a Fell Winter indeed, and I am expecting White Wolves to cross the river.'
> J. R. R. Tolkien, academic and writer, Oxford, 1961

Little has changed, over 250 years later, when Pete McCarthy travels as a child on the Cork steampacket, a cattle boat 'with berths for thirty or forty passengers as a sideline', as he recalls in *McCarthy's Bar* (2000).

'We'd leave from Pier Head at night, in what now seems like a scene from a period movie playing inside my head: men in hats, fog, customs officers wielding pieces of chalk. The crossing would turn rough in the early hours of the morning, as we rounded the southeast corner of Ireland, and the swell of the Atlantic hit the Irish Sea. The seasickness was spectacular. Today's ferries may be sleek and comfortable, but they deny young people the unforgettable experience of witnessing at first hand a cow throwing up. I suppose we can't stand in the way of progress, but holiday travel's a duller business without bovine projectile vomit.'

-27.2C Altnaharra 1995 ❄

16.1C Wistanstow 1925 ☀

31 DECEMBER

'HOME AWAITED THEM WARM' The skies are clear as 284 Hebridean soldiers and sailors board the armed yacht HMS *Iolaire* at Kyle of Lochalsh () tonight in 1918. (This is a tragic story, so don't read on if you're easily depressed.) *Iolaire* has lifeboats and lifejackets enough for only a hundred, but the men are returning from the trenches and the horrors of the First World War: since they are so anxious to get home for the traditional Hogmanay celebrations, the rule-book is ignored. 'Home awaited them warm' the poet Donald MacPhail writes. 'And all was best prepared.'

As they head north across the Minch, bound for Stornaway on the Isle of Lewis, a fresh wind blows up from the south, bringing sleet squalls and heavy seas. Approaching the harbour, *Iolaire* strikes rocks called the Beasts of Holm (). The crush to abandon ship is intense. The lifeboats are dashed against the side or engulfed by the now raging sea. Although *Iolaire* sinks scarcely 20 feet (6 metres) from land, 205 men drown.

It's the worst peacetime maritime disaster in United Kingdom waters. Nearly every family on the island is bereft. The desperate damage to the community is reflected in mass emigrations that take place over the next decade. The *Stornoway Gazette* describes the tragedy as 'the blackest day in the history of the island', plunging 'every home and every heart in Lewis into grief unutterable'.

'A most terrible fog came on, and Lupin would go out in it, but promised to be back to drink out the Old Year – a custom we have always observed. At a quarter to twelve Lupin had not returned, and the fog was fearful. As time was growing close, I got out the spirits. Carrie and I deciding on whisky, I opened a fresh bottle; but Carrie said it smelt like brandy. As I knew it to be whisky, I said there was nothing to discuss. Carrie, evidently vexed that Lupin had not come in, did discuss it all the same, and wanted me to have a small wager with her to decide it by the smell. I said I could decide it by the taste in a moment. A silly and unnecessary argument followed, the result of which was we suddenly saw it was a quarter-past twelve and, for the first time in our married life, we missed welcoming in the New Year. Lupin got home at a quarter-past two, having got lost in the fog – so he said.'
George and Weedon Grossmith, Diary of a Nobody, *1888*

-18.6C Eskdalemuir 1961

15.6C Great Yarmouth 1901

TIMELINE

OTHER KEY DATES IN HISTORY

55BC – Caesar's first invasion (*8 July*)

0AD

409 – Roman legions abandon Britain
450-750 – Angles, Saxons and Jutes settle in Britain

900 – 1300 (roughly) – Medieval Warm Period (*20 February*)

971 – The canonization of St. Swithun (*15 July*)
1066 – Battle of Stamford Bridge (*1 October*)   1066 – Battle of Hastings
1066 – Invasion of William of Normandy (*3 October*)

1086 – Domesday Book

1120 – White Ship disaster (*2 December*)
1174 – Work on Wells Cathedral begins (*20 February*)
1212 – First Great Fire of London (*17 July*)
1215 – Magna Carta (*17 July*)

1250

1263 – Battle of Largs (*9 October*)
1277 – Rev. William Merle begins his daily
  weather record (*17 March*)
1287 – Storm creates the Norfolk Broads
  (*29 December*)
1314 – Battle of Bannockburn (*2 July*)
1315-18 – The Great Famine (*12 May*)

1337 – Hundred Years War with France begins

1348 – The Black Death arrives in Britain
  (*30 June*)

1350 – 1850 (roughly) – Little Ice Age (*3 March*)

1381 – Peasants' Revolt

1387 – Chaucer writes *The Canterbury Tales*
  (*17 April*)
1400 – Rebellion of Owain Glyndwr
  (*10 September*)

1415 – Battle of Agincourt
1455 – Wars of the Roses begin

1461 – Battle of Mortimer's Cross (*11 February*)
1461 – Battle of Towton (*7 April*)
1471 – Battle of Barnet (*23 April*)

1474 – William Caxton prints the first book in English
1485 – Battle of Bosworth Field

1500

1509–47 – Reign of Henry VIII

1513 – Battle of Flodden (*19 September*)

1534 – Papal authority abolished in England
1536 – Political union of Wales and England

1545 – *Mary Rose* sinks (*29 July*)

1549 – First Book of Common Prayer
1558–1603 – Reign of Elizabeth I

1570 – 'All Saint's Flood' (*11 November*)

1577–80 – Drake's first voyage around the world

1588 – Spanish Armada defeated (*17 August*)
1597 – Third Armada (*24 October*)

1600

1600 – East India Company founded
1603–25 Reign of James I (VI of Scotland)

1607 – Three ships sail for Jamestown (*5 January*)
1607 – Great Flood on the River Severn (*30 January*)
1607 – Flight of the Earls (*2 October*)

1616 – Death of Shakespeare
1642 – English Civil War begins

1643 – Robert Hooke develops a domestic barometer (*17 November*)
1649 – King Charles I executed (*9 February*)
1650 – Battle of Dunbar (*13 September*)
1659 – CET monthly temperature series begins

1660 – Royal Society founded

1661 – Yacht racing invented (*10 October*)
1666 – Great Fire of London (*12 September*)
1666 – The Great Plague (*10 November*)
1684 – Great Frost Fair on the Thames (*19 January*)
1688 – The Glorious Revolution (*15 November*)

1689–1702 – Reign of William III & Mary II
1690 – Battle of the Boyne

1692 – Glencoe Massacre (*22 February*)
1695 – A storm inspires the first lighthouse (*3 January*)
1694 – The 'Seven Ill Years' (*7 November*)
1697 – Severest ever hailstorm in Britain (*15 May*)

1700

1701- Jethro Tull invents the seed drill

1703 - The Great Storm (*7 December*)
1707 - *HMS Association* sinks (*2 November*)

1707 - Act of Union between England
and Scotland

1714 - George I arrives in England, in fog
(*29 September*)

1714-27 - Reign of George I

1739-40 - 'Time of the Black Frost' (*13 February,
2 May*)
1736 - Robert Marsham invents 'Phenology'
(*10 April*)

1745 - Second Jacobite Rebellion

1746 - Battle of Culloden (*27 April*)
1764 - A storm leads to lightning conductors
on tall buildings (*18 June*)

1764 - Spinning Jenny invented

1775-81 - American War of Independence

1782 - The 'Black Aughty Twa' (*15 September*)
1784 - First Mail Service coach leaves Bristol
for London (*2 August*)
1789 - 'The Original' - first purpose-built lifeboat
(*15 March*)
1789 - George III visits Weymouth for the
first time (*13 July*)

1793-1802 - War with France

1796 - Attempted French invasion of Ireland
(*21 December*)

1800

1801 - Act of Union creating the United
Kingdom of Great Britain and Ireland
1803-15 - War with France
1805 - Battle of Trafalgar

1805 - Admiral Beaufort devises his 'Wind
Force Scale' (*13 January*)
1814 - Last Frost Fair on the Thames (*4 February*)
1815 - Wordsworth's *Daffodils* published (*15 April*)
1816 - 'Year without a summer' (*13 October*)
1819 - Keats writes 'Ode to Autumn'
(*19 September*)
1821-22 - Constable paints his cloud 'sketches'
(*23 October*)
1822 - Storm inspires Sir William Hillary to found
the RNLI (*6 October*)
1829 - A Hebridean storm inspires Felix
Mendelssohn's most famous overture (*7 August*)

1815 - Battle of Waterloo

1829 - Stephenson's Rocket

1832 - The Great Reform Bill

1836 – Avalanche in Sussex (*27 December*)

1837–1901 – Reign of Queen Victoria
1837 – First telegraphic message sent

1838 – Turner paints *The Fighting Temeraire*
(*6 September*)
1838 – Grace Darling rescues survivors of the
Forfarshire (*7 September*)
1845-9 – Irish potato famine (*8 September*)
1846 – Scottish potato famine (*25 May*)
1846 – Hailstorm destroys 7,000 panes of glass
in the Houses of Parliament (*3 August*)
1848 – Moray fishing disaster (*19 August*)
1854 – Meteorological Department of the Board of
Trade (predecessor of the UK Meteorological
Office) founded (*1 August*)
1854 – First cork and canvas lifebelt designed
(*15 November*)
1858 – Heat-wave in London (the 'Great Stink')
leads to a sewage system (*23 June*)
1859 – Last malaria epidemic in Britain (*28 August*)
1859 – *Royal Charter* disaster leads to 'Storm
Warnings' (*25 October*)
1861 – First published weather forecast in *The
Times* (*1 August*)

1877 – Proclamation of the Empire of India

1879 – Tay Bridge disaster (*28 December*)
1881 – Eyemouth fishing disaster (*14 October*)
1890s – Ice climbing invented in Scotland
(*22 April*)

1899–1902 – Second Boer War

1900

1901 – Monet paints London in fog (*23 February*)
1908 – Windscreen wiper invented (*26 April*)
1909 – Louis Bleriot is the first person to fly
across the Channel (*25 July*)
1911 – 'Perfect Summer' (*22 July*)
1912 – *Titanic* sinks (*14 April*)

1914–18 – First World War

1916 – Battle of Jutland (*1 June*)
1916 – Death of Kitchener (*5 June*)
1922 – First ever radio broadcast weather
forecast (*14 November*)

1922 – Anglo-Irish Treaty: partition of Ireland

1924 – Welsh rain inspires A.A. Milne (*22 August*)
1925 – First BBC Shipping Forecast (*4 July*)
1928 – Penicillin discovered (*3 September*)

1933 – Invention of the 'Catseye' (*3 December*)

1939–45 – Second World War

1940 – Dunkirk (*28 May*)
1944 – D-Day (*4 June, 6 June*)

1946 – National Health Service

1947 – Severe winter (*27 February, 4 March*)
1947 – Keswick Mountain Rescue formed
(*14 March*)
1948 – First Heathrow air crash (*2 March*)
1952 – Lynmouth Flood (*15 August*)

1952 – Coronation of Elizabeth II

1952 – Great London Smog (*6 December*)
1953 – East Coast Flood (*31 January*)
1954 – Roger Bannister runs the 'first four-minute
mile' (*6 May*)
1954 – Highest ever annual rainfall recorded at
Sprinkling Tarn, Lake District (*8 May*)
1954 – First live TV weather forecast on BBC
(*16 May*)
1954 – Britain's record daily rainfall, Martinstown,
Dorset (*18 July*)
1956 – Jim Laker takes 19 wickets in a Test
match (*31 July*)

1956 – Suez Crisis

1956 – Clean Air Acts (*5 October*)
1957 – Explosion at Windscale plutonium plant
(*11 October*)
1957 – Lewisham rail crash (*4 December*)

1959 – M1 opens

1962–3 – Coldest winter in living memory
(*26 January, 6 February*)
1966 – Aberfan disaster (*21 October*)
1971 – Cairngorm tragedy (*22 November*)

1973 – UK joins Common Market

1976 – Hottest summer (*4 July*)
1979 – Winter of Discontent (*10 January*)
1979 – Fastnet race disaster (*11 August*)
1981 – Queens Park Rangers roll out the first
artificial football pitch (*17 January*)

1982 – Falklands War

1987 – Great storm in south east England
(*15 October*) and Stock Market crash (*16 October*)
1991 – 'BRITISH RAIL BLAMES THE WRONG KIND
OF SNOW' (*11 February*)

1991 – World Wide Web invented

1991 – Britain's first wind farm opens (*22
December*)
1999 – North Sea storm reveals Seahenge, Norfolk
(*1 January*)

1999 – Devolution for Scotland and Wales

2000

2003 – Hottest day on record (*10 August*)
2004 – Boscastle Flood (*16 August*)
2006 – Severn Bore surfing record set (*1 April*)
2006 – Warmest July on record for large parts
    of Britain (*19 July*)
2007 – Sunniest April on record for the majority
    of Britain (*6 April*)
2007 – Wettest June on record for E and
    NE England

2003 – Second Gulf War

LOCAL WEATHER EFFECTS

West Shetland Basin, most storm-battered region of the Atlantic (*24 November*)

Haar (*29 August*)

Haar (*29 August*)

Haar (*29 August*)

Fort William (*8 May*)

Houghall (*10 December*)

Seathwaite (*8 May*)

The Helm (*17 may*)

'Cyclonic bomb' (*6 January*)

Blaenau Ffestiniog (*8 May*)

Fohn Effect 27 January, 12 March

'Cheshire Gap' (wind funnel)

Thunderstorm alley

The Roger (*15 July*)

Fen Blow (*4 may*)

Santon Downham (*10 December*)

Rickmansworth (*10 December*)

The Cribbar (*4 September*)

Dartmoor fog (*22 October*)

Weymouth (*13 July*)

Bognor Regis (*28 June, 13 July*)

Brighton (*13 July*)

WIND ROSE (LONDON)

North

West ← → East

South

✳ Major frost hollows (*10 December*)

🌢 Wettest places (*8 May*)

☼ Sunniest places (*28 June, 13 July*)

General effects, such as sea and land breezes or orographic cloud (over hills and mountains), not shown.

KEY FOUL-WEATHER SHIPPING HAZARDS MENTIONED IN THE TEXT

Muckle Flugga (*12 January*)

Cape Wrath
(*12 January*)

Beasts of Holm
(*31 December*)

Skerryvore
(*12 January*)
Dubh Artach
(*12 January*)

Bell Rock
(*12 January*)

Longstone Rock
(*7 September*)

Lacada Point
(*9 November*)

Conister Rocks
(*6 October*)

Whitby Rock
(*30 October, 8 August*)

Aran Islands
(*17 September*)

Skerries
(*25 October*)

Great Bank
(*6 December*)

Blackwater Bank
(*30 April*)

Blasket Sound
(*9 November*)
Fastnet Rock
(*11 August, 13 August*)

Goodwin Sands
(*27 November,
7 December*)

The Needles
(*18 June*)

Carn Boscawen Zawn, Tater Du
(*19 December*)
Scilly Isles
(*2 November*)
Wolf Rock
(*4 January*)

Bolt Head (*15 February*)
Eddystone Rock
(*3 January*)

THE SHIPPING FORECAST

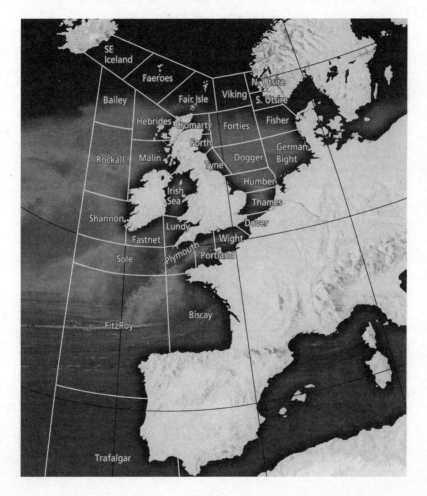

The Shipping Forecast is broadcast four times daily, at 00.48 (FM Radio 4 and LW 198 kHz), at 05.20 (FM and LW), 12.01 (LW only), 17.54 (LW only weekdays, FM and LW at weekends), and read at dictation speed. The forecast begins with any Gale Warnings (winds of Beaufort Force 8 or more), followed by a general synopsis, giving position, pressure (in millibars) and track of pressure areas, followed by each area's forecast. Wind direction is given first, then strength (Beaufort Scale), then precipitation, if any, and, finally, visibility. All terms carry precise definitions:

Gale warnings:
Gales are categorised as *gale, severe gale, storm, violent storm* or
   hurricane force as defined by the Beaufort Scale.
Imminent—expected within 6 hours.
Expected soon—expected within 6-12 hours.
Later—in more than 12 hours' time.

Variation with time:
Then—if a single type of weather (e.g. rain) persists for more than half the
   period, then changes, this is described as that weather followed by the next
   weather, e.g. '*rain then fair*' or '*rain then snow.*'
For a time—only happening once, at neither the start nor end of the period,
   for no more than half the total period.
At times—happening more than once but for no more than half the total
   period, at neither the start nor end of the period.
At first—occurring at the start of the period and ceasing before the middle.
Later—starting more than halfway through the period and continuing to the end.
Soon—expected within 6-12 hours.
Imminent—expected within six hours.

Wind:
North, south, north-west, south-west, etc implies a 45° range centred on the direction
   stated. *Northerly, southerly, north-westerly, south-westerly*, etc implies a 90° range.
Veering—changing of the wind direction clockwise, e.g. SW to W.
Backing—changing of the wind in the opposite direction to veering
   (anti-clockwise), e.g. SE to NE.
Becoming—wind direction changing to or from a variable or cyclonic state.
Variable—wind direction varies by more than 90°, for winds of Force 4
   or less.

Weather:
Snow—solid precipitation.
Rain—liquid precipitation from layer cloud (of size over 500 microns).
Sleet—snow and rain falling together or snow melting as it falls.
Showers—solid or liquid precipitation from a convection (as opposed to layer) cloud.
Squally showers—a squall is defined as 'a sudden increase in wind by at least 8 metres per second,' lasting for at least one minute.
Drizzle—liquid precipitation from layer cloud (of size up to 500 microns).
Freezing rain—liquid precipitation that freezes on contact with a very cold surface.
Hail showers—hail is defined as solid precipitation in the form of balls of ice falling from convective clouds.

Visibility:
Good—visibility more than 5 nautical miles.
Moderate—visibility between 2-5 nautical miles.
Poor—visibility between 1,000 metres and 2 nautical miles.
Fog—visibility less than 1,000 metres.
Fog patches—less than 40% sea area coverage.
Fog banks—40-50% sea area coverage.
Extensive fog—more than 50% sea area coverage.

Icing:
In severe winter cold, combined with strong winds and a cold sea, superstructure icing can occur (usually only in sea area Southeast Iceland). *Light, moderate, severe* or *very severe* (denoting accumulations of 1-14 cm per 24 hours). This is the last item of each sea area forecast.

Movement of pressure systems:
Slowly—less than 15 knots.
Steadily—15-25 knots.
Rather quickly—25-35 knots.
Rapidly—35-45 knots.
Very rapidly—more than 45 knots.

BEAUFORT WIND SCALE

Beaufort Force	Knots	MPH	Description	Sea Condition	Land Condition
0	0	0	Calm	Sea like a mirror	Smoke rises vertically
1	1-3	1-3	Light Air	Ripples without foam crests	Wind motion visible in smoke
2	4-6	4-7	Light Breeze	Small wavelets; crests do not break	Wind felt on exposed skin. Leaves rustle
3	7-10	8-12	Gentle Breeze	Large wavelets; perhaps scattered white horses	Leaves and smaller twigs in constant motion
4	11-16	13-18	Moderate Breeze	Small waves; fairly frequent white horses	Dust and loose paper raised. Small branches begin to move
5	17-21	19-24	Fresh Breeze	Moderate waves; many white horses	Smaller trees sway
6	22-27	25-31	Strong Breeze	Large waves begin to form; white foam crests, probably spray	Large branches in motion. Whistling heard in overhead wires. Umbrellas difficult to handle
7	28-33	32-38	Near gale	Sea heaps up and white foam blown in streaks along the direction of the wind	Whole trees in motion. Effort needed to walk against the wind
8	34-40	39-46	Gale	Moderately high waves; crests begin to break into spindrift	Twigs broken from trees. Cars veer on road
9	41-47	47-54	Strong Gale	High waves. Dense foam along the direction of the wind. Crests of waves begin to roll over. Spray may affect visibility	Light structure damage
10	48-55	55-63	Storm	Very high waves with long overhanging crests. The surface of the sea appears white; tumbling of the sea becomes heavy and shock like. Visibility affected	Trees uprooted. Considerable structural damage
11	56-63	64-73	Violent Storm	Exceptionally high waves. The sea is completely covered with long white patches of foam lying in the direction of the wind. Visibility affected	Widespread structural damage
12	64-80	74-95	Hurricane	The air is filled with foam and spray. Sea completely white with driving spray. Visibility very seriously affected	Considerable and widespread damage to structures

THE TORRO TORNADO INTENSITY SCALE

Tornado Intensity	Description & Windspeeds (mph)	Damage (guidance only)
T0	Light Tornado 39 - 54	Light litter raised from ground level in spirals. Tents, marquees seriously disturbed. Some exposed tiles, slates on roofs dislodged. Trail visible through crops. Wheelie bins tipped and rolled.
T1	Mild Tornado 55 - 72	Deckchairs, small plants, heavy litter become airborne. Minor damage to sheds. More serious dislodging of tiles, slates. Chimney pots dislodged. Wooden fences flattened. Slight damage to hedges and trees.
T2	Moderate Tornado 73 - 92	Light caravans blown over. Garden sheds destroyed. Garage roofs torn away and doors imploded. Much damage to tiled roofs and chimneys. Ridge tiles missing. General damage to trees; small trees uprooted. Bonnets blown open on cars. Windows blown open or glazing sucked out of frames.
T3	Strong Tornado 93 - 114	Mobile homes overturned. Light caravans destroyed. Garages and weak outbuildings destroyed. House roof timbers exposed. Bigger trees snapped or uprooted. Debris carried considerable distances. Garden walls blown over. Buildings shaking.
T4	Severe Tornado 115 - 136	Mobile homes airborne. Sheds airborne for considerable distances. Entire roofs removed. Roof timbers of stronger brick or stone houses completely exposed. Gable ends torn away. Trees uprooted or snapped. Traffic signs folded. Debris carried up to 2km leaving an obvious trail.
T5	Intense Tornado 137 - 160	Wall plates, entire roofs and several rows of bricks on top floors removed. Items sucked out from inside houses including partition walls and furniture. Older, weaker buildings collapse.
T6	Moderately-Devastating Tornado 161 - 186	Strongly built houses suffer major damage or demolished. Bricks and blocks etc. become dangerous airborne debris. National grid pylons damaged or twisted. Exceptional or unusual damage found, e.g. objects embedded in walls or small structures elevated and landed with no obvious damage.
T7	Strongly-Devastating Tornado 187 - 212	Brick and wooden-frame houses demolished. Steel-framed warehouse-type constructions destroyed or seriously damaged. Locomotives thrown over. Noticeable de-barking of trees by flying debris.

Tornado Intensity	Description & Windspeeds (mph)	Damage (guidance only)
T8	Severely-Devastating Tornado 213 - 240	Motorcars carried great distances. Steel and other heavy debris strewn over great distances. A high level of damage within the periphery of the tornado path.
T9	Intensely-Devastating Tornado 241 - 269	Many steel-framed buildings demolished. Locomotives or trains hurled some distances. Complete debarking of any standing tree-trunks. Inhabitants' survival reliant on shelter below ground level.
T10	Super Tornado 270 - 299	Entire frame houses and similar buildings lifted bodily from foundations and carried some distances. Destruction of a severe nature, rendering a broad linear track largely devoid of vegetation, trees and man-made structures.

This scale is primarily used in the United Kingdom, whereas the Fujita scale is used in North America, Europe, and the rest of the world.

DEFINITIONS

(A selection of terms mentioned but not fully explained in the text.)

AIR PRESSURE: The force of the atmosphere pressing down on a unit area of the Earth's surface. The effect is caused by gravity. Also called 'atmospheric pressure', it's about 100,000 newtons per square metre or 1,000 millibars at sea-level.

ANEMOMETER: An instrument for measuring wind speed.

ANTICYCLONE: A large area of high atmospheric pressure characterised by winds circulating in a clockwise fashion in the Northern Hemisphere – a 'high' is usually accompanied by settled weather.

BLACK ICE: Transparent ice that forms when liquid water on the ground freezes – for instance when the temperature falls sharply after rain.

BLIZZARD: Snow accompanied by winds with an average speed of at least 32 mph, leading to significantly reduced visibility.

CLIMATE: The long-term average weather of a region (sometimes also including information about climatic extremes).

CONVECTION: Heat transfer through fluids via rising currents. Usually used to mean rising air motion and occurring when less dense, warmer air lies beneath denser, cooler air.

DEPRESSION: A low-pressure weather system (also called a 'low' or mid-latitude 'cyclone'). Generally, depressions move across mid latitudes relatively quickly bringing strong winds, cloud and rain or snow. Winds circulate in an anti-clockwise direction around a depression in the Northern Hemisphere.

DEW: When the temperature falls at night, water vapour in the air may condense into small water droplets on to objects at or near the ground. The 'Dew Point' is the temperature at which air becomes saturated with water vapour and below this temperature dew (on the ground) or fog (in the air) forms.

FOG: Water droplets in the air, reducing visibility to less than 1 km.

FRONT: The boundary between two air masses, often accompanied by extensive bands of cloud and precipitation.

GROUND FROST: White ice crystals deposited on the surface of objects when the ground temperature drops below 0C. Not to be confused with an air frost which occurs when the air temperature at screen height falls to 0C. Ground frosts are more common than air frosts.

GLAZE: Clear ice, usually formed when rain falls on a frozen surface.

HAZE: Reduced visibility due to smoke or dust.

HYGROMETER: An instrument used for measuring the humidity of air.

JET STREAM (Polar Front): A strong, narrow current of high-level wind that can reach speeds in excess of 200 mph, usually around 5-10 km up, in the atmosphere.

KNOT: One knot = 1.152 statute mph = 1.853 kph.

MILLIBAR: International unit for measuring air pressure, also called a hectopascal.

MIST: Water droplets suspended in the air causing an impairment of visibility to between 1 and 10km (cf. Fog).

LA NIÑA: A periodic cooling of the tropical Pacific Ocean, which affects weather in some parts of the world.

EL NIÑO: A periodic warming of the tropical Pacific Ocean, which affects weather in some parts of the world.

RAIN SHADOW: An area of decreased rainfall on the lee side of a hill or mountain.

SEA BREEZE: A cool breeze off the sea, felt at the coast, usually on sunny days. Land breezes occur, on clear nights, in the opposite direction.

SEA-LEVEL: The normal level of high tide, used for measuring height or depth.

SLEET: A mixture of rain and snow.

SQUALL: A quickly strengthening wind that lasts for a few minutes. Longer than a gust, squalls can become violent.

SYNOPTIC CHART: A weather map showing the key weather features over a geographical area at a particular time.

TEMPERATURE INVERSIONS: Warm air overlaying cold air (a reversal of the normal lapse rate of temperature with increasing altitude).

WIND CHILL: The effect of wind making it feel colder than it actually is.

WIND DIRECTION: Wind direction is given as the direction from which it originates, i.e. an easterly blows *from* the east to the west.

SELECT BIBLIOGRAPHY

We have consulted a vast range of sources. Some of the most evocative, accessible and reliable are included here.

Meteorology and related
Courtney, Nicholas. *Gale Force 10: The Life and Legacy of Admiral Beaufort* (2003)
Defoe, Daniel. *The Storm* (1704)
Eden, Philip. *Book of the Weather* (2003)
Fagan, Brian. *The Little Ice Age* (2000)
Gribbin, John and Mary. *Fitzroy: The Remarkable Story of Darwin's Captain and the Invention of the Weather Forecast* (2003)
Hamblyn, Richard. *The Invention of Clouds: How an Amateur Meteorologist Forged the Language of the Skies* (2001)
Hulme, M. and Barrow, E. (eds) - *Climates of the British Isles: Present, Past and Future* (1997)
Inwards, Richard. *Weather Lore: A Collection of Proverbs, Sayings and Rules Concerning the Weather* (1893)
Lamb, H.H. *The English Climate* (1964)
Manley, Gordon. *Climate and the British Scene* (1952)
Stirling, Robin. *The Weather of Britain* (1997)
Wheeler, D. and Mayes, J. (eds). Regional Climates of the British Isles (1997)
White, Gilbert. *The Natural History of Selborne* (1789)

General
Austen, Jane. *Pride and Prejudice* (1813)
Baker, J.A. *The Peregrine* (1967)
Bathurst, Bella. *The Lighthouse Stevensons* (1999)
Bell, Adrian. *Corduroy* (1930)
Blackmore, R.D. *Lorna Doone* (1869)
Brontë, Emily. *Wuthering Heights* (1847)
Card, Frank. *Whensoever: Fifty years of the RAF Mountain Rescue Service (1943-1993)* (1993)
Carroll, Lewis. *Alice in Wonderland* (1865)

Chouinard, Yvon. *Climbing Ice* (1978)

Cobbett, William. *Rural Rides* (1830)

Conrad, Joseph. *Youth* (1902)

Chatwin, Bruce. *On the Black Hill* (1982)

David, Elizabeth. *A Book of Mediterranean Food* (1950)

Deakin, Roger. *Waterlog* (1999)

de Havilland, Geoffrey. *Sky Fever* (1979)

de Quincey, Thomas. *Confessions of an English Opium Eater* (1804)

Dickens, Charles. *Bleak House* (1853)

Dickens, Charles. *The Pickwick Papers* (1836)

du Maurier, Daphne. *Rebecca* (1938)

Evelyn, John. *The Diary of John Evelyn*

Grossmith, George and Weedon. *Diary of a Nobody* (1888)

Hardy, Thomas. *Far From the Madding Crowd (1874)*

Hartley, L.P. *The Go-Between* (1952)

Harvey, Graham. *The Forgiveness of Nature: The Story of Grass* (2002)

Herriot, James. *It Shouldn't Happen to a Vet* (1972)

Jerome, Jerome K. *Three Men in a Boat* (1889)

Johnson, Samuel and Boswell, James. *Journey to the Western Islands of Scotland and the Journal of a Tour to the Hebrides* (1775)

Kilvert, Rev. Francis. *Kilvert's Diary 1870-79*

Lawrence, D.H. *Lady Chatterley's Lover* (1928)

Lee, Laurie. *Cider With Rosie* (1959)

Mabey, Richard. *Gilbert White: A Biography* (1986)

Mackenzie, Compton. *Whisky Galore!* (1947)

Maxwell, Gavin. *Ring of Bright Water* (1960)

Pepys, Samuel. *The Diaries of Samuel Pepys*

Raban, Jonathan. *Coasting* (1987)

Ransome, Arthur. *Winter Holiday* (1933)

Scott, Peter. *The Eye of the Wind* (1961)

Stevenson, Robert Louis. *Kidnapped* (1886)

Swift, Graham. *Waterland* (1983)

Thompson, Flora. *Lark Rise to Candleford* (1945)

Thomson, I.D.S. *The Black Cloud: Scottish Mountaineering Misadventures (1928-1966)* (1993)

White, T.H. *England Have My Bones* (1936)

Woodforde, James. *The Diary of a Country Parson 1758-1802* (1924-31)

Woolf, Virginia. *Orlando* (1928)

Websites

Sun rise and set times: aa.usno.navy.mil/data/docs/rs_oneyear.php

Latest satellite imagery: www.metoffice.gov.uk/education/data/satellite.html

RNLI: www.rnli.org.uk/

TORRO: www.torro.org.uk/

For historical information on extreme weather:

booty.org.uk/booty.weather/metindex.htm

www.personal.dundee.ac.uk/~taharley/britweather.htm

The Met Office: www.metoffice.gov.uk/

Atmospheric Optics: www.atoptics.co.uk/

Royal Meteorological Society (RMS:) www.rmets.org/index.php

RMS Education Activities: www.rmets.org/education/index.php

American Meteorological Society Glossary of Meteorology:

amsglossary.allenpress.com/glossary

ACKNOWLEDGEMENTS

A book where the research remit effectively includes the full canon of English language non-fiction writing—and plenty of fiction besides—presents a daunting task. We could not have contemplated it without the help of many friends and experts.

For general meteorological expertise, explanation of processes, data-checking and proof-reading, we are indebted to Dr Steve Dorling of the University of East Anglia. Jonathan Webb, of the Tornado and Storm Research Institute, directed us to published sources and generously supplied us with his own pre-publication research data as well as answering numerous questions as and when they occurred to us. The Meteorological Office, too, has been generous both with advice and their weather data sets. Thanks especially to Abigail Barbour, Graham Bartlett, Mark Beswick, Neil Davis, Barry Gromett, Jen Hardwick and Sally Molloy. Thank you also to Niall Brooks at Met Eireann.

Numerous researchers have assisted us, energetically and resourcefully chasing often obscure, abstruse or apocryphal stories down ever narrower and dustier cul-de-sacs—such as on exactly which wet day did Sir Walter Raleigh lay down his cloak for Elizabeth I, when did the headline FOG IN CHANNEL: CONTINENT CUT OFF first appear, or when was the last person killed by a falling icicle? Special thanks are due to Greg Fedorenko, Colin Morris, Andrew Warburton and Edward Wilford. Other researchers who assisted us include Ruth Andrzejewska, Peter Auger, Guy Brandon, Alexander Carroll, Marina Falbo, Helen Graham-Matheson, Sammi Luxa, John McHugh, Peter Reid, Joel Sam, Clare Stansfeld and Sarah Thacker. We are also grateful to Matt Worgan for design work.

Of the many experts and representatives of learned bodies or other institutions we have spoken to, particular thanks are due to Clara Anderson, assistant archivist at the Royal Society; David Barber, Football Association historian; Barry Cox at the RNLI archives; Giles Coren; Brent Elliot of The Royal Horticultural Society; Hattie Ellis; William Farara; Robert Hardman; Caroline Johnson of The National Motor Museum, Beaulieu; Christiane Kroebel, Whitby Literary and Philosophical Society; Sean Magee; Gordon Macey,

QPR club historian; Sue Mallinson of The Woodland Trust; The Daphne du Maurier Museum; Tomás Ó Maolalaidh, Foras na Gaeilge, Dublin; Robert Nicholson, curator, James Joyce Museum; Carol Palfrey of The Broads Society; John Prince at the House of Commons Library; Laurie Rae, curator, British Golf Museum; Neil Robinson, MCC museum; Yvonne Sibbald, Alpine Club Library; Dr Bob Sharp of Lomond Mountain Rescue; Graham Sharpe, William Hill Press Office; David B. Smith, historian of The Royal Caledonian Curling Club; Jed Smith, curator, Museum of Rugby, Twickenham; Dr Tim Sparks of The National Environmental Research Council; Erin O'Neill at the BBC Written Archives Centre; and Christopher Woodward of The National Garden History Museum.

We have drawn limitlessly and shamelessly on the help of family, friends and others who have dispensed time and wisdom, including Paul Aked, Alf Alderson, Captain 'Sydney' Allan, Tim Baker, Max Baring, Dr Mark Bateman, Philippa Beale, Natasha Burton, Sam Carter, Peter Chilvers, Dennis Clareboro, Sue Clifford, Roderick Corrie, Paddy Cramsie, Mike Creswell, Professor Stephen Daniels, Neil Davis, Richard Dodd, Dr Robert Doe, Mark Egerton, Matthew Engel, Simon Evans, Francesca Fairbairn, Peter and Elizabeth Freeman, Jake Gavin, Tony and Liz Gentil, Dr Martin George, Reg Green, Ed Gretton, Richard and Pat Harrad, Inger Hatloy, Professor Ronald Hutton, Rodney and Jill Hill, Nick Holmes, Joanna Kerr, Oliver James, Ann Lees, Keith Lewis and Frank O'Leary of Kolvox Computer Products, Mark Lloyd, Nigel McMorris, Harry Marshall, Geraldine Mathieson, Arne Maynard, Lucy Moore, Gerald Morris, Robert Noel, John O'Donnell, Eck Ogilvie-Grant, Jonny Ohlson, Lucy Parham, John Phibbs, James Penn, Richard Penn, Rosemary Penn, Arabella Pike, Nick Powell, Jane Raven, Bryan Read, Polly Robertson, Sally and Steven Thomas, Mike Town, Hank Wallace, Christine Watson, Molly Watson, Ben Weatherall, Charlie Williams, Roger Williams, Stephen Williams, Zac Withers, Gay and Paul Woodley, Peter Woodward. Also, to every builder, farmer, lorry driver, mountaineer, plumber or telephone engineer we have encountered in the last eighteen months who has been grilled for stories. We apologise to anyone we have left out.

Finally, and most of all, thanks to our editor, Nick Davies at Hodder & Stoughton, for commissioning the book, our agents, Camilla Hornby of Curtis Brown and Felicity Rubinstein of Lutyens Rubinstein, and everyone at Hodder who has helped guide such a complicated project to completion in so short a time, especially Rupert Lancaster, Nicola Doherty and Eleni Fostiropoulos. Finally, finally, to our longsuffering wives, the two Vs — for the many, many missed bath times.

PERMISSIONS

The following are reproduced by permission of: The Random House Group Ltd, for Martin Amis, *Experience: A Memoir.* Faber & Faber, for Alan Bennett, *The Alan Bennett Diaries 1980-1990.* Pollinger Limited, for Wilfred Blunt, *Of Flowers and a Village.* The Estate of Lord Bonham-Carter, for Violet Bonham-Carter, *'Champion Redoubtable', The Diaries and Letters of Violet Bonham Carter 1914-1945.* Greene & Heaton Ltd, for Bill Bryson, *Notes from a Small Island.* Johnson & Alcock Ltd, for Sir Hugh Casson, *Diary.* Clarebooks, for Barbara Castle, *The Castle Diaries 1964-70.* The University of Birmingham, for Neville Chamberlain, *The Neville Chamberlain Diary Letters: The Downing Street Years, 1934-1940.* Weidenfeld & Nicolson, for Sir Henry Channon, *'Chips': The Diaries of Sir Henry Channon.* Curtis Brown Group Limited, for Clementine Churchill, *Speaking for Themselves: The Personal Letters of Winston and Clementine Churchill.* Curtis Brown Group Limited, for Winston Churchill, *Speaking for Themselves: The Personal Letters of Winston and Clementine Churchill.* Gomer Press, for William Condry, *A Year in a Small Country.* A. & C. Black Publishers Ltd, for Noel Coward, *The Noel Coward Diaries.* David Higham Associates Limited, for Roald Dahl, *Charlie and the Chocolate Factory.* *Waterlog* by Roger Deakin, published by Chatto & Windus. Reprinted by permission of The Random House Group Ltd. Virago Press for E.M. Delafield, *Diary of a Provincial Lady.* Chichester Partnership, for Daphne du Maurier, *Rebecca.* Judy Daish Associates Ltd for Sir Richard Eyre, *National Service: Diary of a Decade at the National Theatre.* Gillon Aitken Associates Ltd, for Helen Fielding, *Bridget Jones's Diary.* Faber & Faber, for Marjorie Fish, *A Flower for Every Day.* Penguin Group, for Ian Fleming, *Chitty Chitty Bang Bang.* Curtis Brown Group Limited, for Stella Gibbons, *Cold Comfort Farm.* Sheil Land Associates Ltd, for Joyce Grenfell, *The Time of My Life.* Christopher Sinclair-Stevenson, for Alec Guinness, *My Name Escapes Me* © the estate of Alec Guinness. Crowood Press, for Geoffrey de Havilland, *Sky Fever.* David Higham Associates Limited, for James Herriot, *It Shouldn't Happen to a Vet.* Transworld Ltd., for Professor John Hull, *Touching the Rock: An Experience of Blindness.* Curtis Brown Group Ltd, for Christopher Isherwood, *Christopher Isherwood Diaries, Vol 1: 1939-1960.* Tony Peake, for Derek Jarman, *Smiling in Slow Motion.* Pan Macmillan Publishers, for Gertrude

PICTURE ACKNOWLEDGEMENTS

Page ii © PA Photos (Queen's Birthday Parade, 12th June 1982). January 3rd: Trinity House Collection. February 6th: © Bettmann/Corbis. February 12th: © Brockbank Partnership/ www.russellbrockbank.co.uk. March 15th: By permission of the Special Collections and Archives Librarian, Robinson Library, Newcastle University. 20th March: © The Estate of E.H. Shepard reproduced with permission of Curtis Brown Limited London. April 13th: Edmund H. New from *The Natural History of Selborne* by Gilbert White (1937 Edition). May 6th: Hulton Archive/Getty Images. May 14th: Illustration by Peter Scott from his autobiography *The Eye of the Wind*, 1961. June 1st: Neil Gower. June 18th: Courtesy www.needles.shalfleet.net. July 13th: Neil Gower. July 18th: National Meteorological Library and Archive, the Met Office. July 30th: Colen Campbell *Vitruvius Britannicus, The British Architect*, 1717. August 7th: Mendelssohn Archives, Berlin. September 4th: Kerstin Finke/www.kfishsurf.com. September 5th: Neil Gower. September 19th: Horse Chestnut from *Flora Nonacensis,* 1811/Bridgeman Art Library. October 5th:*Punch* 29th September 1888. October 6th: By permission of the Royal National Lifeboat Institution. October 17th: Neil Gower. October 25th: Rear Admiral Robert Fitzroy, *The Weather Book: A Manual of Practical Meteorology,* 1863/National Meteorological Library and Archive, the Met Office. October 29th: Hulton Archive/Getty Images. November 1: © PA Photos. November 6th: Neil Gower. November 15th: Guildhall Library, City of London. December 3rd: © Hulton Archive/Getty Images. December 10th: © NI Syndication. Page 423: The Shipping Forecast (map of sea areas since 2002) © Crown copyright 2007, the Met Office. Lettering for date headings by Edward Bettison. Maps of the British Isles, weather and map symbols, by Neil Gower.

I'm leaving because the weather is too good. I hate London when it's not raining.

Groucho Marx

HAVE YOU GOT A GREAT BRITISH WEATHER STORY?

Something funny, moving, charming, telling, historically significant and, above all, TRUE? Do you know the exact date? Did it have consequences? Did it happen within the British Isles or British waters, as defined in the Introduction?

If so, tell us about it at www.wrongkindofsnow.co.uk or post a brief account to The Wrong Kind of Snow, Hodder & Stoughton, 338 Euston Road, London NW1 3BH